Stable Isotopes and Biosphere–Atmosphere Interactions:

Processes and Biological Controls

This is a volume in the

PHYSIOLOGICAL ECOLOGY Series
Edited by Harold A. Mooney

A complete list of books in this series appears at the end of the volume.

Stable Isotopes and Biosphere–Atmosphere Interactions:

Processes and Biological Controls

L. B. Flanagan
Department of Biological Sciences
University of Lethbridge
Lethbridge, Alberta, Canada

J. R. Ehleringer
Department of Biology
University of Utah
Salt Lake City, USA

D. E. Pataki
Department of Biology
University of Utah
Salt Lake City, USA

ELSEVIER
ACADEMIC
PRESS

Amsterdam • Boston • Heidelberg • London • New York • Oxford
Paris • San Diego • San Francisco • Singapore • Sydney • Tokyo

Elsevier Academic Press
525 B Street, Suite 1900, San Diego, California 92101-4495, USA
http://www.elsevier.com

Elsevier Academic Press
84 Theobald's Road, London WC1X BRR, UK
http://www.elsevier.com

Library of Congress Control Number: 2004116780

British Library Cataloguing in Publication Data
A catalogue record for this book is available from the British Library

ISBN 0-12-088447-X

Printed and bound in Great Britain
05 06 07 08 9 8 7 6 5 4 3 2 1

Contents

Contributors

Number in parentheses after each name indicates the chapter number for the author's contribution

Dennis D. Baldocchi (7)
Ecosystem Science Division
Department of Environmental Science
Policy and Management
151 Hilgard Hall
University of California, Berkeley
Berkeley, CA 94720, USA

Margaret M. Barbour (2)
Landcare Research
Gerald St
PO Box 69
Lincoln, New Zealand

Joseph A. Berry (9)
Department of Plant Biology
Carnegie Institution of Washington
290 Panama Street
Stanford, CA 94305, USA

David R. Bowling (7)
Department of Biology
University of Utah
257 S 1400 E
Salt Lake City, UT 84112-0840
USA

Susan E. Bush (12)
Department of Biology
University of Utah
257 S 1400 E
Salt Lake City, UT 84112-0840
USA

Lucas A. Cernusak (2)
Environmental Biology Group
Research School of Biological Sciences
Australian National University
GPO Box 475
Canberra, ACT 2601, Australia

Jeffrey Chanton (6)
Department of Oceanography
Florida State University
Tallahassee, FL 32306-4320
USA

Lia Chasar (6)
US Geological Survey WRD
Tallahassee, FL 32301-1372
USA

Philippe Ciais (14)
Laboratoire des Sciences du Climat et de l'Environnement
Unité mixte CEA-CNRS
Bat 709, CE l'Orme
des Merisiers
91191 Gif sur Yvette, France

Matthias Cuntz (14)
Laboratoire des Sciences du Climat et de l'Environnement
Unité mixte CEA-CNRS
Bat 709, CE l'Orme
des Merisiers
91191 Gif sur Yvette, France

Todd Dawson (8)
Center for Stable Isotope Biogeochemistry
Department of Integrative Biology
University of California, Berkeley
3060 Valley Life Sciences Building #3140
Berkeley, CA 94720-3140, USA

James R. Ehleringer (12)
Department of Biology
University of Utah
257 S 1400 E
Salt Lake City, UT 84112-0840
USA

Alf Ekblad (4)
Section of Biology
Department of Natural Sciences
Örebro University
SE-701 82, Örebro, Sweden

Graham D. Farquhar (2)
Environmental Biology Group
Research School of Biological Sciences
GPO Box 475, Canberra
ACT 2601, Australia

Lawrence B. Flanagan (1, 10)
Department of Biological Sciences
University of Lethbridge
4401 University Drive
Lethbridge, Alberta, T1K 3M4
Canada

Vincent Gitz (14)
Centre de International Recherche sur l'Environnement et le Développement
45 bis avenue de la Belle Gabrielle
94736 Nogent sur Marne
France

Paul Glaser (6)
Department of Geology
University of Minnesota
Minneapolis, MN 55455
USA

Gerd Gleixner (3)
Max-Planck-Institut für Biogeochemistry
PO Box 10 01 64-07701
Jena, Germany

Brent Helliker (9)
Department of Plant Biology
Carnegie Institution of Washington
290 Panama Street
Stanford, CA 94305, USA

Mona N. Högberg (4)
Department of Forest Ecology
SLU
SE-901 83, Umeå, Sweden

Peter Högberg (4)
Department of Forest Ecology
SLU
SE-901 83, Umeå, Sweden

Jennifer Y. King (11)
Department of Ecology, Evolution and Behavior
University of Minnesota
St Paul, MN 55108, USA

Daniel Milchunas (11)
Natural Resources Ecology Laboratory
Colorado State University
Fort Collins, CO, USA

John B. Miller (16)
NOAA/CMDL
R/CMDL1
325 Broadway
Boulder, CO 80303, USA

Jack Morgan (11)
USDA/ARS
Crops Research Laboratory
1701 Center Ave
Fort Collins, CO 80526, USA

Arvin R. Mosier (11)
USDA/ARS
PO Box E
301 S. Howes
Fort Collins, CO 80522-0470
USA

Florent Mouillot (14)
Department of Global Ecology
Carnegie Institution of
Washington
290 Panama Street
Stanford, CA 94305, USA

Anders Nordgren (4)
Department of Forest Ecology
SLU
SE-901 83, Umeå, Sweden

Anders Ohlsson (4)
Department of Forest Ecology
SLU
SE-901 83, Umeå, Sweden

Diane E. Pataki (12)
Department of Biology
University of Utah
257 S 1400 E
Salt Lake City, UT 84112-0840
USA

Present Address for Diane E. Pataki:
Department of Earth System
Science
3313 Croul Hall
University of California, Irvine
Irvine, CA 92697-3100
USA

Elise Pendall (11)
Department of Botany
University of Wyoming
Laramie, WY 82071, USA

Tibisay Pérez (5)
Department of Earth System
Science
University of California, Irvine
Irvine, CA 92697-3100, USA

Present Address for Tibisay Pérez:
Centro de Quimica
Instituto Venezolano de
Investigaciones Cientificas
Aptdo 21827 Caracas, 1020-A
Venezuela

Philippe Peylin (14)
UMR BioMCo, Université Paris 6,
INRA-CNRS INRA-INAPG
Bâtiment EGER; Aile B/C 1 er
étage 78850
THIVERVAL-GRIGNON, France

Agneta H. Plamboeck (4)
Swedish Defence Research
Agency
Division of NBC-Defence
SE-901 82, Umeå, Sweden

Thom Rahn (15)
Los Alamos National Laboratory
EES-6, MS-D462
Los Alamos, New Mexico 87545
USA

James T. Randerson (13)
Department of Earth System
Science
3212 Croul Hall
University of California, Irvine
Irvine, CA 92697-3100, USA

William J. Riley (9)
Earth Sciences Division
Lawrence Berkeley National Laboratory
1 Cyclotron Road (90-102A)
Berkeley, CA 94720, USA

Marko Scholze (14)
Max-Planck-Institut für Meteorologie
Bundestraße 53
20146 Hamburg, Germany

Don Siegel (6)
Department of Geology
Syracuse University
Syracuse, NY 13244, USA

Bhupinderpal-Singh (4)
Department of Forest Ecology
SLU
SE-901 83, Umeå, Sweden

Christopher Still (9)
Department of Geography
3611 Ellison Hall
University of California
Santa Barbara
Santa Barbara, CA 93106
USA

Kevin Tu (8)
Center for Stable Isotope Biogeochemistry
Department of Integrative Biology
University of California
Berkeley
3060 Valley Life Sciences Building #3140
Berkeley, CA 94720-3140
USA

1

Introduction: Stable Isotopes and Earth System Science

Lawrence B. Flanagan

Global Change and Earth System Science

The globe is being dramatically affected by environmental changes such as alterations to the composition of the atmosphere (e.g., carbon dioxide, methane, and nitrous oxide concentrations), associated shifts in climate and reductions in biological diversity (Vitousek, 1994). Earth system science, a multidisciplinary field of study, has become established to improve our understanding of the functioning of ecosystems across the entire planet and to train a new generation of scientists with the necessary skills to tackle the complex issues associated with global change. Some have argued that most universities have been slow to embrace the challenge of global change science and have not moved quickly enough to set up an effective framework or programs for fostering interdisciplinary and multidisciplinary research and teaching (Lawton, 2001). In too many institutions traditional academic departments and subject areas are the norm and this traditional organizational framework can often limit the interactions necessary for fostering truly multidisciplinary research like that needed for earth system science.

Physiological Ecology and Global Gas Exchange

Physiological ecology is a traditional academic discipline with a long history (Billings, 1985) and one that is contributing substantially to the development of global change studies. As a discipline with an early focus on the fluxes of carbon, water, and energy at the organism level, ecophysiology as a science was 'pre-adapted' to expand into broader-scale ecosystem and global biogeochemical studies (Field and Ehleringer, 1993). Physiological ecologists have embraced several new technologies, particularly eddy covariance flux measurements, remote sensing and stable isotope analyses, for scaling their detailed understanding of the mechanisms controlling gas exchange (photosynthesis, respiration, transpiration) at the leaf and whole plant level to larger spatial and longer temporal scales. This book brings together much of the recent progress in the application of stable isotope analyses to understanding biosphere–atmosphere exchange of carbon dioxide, methane, and nitrous oxide.

Physiological ecologists have been able to effectively contribute to global studies because they have a strong understanding of the mechanistic processes that control stable isotope fractionation during important enzymatic reactions. For example, the isotope effects caused by enzymes such as ribulose bisphosphate carboxylase and carbonic anhydrase during photosynthesis or cytochrome oxidase during respiration, have been studied in test tubes or closed gas exchange chambers under controlled conditions and mechanistic models of these processes have been developed (Farquhar *et al.*, 1982; Guy *et al.*, 1989; Farquhar *et al.*, 1993; Gillon and Yakir, 2001). These models can be used on a global basis because of the huge magnitude of the CO_2 and O_2 fluxes through the enzymatic reactions (Yakir, 2002). Rules of mass balance apply in the relatively closed atmosphere–biosphere system and allow the operation of enzymatic processes to be observed through the seasonal and interannual shifts in the isotopic composition of atmospheric CO_2 in a manner similar to what can be observed in leaf chamber and terrarium experiments (Flanagan and Ehleringer, 1998). Berry (1999) captured the essence of this approach in a diagram that depicted the ocean–land–atmosphere system as a bathtub that supplies water to two organelles (chloroplast and mitochondrion) which, in turn, exchange carbon dioxide and oxygen with the atmosphere.

Historical Background and the Contents of this Book

This book is a product of the fourth conference in a series of meetings held over the last 18 years on the application of stable isotope techniques in the ecological sciences. The first meeting was held at the UCLA Lake Arrowhead Conference Center in April 1986 and focused on a broad range of applications using stable isotope techniques to study questions in plant and animal ecology (Rundel *et al.*, 1988). The book that resulted from the Lake Arrowhead meeting was instrumental in promoting many exciting opportunities and illustrated that stable isotope techniques opened up new fields of enquiry in ecological science. A large number of graduate students and postdoctoral fellows, who now dominate the forefront of current ecological stable isotope research, were strongly influenced by the Lake Arrowhead volume. The focus of the next meeting, held at University of California, Riverside, in January 1992, was on plant physiological ecology, particularly the application of carbon isotope techniques to studies of water-use efficiency (Ehleringer *et al.*, 1993). Other topics discussed at the Riverside meeting included the application of hydrogen and oxygen isotopes in studies of plant water use, and the first theoretical and experimental studies of the mechanisms of oxygen isotope discrimination during photosynthetic CO_2 exchange. In addition, a few presentations at the Riverside meeting highlighted the potential of applying Keeling's approach to estimate the isotope ratio of carbon dioxide respired by an ecosystem (Keeling, 1961; Pataki *et al.*, 2003). This provided an opportunity to extend the scale of ecophysiological research, which had previously been focused on the leaf or whole plant scale, to the ecosystem-level. These conceptual advances coincided with the development and expansion of earth system science and funding of new multidisciplinary programs fostered by the International Geosphere–Biosphere Program (IGBP). The Boreal Ecosystem–Atmosphere Study (BOREAS) program in Canada and similar studies in the Amazon Basin of Brazil (LBA) provided new opportunities for including stable isotope measurements in large-scale ecosystem studies. The third conference held in Newcastle-upon-Tyne, UK during August 1996 followed on these developments and included presentations on studies that were beginning to forge a greater integration among biological, ecological, and geochemical processes (Griffiths, 1998). The fourth meeting, which was held in Banff, Canada in May 2002, focused on biosphere–atmosphere interactions and the role that stable isotope measurements play in providing mechanistic insights about physiological processes operating at large spatial scales.

While the range of topics covered in the Banff meeting was reduced compared to the Newcastle meeting, there was an increased refinement and improved sophistication in application of stable isotope techniques to biological and ecological studies. The first part of this book includes chapters that highlight advances recently made in the understanding of stable isotope effects that occur during the acquisition of carbon and water by plants, and during soil processes that control the breakdown of organic matter and the production of carbon dioxide, methane, and nitrous oxide. The second part presents studies of the application of stable isotope techniques and models to understand processes operating during ecosystem CO_2 and H_2O exchange in natural, experimental (elevated CO_2), and urban environments. The final part of the book examines the application and interpretation of global-scale processes and budgets for atmospheric CO_2, CH_4, and N_2O. The organizational committee of the Banff meeting promoted the involvement of many young investigators (post-doctoral fellows and individuals who had recently been appointed as university professors) as speakers at the meeting, in addition to a number of established and more experienced researchers. The excellent presentations made by these young researchers in Banff and the chapters of this book indicate that the field has an exciting future in helping to solve many open questions in earth system science.

Acknowledgments

The idea for the Banff conference was conceived by the steering committee of the Biosphere–Atmosphere Stable Isotope Network (BASIN), a core project of Focus 1 (Ecosystem Physiology) of the Global Change and Terrestrial Ecosystems project within the IGBP program. The BASIN steering committee members were: J. R. Ehleringer (chair), J. A. Berry, N. Buchmann, L. B. Flanagan, D. E. Pataki, and D. Yakir. Financial support for the BASIN program and the Banff meeting was provided by a grant from the National Science Foundation (USA) to J. R. Ehleringer (University of Utah).

References

Berry J. A. (1999) Ghosts of biospheres past. *Nature* **400**: 509–510.

Billings W. D. (1985) The historical development of physiological plant ecology. In *Physiological Ecology of North American Plant Communities* (B. Chabot and H. A. Mooney, eds) pp. 1–15. Chapman Hall, New York.

Ehleringer J. R., Hall A. E. and Farquhar G. D. (1993) *Stable Isotopes and Plant Carbon–Water Relationships.* Academic Press, San Diego.

Farquhar G. D., O'Leary M. H. and Berry J. A. (1982) On the relationship between carbon isotope discrimination and the intercellular carbon dioxide concentration in leaves. *Aust J Plant Physiol* **9**: 121–137.

Farquhar G. D., Lloyd J., Taylor J. A., Flanagan L. B., Syvertsen J. P., Hubick K. T., Wong S. C. and Ehleringer J. R. (1993) Vegetation effects on the isotope composition of oxygen in atmospheric CO_2. *Nature* **363**: 439–443.

Field C. B. and Ehleringer J. R. (1993) Introduction: questions of scale. In *Scaling Physiological Processes: Leaf to Globe* (J. R. Ehleringer and C. B. Field, eds) pp. 1–4. Academic Press, San Diego.

Flanagan L. B. and Ehleringer J. R. (1998) Ecosystem–atmosphere CO_2 exchange: interpreting signals of change using stable isotope ratios. *Trends Ecol Evol* **13**: 10–14.

Gillon J. and Yakir D. (2001) Influence of carbonic anhydrase activity in terrestrial vegetation on the ^{18}O content of atmospheric CO_2. *Science* **291**: 2584–2587.

Griffiths H. (1998) *Stable Isotopes: Integration of Biological, Ecological and Geochemical Processes.* BIOS Scientific Publishers Ltd., Oxford, UK.

Guy R. D., Berry J. A., Fogel M. L. and Hoering T. C. (1989) Differential fractionation of oxygen isotopes by cyanide-resistant and cyanide-sensitive respiration in plants. *Planta* **177**: 483–491.

Keeling C. D. (1961) The concentration and isotopic abundance of carbon dioxide in rural and marine air. *Geochim Cosmochim Acta* **24**: 277–298.

Lawton J. (2001) Earth system science. *Science* **292**: 1965.

Pataki D. E., Ehleringer J. R., Flanagan L. B., Yakir D., Bowling D. R., Still C. J., Buchmann N., Kaplan J. O. and Berry J. A. (2003) The application and interpretation of Keeling plots in terrestrial carbon cycle research. *Global Biogeochem Cycles* **17**: doi: 10.1029/2001GB001850.

Rundel P. W., Ehleringer J. R. and Nagy K. A. (1988) *Stable Isotopes in Ecological Research. Ecological Studies 68*. Springer-Verlag, New York.

Vitousek P. M. (1994) Beyond global warming: ecology and global change. *Ecology* **75**: 1861–1876.

Yakir D. (2002) Sphere of influence. *Nature* **416**: 795.

Part I

Stable Isotopes and Physiological Processes

2

Factors Affecting the Oxygen Isotope Ratio of Plant Organic Material

Margaret M. Barbour, Lucas A. Cernusak, Graham D. Farguhar

Early work by Epstein *et al.* (1977) showed that cellulose from aquatic plants was 27 ± 4‰ more enriched in ^{18}O than the water in which it grew. They explained the observed 27‰ fractionation by suggesting that CO_2 had been shown to be 41‰ more enriched than water at equilibrium, and that fixation of CO_2 by Rubisco required one CO_2 molecule and one H_2O molecule (2/3 of 41‰ is about 27‰). However, DeNiro and Epstein (1979) later ruled out the 2/3 CO_2 hypothesis by pointing out that one of the oxygen atoms added by CO_2 fixation is lost in the photosynthetic carbon reduction cycle (PCR cycle). An alternate hypothesis was proposed by Sternberg and DeNiro (1983), who suggested that the oxygen atoms in water and those in organic molecules exchanged isotopically, allowing organic material (including cellulose) to reflect variation in plant water. Using acetone as a model molecule, they demonstrated that at equilibrium the organic molecule is 28‰ more enriched in ^{18}O than the water with which it exchanged. Hence, the oxygen isotope composition of plant organic material ($\delta^{18}O_p$) should largely reflect the oxygen isotope composition of plant water.

$\delta^{18}O$ of water in plants changes as a result of: (1) variations in $\delta^{18}O$ of water taken up by plants; (2) leaf water enrichment in ^{18}O during transpiration, the extent to which is dependent on the atmospheric conditions (relative humidity, and $\delta^{18}O$ of water vapor in the atmosphere) and the stomatal regulation of water loss; and (3) variation in $\delta^{18}O$ of water in cells forming organic material (e.g., cellulose), which may be a mixture of unenriched source water and enriched leaf water. Variation in $\delta^{18}O$ of plant organic material may also occur as a result of variation in the extent of isotopic exchange between water and organic molecules during biosynthesis.

This chapter will outline current understanding of the factors affecting $\delta^{18}O_p$, formalize this understanding in mathematical models, and then describe potential applications of the technique.

Definitions

As the absolute isotopic composition of a material is difficult to measure directly, isotope ratios are generally compared to that of a standard. In the case of $^{18}O/^{16}O$, the standard is commonly Vienna-Standard Mean Oceanic Water (VSMOW), with an isotope ratio of 2.0052×10^{-3} (Gonfiantini, 1984). Isotopic compositions are expressed as relative deviations from VSMOW, and denoted $\delta^{18}O = R_p/R_{st} - 1$, where R_p and R_{st} are the isotope ratios of the substance of interest and the standard, respectively. Variation in the $\delta^{18}O$ of source (soil) water may be removed from $\delta^{18}O_p$ by presenting the composition as an enrichment above source water ($\Delta^{18}O_p$) given by: $\Delta^{18}O_p = R_p/R_s - 1$, where R_s is the $^{18}O/^{16}O$ ratio of source water. To a close approximation, $\Delta^{18}O_p$ may be estimated by $\Delta^{18}O_p = \delta^{18}O_p - \delta^{18}O_s$, where $\delta^{18}O_s$ is the isotopic composition of source water (relative to VSMOW).

Factors Contributing to Variation in $\delta^{18}O_p$

Variation in Source Water $\delta^{18}O$

The oxygen isotope ratio of soil water taken up by a plant depends, to a first approximation, on the temperature of droplet formation for rain and snow falling at the site. Precipitation becomes more depleted in ^{18}O as temperature decreases at higher latitudes and altitudes. There is also an amount effect, with high precipitation sites having more depleted water than would be expected from site temperatures. The temperature dependence of $\delta^{18}O$ of precipitation ($\delta^{18}O_\tau$), and the record of variation in $\delta^{18}O_\tau$ in plant tissue has been recognized for some time, and prompted Libby *et al.* (1976) to suggest that it may be possible to use cellulose from annual tree rings as 'isotopic thermometers'.

The mean annual $\delta^{18}O$ of precipitation ($\delta^{18}O_R$) has been related to mean annual temperature (T, in °C), precipitation amount (P, in m), and site elevation (E_v, in m) by Barbour *et al.* (2001):

$$\delta^{18}O_R = 0.52T - 0.006T^2 + 2.42P - 1.43P^2 - 0.046\sqrt{E_v} - 13.0 \quad (2.1)$$

Equation 2.1 was generated by a multiple regression of data collected at International Atomic Energy Agency (IAEA) sites around the world, and presented in IAEA (1992). Figure 2.1 shows the strong temperature dependence of $\delta^{18}O_R$ at lower temperature sites, and that warmer sites with high annual precipitation diverge from this relationship.

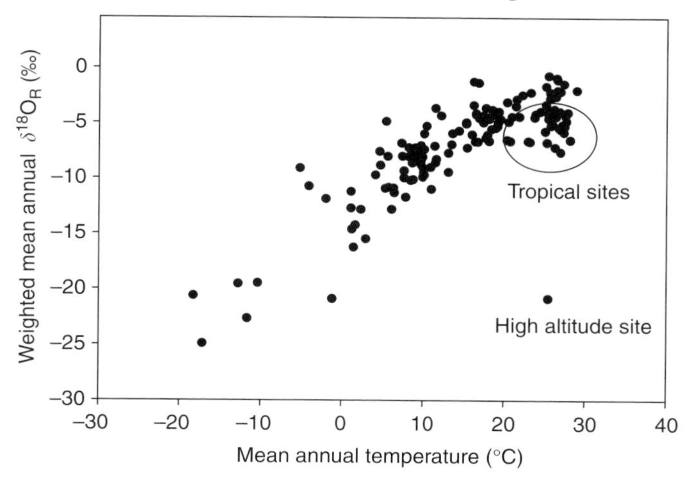

Figure 2.1 The relationship between site mean annual temperature and mean annual $\delta^{18}O$ of rainfall, weighted by rainfall amount, for IAEA sampling sites around the world. Data from IAEA (1992).

Using Eq. 2.1, Barbour *et al.* (2001) explained 65% of variation in $\delta^{18}O$ of wood from *Pinus* species collected around the world. When trees are grown in a constant environment, with $\delta^{18}O$ of source water carefully controlled and measured, much closer relationships are found. For example, Roden and Ehleringer (1999) found that 96 to 99% of variation in $\delta^{18}O$ of wood cellulose from three temperate tree species grown in a closely controlled hydroponic system could be explained by variation in $\delta^{18}O$ of the water in the hydroponic tanks. However, it was clear from Roden and Ehleringer's work that relative humidity of the growth environment also had strong effects on $\delta^{18}O$ of cellulose. This effect will be discussed in the next section. $\delta^{18}O$ of water taken up from the soil may vary considerably from $\delta^{18}O_R$. While variation in $\delta^{18}O$ of source water has not often been utilized in studies to date, variation in the hydrogen isotope ratio (δD) of source water has been studied in a number of experimental systems. Parallels between $\delta^{18}O$ and δD of source water may be drawn because the two isotopes share a unique relationship in precipitation: $\delta D = 8\delta^{18}O + 10‰$ (Craig, 1961). Several studies have exploited large differences between δD of surface soil water and that of groundwater, or seasonal variation in δD of rainfall, to determine the source of plant water within the soil profile (Dawson *et al.*, 1998, and references therein).

As demonstrated by Barbour *et al.* (2004), removing spatial and temporal variation in source water $\delta^{18}O$ by presenting plant organic matter $^{18}O/^{16}O$ as an enrichment above source water (i.e., $\Delta^{18}O_p$, as described earlier) will simplify interpretation and enable variation due to leaf water enrichment and isotopic exchange to be identified.

Leaf Water Evaporative Enrichment

Few studies have compared $\delta^{18}O$ of leaf water with that of organic material formed at the same time. Wang *et al.* (1998) presented leaf water and cellulose oxygen isotope compositions from a wide range of plant forms grown in a common garden. Re-plotting their data reveals that cellulose was, on average, 19‰ more enriched than leaf water, with a slope for the fitted regression line of 0.48. That is, the range in leaf water $\delta^{18}O$ was reduced by about half in cellulose. Their data also show a wide variation about the fitted line (Fig. 2.2). A much closer fit was found by Helliker and Ehleringer (2002a) for leaf water and cellulose of a range of grass species grown under constant conditions and a range of relative humidities. They found that 97% of variation in cellulose $\delta^{18}O$ was explained by variation in leaf water $\delta^{18}O$. Less of the leaf water signal was dampened in the well-controlled environment, with a slope between the two of 0.72. Dampening of the leaf water signal in cellulose will be explored in more detail later.

Enrichment of leaf water relative to soil water was first demonstrated by Gonfiantini *et al.* (1965). A model of evaporative enrichment developed by Craig and Gordon (1965) for enrichment of a free water surface is commonly applied to leaf water, with modifications. This model relates enrichment of leaf water above source water ($\Delta^{18}O_e$) to the kinetic fractionation during diffusion through the stomata and leaf boundary layer (ε_k),

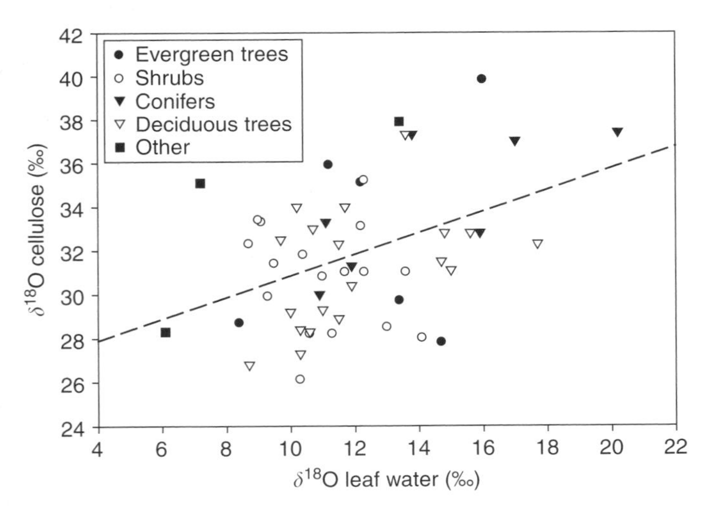

Figure 2.2 The relationship between measured $\delta^{18}O$ of leaf water and $\delta^{18}O$ of cellulose for a range of plants species grown in a common garden. The fitted line has a slope of 0.48, and when $\delta^{18}O$ of leaf water is zero, cellulose $\delta^{18}O$ is fitted to be 26.1‰ ($r = 0.405$; $P = 0.003$). Data from Wang *et al.* (1998).

the proportional depression of water vapor pressure by the heavier $H_2{}^{18}O$ molecule (ε^*), the oxygen isotope composition of water vapor relative to source water ($\Delta^{18}O_v$) and scaled by the ratio of ambient to intercellular water vapor pressure (e_a/e_i) (Craig and Gordon, 1965; Dongmann *et al.*, 1974; Farquhar and Lloyd, 1993) by:

$$\Delta^{18}O_e = \varepsilon^* + \varepsilon_k + (\Delta^{18}O_v - \varepsilon_k)e_a/e_i \tag{2.2}$$

In well mixed conditions, $\Delta^{18}O_v$ is often close to $-\varepsilon^*$, so that $\Delta^{18}O_e$, and therefore $\Delta^{18}O_p$ (to some extent) are proportional to $1 - e_a/e_i$. As a result, $\Delta^{18}O_p$ should be negatively related to relative humidity (RH). Equation 2.2 also predicts that at constant e_a, increasing stomatal conductance will result in less enrichment at the sites of evaporation within leaves. This is because as transpiration rate increases with higher stomatal conductance, both leaf temperature and e_i decrease.

The negative relationship predicted between RH and $\Delta^{18}O_p$ (or between RH and $\delta^{18}O_p$, if source water $\delta^{18}O$ is constant) has been observed in a number of studies (e.g., Edwards and Fritz, 1986; Saurer *et al.*, 1997; Roden and Ehleringer, 1999; Barbour and Farquhar, 2000). However, other studies report no evidence of a humidity signal in $\delta^{18}O_p$ (DeNiro and Cooper, 1989).

While Eq. 2.2 predicts general trends in leaf water enrichment quite well, in some cases measured leaf water $\delta^{18}O$ was found to be less enriched than that predicted (e.g., Yakir *et al.*, 1989; Flanagan *et al.*, 1994; Wang *et al.*, 1998), while in others leaf water was more enriched than predicted (Bariac *et al.*, 1994a; Wang and Yakir, 1995; Helliker and Ehleringer, 2000). A number of approaches have been taken to address these discrepancies, including: (1) pools of water within a leaf (e.g., Yakir *et al.*, 1990); (2) unenriched water within veins lowering the bulk leaf water enrichment (e.g., Roden and Ehleringer, 1999); (3) a string of interconnected pools of water within the leaf (e.g., Helliker and Ehleringer, 2000); (4) diurnal changes in the evaporative environment and water content of the leaf (e.g., Cernusak *et al.*, 2002); and (5) the ratio of convection of unenriched water towards the sites of evaporation to back diffusion of enrichment from those sites (e.g., Farquhar and Lloyd, 1993). These different approaches are outside the scope of this chapter, so here we consider only the broad implications of the different treatments on $\Delta^{18}O_p$. Effects on $\Delta^{18}O_p$ additional to those outlined by Eq. 2.2 are seen in the final two treatments.

Diurnal variation in leaf water enrichment is expected to occur largely as a result of diurnal variation in temperature. Ambient vapor pressure usually remains nearly constant over a diurnal cycle, such that variation in the term e_a/e_i will be driven mostly by temporal changes in the saturation vapor pressure within the leaf, which will vary exponentially as a function of leaf temperature. Such diurnal variation in leaf water enrichment is commonly observed, with the extent of variation depending on environmental conditions (Dongman *et al.*, 1974; Zundel *et al.*, 1978; Yakir *et al.*, 1990; Walker and Lance, 1991; Bariac *et al.*, 1994b; Cernusak *et al.*, 2002), and is likely to affect $\Delta^{18}O_p$. In expanding leaves, it will cause diurnal variation in the isotopic composition of the water in which new leaf material is forming. In mature leaves acting as carbohydrate sources, it will cause diurnal variation in the $\Delta^{18}O$ of the carbohydrates exported from the leaf. The $\Delta^{18}O_p$ in either case should therefore reflect a photosynthesis-weighted average of the diurnal variation in leaf water enrichment. Because both tissue synthesis and carbohydrate export can continue at night, the net effect is likely to be a reduction in $\Delta^{18}O_p$ relative to that which would be predicted if only the midday evaporative conditions were taken into account. Further research is necessary to accurately quantify the effects of diurnal variation in leaf water enrichment on $\Delta^{18}O_p$.

A Péclet effect, where the convection of unenriched water to the evaporating sites is opposed by backward diffusion of $H_2{}^{18}O$, will have important and testable effects on $\Delta^{18}O_p$. The Péclet effect (Farquhar and Lloyd, 1993) predicts a somewhat reduced response of $\Delta^{18}O_p$ to the changes in the external evaporative environment (i.e., changes in RH), but also a somewhat enhanced response to leaf-driven changes in evaporation (i.e., changes in stomatal conductance). In this model the organic molecules exchange with leaf water somewhat less enriched than predicted by Eq. 2.2, and the extent of the difference between the Craig and Gordon-predicted and Péclet-predicted leaf water enrichments increases with increasing transpiration. Under constant humidity, plants control transpiration rate by stomatal aperture. The Craig and Gordon model predicts that $\Delta^{18}O$ of leaf water (and so $\Delta^{18}O_p$) should decrease slightly with increasing stomatal conductance. Inclusion of a Péclet effect significantly enhances the dependence of $\Delta^{18}O$ of leaf water on stomatal conductance, as shown in Fig. 2.3. Strong relationships between stomatal conductance and $\Delta^{18}O_p$ have been found for cotton leaves grown in a humidity-controlled glasshouse (Barbour and Farquhar, 2000), and for field-grown wheat (Barbour *et al.*, 2000a).

Some indirect evidence in support of the relevance of a Péclet effect to $\Delta^{18}O_p$ is presented by Barbour *et al.* (2000b), who found that $\Delta^{18}O$ of sucrose was negatively related to e_a/e_i (as predicted by both the simple Craig and Gordon model and the Péclet effect extension), but that the Craig and Gordon prediction of $\Delta^{18}O$ of sucrose was significantly more sensitive to changes in e_a/e_i than was observed (as would be the case with a Péclet effect). In this analysis the Craig and Gordon predicted $\Delta^{18}O$ of sucrose was calculated from Eq. 2.2, plus the fractionation factor between carbonyl oxygen and water (taken to be 27‰).

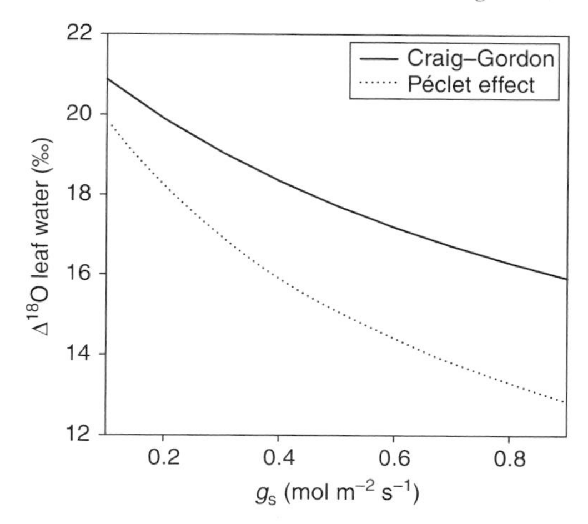

Figure 2.3 The predicted dependence of $\Delta^{18}O$ of leaf water on stomatal conductance (g_s) using the Craig–Gordon and Péclet models of leaf water enrichment.

The leaf water oxygen isotope signal is dampened in organic material formed from exported sucrose in other plant parts. This dilution is mainly a result of isotopic exchange of organic molecules with water in the sink cells forming new organic material. This 'sink cell water' may be isotopically rather different to source leaf water. For example, Adar *et al.* (1995) found that water in the new xylem cells of *Tamarix jordanis* tree stems was about −2.4‰, reflecting soil water, while leaf water of the same plant was +25.2‰. Barbour and Farquhar (2000) suggested that while tree stem xylem cell water may reflect source water $\delta^{18}O$, other tissue (particularly those closer to the source leaves) could be more like leaf water $\Delta^{18}O$, due to unloading of phloem water (at $\Delta^{18}O$ of leaf water) with phloem sugar. This idea is supported by calculations (Bret-Harte and Silk, 1994) suggesting phloem water could potentially supply 80% of the water required for cell expansion in corn root tips. Bret-Harte and Silk's (1994) calculations mean that the proportion of water in developing cells sourced from the xylem (p_x) could be as low as 0.2. Barbour and Farquhar (2000) also included in the parameter p_x the possibilities that water in phloem and xylem may exchange during sucrose transport, and that water in the developing cells may become enriched by transpiration from these cells.

Evidence of a mixture of phloem and xylem water in leaf sink cells was recently presented by Helliker and Ehleringer (2002b), who show that water in the intercalary meristems of *Lolium multiflorum* leaves (−13 to −9.6‰) was intermediate between source water (−16.3‰) and leaf water (−7.3 to 16.3‰). These authors calculated a value for p_x of 0.62 for

Lolium multiflorum, and a range in p_x of between 0.50 and 0.62 for ten grass species in a previous study (Helliker and Ehleringer, 2002a).

Further, Cernusak *et al.* (2002) show that *Lupinus angustifolius* pod and seed water ($\delta^{18}O$ = 5‰) was somewhat more enriched than stem xylem water ($\delta^{18}O$ = −3‰), but less enriched than leaf water ($\delta^{18}O$ of between 0 and 23‰). Water bled from the phloem of pods was between pod and leaf water ($\delta^{18}O$ between 7 and 15‰) and tended to follow diurnal patterns of leaf water enrichment. Recalculating data from Cernusak *et al.* (2002) gives p_x values of 0.31 for phloem water and 0.69 for pod and seed water at midday. The observation that phloem water is less enriched than leaf water supports the suggestion (Barbour and Farquhar, 2000) that some exchange between phloem and xylem water has occurred.

Isotopic Exchange Between Water and Organic Oxygen

As described above, organic molecules reflect the water in which they formed due to isotopic exchange between carbonyl oxygen and water (Sternberg *et al.*, 1986). Oxygen atoms in other functional groups, such as hydroxyl, carboxyl, and phosphate groups, are not exchangeable at normal cellular temperature and pH. The exchange of oxygen atoms between water and carbonyl groups is possible due to the formation of a short-lived gem–diol intermediate (Samuel and Silver, 1965), as shown in Fig. 2.4.

At equilibrium oxygen atoms in carbonyl groups are between 25 and 30‰ more enriched than the water in which they formed (Sternberg and DeNiro, 1983). Many intermediates in the biochemical pathways leading to synthesis of structural and non-structural carbohydrates contain carbonyl oxygen groups, so the exchange reaction becomes important in determining the $\delta^{18}O$ of plant tissue as a whole. Acetone, with a single exchangeable oxygen, was found to be 28‰ more enriched than the water with which it exchanged (Sternberg and DeNiro, 1983). If a substance contains more than one oxygen atom that has gone through a carbonyl group, an average fractionation factor (ε_{wc}) is applicable, even though slight differences in fractionation may occur for different oxygen atoms, depending on the proximity of other atoms (Schmidt *et al.*, 2001).

The rate of exchange of carbonyl oxygen varies considerably between molecules, with larger molecules being much slower to reach equilibrium

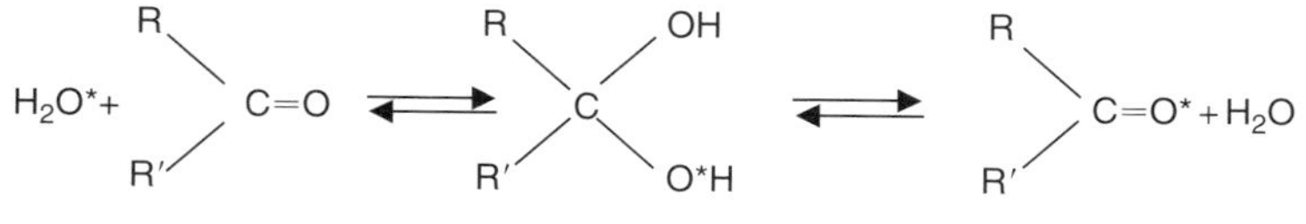

$$H_2O^* + (R)(R')C{=}O \rightleftharpoons (R)(R')C(OH)(O^*H) \rightleftharpoons (R)(R')C{=}O^* + H_2O$$

Figure 2.4 The exchange of oxygen atoms between carbonyl groups and water via a gem–diol intermediate.

than small molecules. Acetone has a half-time to equilibration with water of about 10 min, while fructose 6-phosphate takes 166 min and fructose 1,6-bisphosphate 29 min (Model *et al.*, 1968). These rates are likely to be considerably faster *in vivo*, when enzymes such as aldolase would catalyze the reactions (Model *et al.*, 1968). The most important exchange occurs in triose phosphates, as two of the three oxygen atoms are in carbonyl groups and the half-time to equilibration is known to be rapid (Sternberg *et al.*, 1986, 1989; Farquhar *et al.*, 1998). By following the $\Delta^{18}O$ of sucrose after a step change in vapor pressure deficit (VPD), Barbour *et al.* (2000b) clearly demonstrated that the $\Delta^{18}O$ of sucrose reflects the leaf evaporative environment. Considering the rapid exchange expected in triose phosphate, Barbour *et al.* (2000b) suggested that sucrose synthesized and immediately exported from a leaf should be in full isotopic equilibrium with average leaf water.

However, the possibility that sucrose may depart from full equilibrium has also been investigated. The approach to isotopic equilibrium of the molecule of interest ($\Delta^{18}O$) is given by first-order kinetics by (Barbour, 1999):

$$\Delta^{18}O = \Delta_{eq} - (\Delta_{eq} - \Delta_{ini})e^{-k\tau} \tag{2.3}$$

where Δ_{eq} is the isotopic composition at equilibrium (i.e., water plus 27‰), Δ_{ini} is the initial composition before exchange, k is the natural log of 2 divided by the half-time to equilibration of the molecule, and τ is the residence time. If intermediates within biochemical pathways have short residence times compared to the half-times to equilibration of their carbonyl oxygen, it is possible that end-products, like sucrose, may not be in full isotopic equilibrium with leaf water. Following the history of each oxygen atom in sucrose, Barbour (1999) was able to predict that if full equilibrium was not reached, changes in the ratio of carboxylation to oxygenation by Rubisco would be reflected in $\Delta^{18}O$ of sucrose. This is because during photorespiration an oxygen atom from water is added to glycine by serine hydroxymethyl transferase to make serine. This oxygen forms a phosphate-bridging oxygen in triose phosphate, so does not go through a carbonyl group in the photorespiratory cycle, meaning that a proportion of dihydroxyacetone 3-phosphate leaving the chloroplast to make sucrose may have an oxygen atom in this position that has not been enriched by exchange with chloroplastic water. That is, a proportion of atoms in this position will be about 27‰ less enriched than other, exchangeable oxygen. If full equilibration of all oxygen atoms in sucrose does not occur during sucrose synthesis, then changes in the CO_2 and O_2 concentrations will be reflected in sucrose $\Delta^{18}O$ (via changes in the ratio of carboxylation to oxygenation).

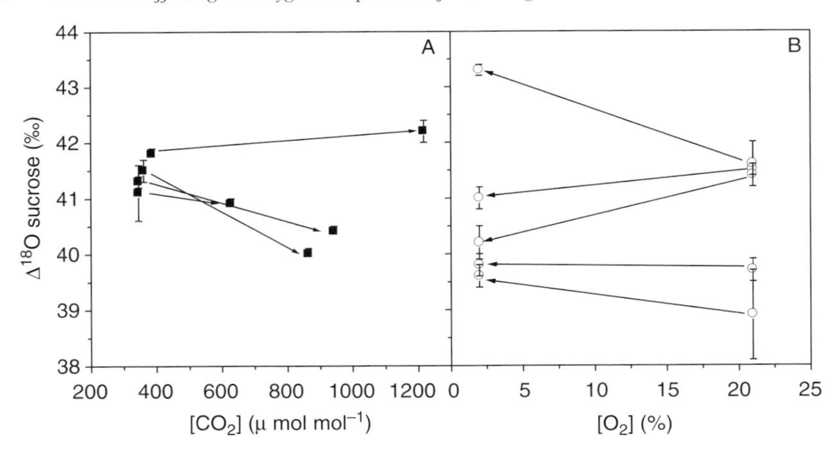

Figure 2.5 The effect of a step change in CO_2 (A) and O_2 (B) concentration on the $\Delta^{18}O$ of sucrose from phloem sap bled from castor bean leaves in a controlled-environment cuvette. Lines joining points indicate the change for a single leaf as a result of the step change. From Barbour (1999).

Barbour (1999) tested this prediction by sampling phloem sap sucrose from castor bean before and after a step change in either CO_2 or O_2 concentration. No consistent response of $\Delta^{18}O$ of sucrose to a change in either was found (Fig. 2.5). The best predictions of $\Delta^{18}O$ of sucrose were found when full equilibration of all carbonyl oxygen was allowed, leading the author to conclude that, within measurement error, sucrose is fully equilibrated with leaf water.

However, cellulose is known to be some way from full isotopic equilibrium with sink cell water, because cellulose from tree rings reflects leaf water $\Delta^{18}O$ (to some extent) rather than source water (Roden *et al.*, 2000). To form cellulose, or other carbohydrate or secondary metabolite, sucrose exported from the source leaf must be broken down into hexose phosphates. This means that some oxygen atoms must go through carbonyl groups, and therefore become exchangeable with local water. Of the five oxygen atoms in the repeating unit of cellulose, only one goes through a carbonyl group as sucrose is broken down into hexose phosphates for cellulose synthesis. Further, Hill *et al.* (1995) showed that a proportion (y) of hexose phosphates also go through a futile cycle to triose phosphates. Such a cycle would allow a further three out of five oxygen atoms in the cellulose unit to exchange with local water. When hexose phosphates re-form from triose phosphates there is an equal chance that the previously non-exchangeable (i.e., phosphate-bridging) oxygen atom will be in the exchangeable position in the new hexose phosphate. This means that with every turn of the futile cycle the probability of an oxygen atom going through an exchangeable carbonyl group increases. This process has been modeled by Farquhar *et al.* (1998) and Barbour and Farquhar (2000):

$$p_{ex} = 0.2 + \left(0.6 + \frac{0.2}{2-y}\right)y \tag{2.4}$$

where p_{ex} is the proportion of exchangeable oxygen in cellulose formed from sucrose, and $(1 - y)$ is the proportion of hexose phosphate molecules used immediately (i.e., no futile cycling through triose phosphates). Recalculation of published data shows p_{ex} to vary between 0.49 (recalculated from data on *Lemna gibba*; Yakir and DeNiro, 1990) and 0.57 (recalculated from data on carrot; Sternberg *et al.*, 1986).

Δ^{18}O of Plant Material Other Than Sucrose and Cellulose

The oxygen isotopic composition of molecules other than sucrose and cellulose has not often been measured (but see Schmidt *et al.*, 2001). However, it is known that whole leaf tissue is significantly less enriched than cellulose (Barbour and Farquhar, 2000; Barbour *et al.*, 2000a). Of the secondary metabolites, the δ^{18}O of lignin has been studied in more detail.

Lignin forms the second most abundant component of many plant tissues, notably wood, and is known to be isotopically rather different to cellulose (Gray and Thompson, 1977). Lignin is formed from three precursors (monolignols), which differ in the degree of methoxylation of the aromatic ring. The oxygen atoms in the methoxyl groups are added by cytochrome P-450-linked monooxygenases, which cleave molecular oxygen and add one oxygen atom to the aromatic ring, the other being reduced to water. At least initially, these oxygen atoms should retain the isotopic composition of molecular oxygen, minus any fractionation that occurs during the reaction (Barbour *et al.*, 2001). Some exchange of these oxygen atoms with sink cell water during subsequent biochemical steps seems likely (Schmidt *et al.*, 2001), and indeed Barbour *et al.* (2001) found that lignin δ^{18}O was positively correlated with modeled mean annual δ^{18}O of rain for wood samples from *Pinus* species from around the world.

Chloroplastic water, enriched by evaporation, provides the substrate for photosynthetically produced O_2 (Guy *et al.*, 1987). On regional and global scales this enriched O_2 represents the terrestrial contribution to the 23.5‰ enrichment of atmospheric molecular oxygen above mean ocean water (i.e., the Dole effect; Bender *et al.*, 1985). While δ^{18}O of O_2 in the atmosphere is fairly constant around the world, $\delta^{18}O_R$ varies so that Δ^{18}O of O_2 will be variable.

When source-water-related variation in δ^{18}O of lignin was removed by considering Δ^{18}O, 37% of variation in Δ^{18}O of lignin was explained by variation in $\Delta^{18}O$ of molecular oxygen in the atmosphere. Using a fitting procedure to a model of the isotopic history of oxygen in lignin, Barbour *et al.* (2001) suggested that about 60% of the oxygen in lignin from monooxygenase reactions (and hence from O_2) had exchanged with local water during subsequent reactions, most likely during monolignol polymerization.

Modeling $\Delta^{18}O$ of Plant Material

Saurer *et al.* (1997) modified Eq. 2.2 to allow interpretation of observed variation in $\delta^{18}O$ of cellulose ($\delta^{18}O_c$) from three tree species. This expression incorporated a 'dampening factor' (f) to summarize the effects of deviation in leaf water enrichment from Eq. 2.2, and the exchange of oxygen atoms with local cellular water (i.e., full dampening when $f = 0$). The expression was (Saurer *et al.*, 1997):

$$\delta^{18}O_c = \delta^{18}O_s + f \cdot [\varepsilon^* + \varepsilon_k + (\delta^{18}O_v - \delta^{18}O_s - \varepsilon_k)e_a/e_i] + \varepsilon_{wc} \quad (2.5)$$

where $\delta^{18}O_s$ and $\delta^{18}O_v$ are the isotopic composition of source water and atmospheric water vapor, respectively, and ε_{wc} is the fractionation factor between carbonyl oxygen and water (27‰). Roden and Ehleringer (1999; and see also Roden and Ehleringer 2000; and Roden *et al.*, 2000) suggested that $\delta^{18}O_c$ of wood should be a function of the isotopic composition of sucrose imported from the leaf, and of local water in the developing cell, such that:

$$\delta^{18}O_c = p_{ex} \cdot (\delta^{18}O_{wm} + \varepsilon_{wc}) + (1 - p_{ex}) \cdot (\delta^{18}O_{wl} + \varepsilon_{wc}) \quad (2.6)$$

where p_{ex} is the proportion of oxygen that exchanges with local water, $\delta^{18}O_{wm}$ is the isotopic composition of sink cell water, and $\delta^{18}O_{wl}$ is the isotopic composition of leaf water. Equation 2.6 can be simplified by expressing compositions in terms of enrichments above source water, by assuming that sucrose exported from the source leaf is in full isotopic equilibrium with average leaf water enrichment ($\Delta^{18}O_L$, as demonstrated by Barbour, 1999), and by including the term p_x, the proportion of xylem-sourced water in the developing cell. Barbour and Farquhar (2000) presented the simplified form:

$$\Delta^{18}O_c = \Delta^{18}O_L(1 - p_{ex} \cdot p_x) + \varepsilon_{wc} \quad (2.7)$$

where $\Delta^{18}O_c$ is the enrichment of ^{18}O above source water. The oxygen isotope composition of whole leaf tissue ($\Delta^{18}O_l$) has been found to be significantly less enriched than its cellulose, by 7.5‰ for cotton leaves (Barbour and Farquhar, 2000), and by 9.1‰ in wheat leaves (Barbour *et al.*, 2000a).

Equation 2.7 may be rewritten to include the term ε_{cp}, the difference in enrichment between cellulose and whole leaf tissue by:

$$\Delta^{18}O_l = \Delta^{18}O_L(1 - p_{ex} \cdot p_x) + \varepsilon_{wc} + \varepsilon_{cp} \tag{2.8}$$

Note that ε_{cp} is equal to $\Delta^{18}O_l - \Delta^{18}O_c$, so that ε_{cp} is negative. However, a significant weakness in Eq. 2.8 is that ε_{cp} is likely to be variable over time. Cernusak *et al.* (2002) have shown that $\Delta^{18}O_l$ of *Lupinus angustifolius* leaves varied considerably over a diurnal period, suggesting that diurnal variation in $\Delta^{18}O$ of non-structural carbohydrates contributed significantly to the whole leaf $\Delta^{18}O$. Substantial temporal variation in $\Delta^{18}O$ of non-structural carbohydrates is expected, based on the measured variation in $\Delta^{18}O$ of phloem sap sucrose (Barbour *et al.*, 2000b; Cernusak *et al.*, 2002).

Whole wood $\Delta^{18}O$ should be much less variable in time than leaf tissue, due to the lower concentrations of non-structural carbohydrates, and the slower rate of metabolic activity in these cells. However, because lignin can form a large portion of whole wood by weight (about 40%; Barbour *et al.*, 2001), and its isotopic history is rather different (Schmidt *et al.*, 2001), the contribution of lignin to whole wood $\Delta^{18}O$ should be considered separately. Enrichment of lignin above source water ($\Delta^{18}O_{lig}$) may be modeled in a similar form to Eq. 2.7 by including molecular oxygen as a source (Barbour *et al.*, 2001):

$$\Delta^{18}O_{lig} = (1 - p_{O_2}) \times [\Delta^{18}O_L(1 - p_{ex}^{lL} \cdot p_x) + \varepsilon_{wc}] + p_{O_2} \times [(1 - p_{ex}^{lO}) \times (\Delta^{18}O_{O_2} - \varepsilon_{O_2}) + p_{ex}^{lO} \times (\Delta^{18}O_L(1 - p_x) + \varepsilon_{wl})] \tag{2.9}$$

where p_{O_2} is the proportion of oxygen atoms in lignin from molecular oxygen (at least initially), p_{ex}^{lL} is the proportion of oxygen atoms in lignin from leaf water but exchanged with sink cell water during lignin synthesis, p_{ex}^{lO} is the proportion of oxygen atoms in lignin from molecular oxygen but later exchanged with sink cell water, $\Delta^{18}O_{O_2}$ is the enrichment of molecular oxygen over source water, ε_{O_2} is the fractionation associated with monooxygenase reactions, and ε_{wl} is the fractionation associated with exchange of oxygen with water in polymerization intermediates (i.e., may differ from ε_{wc}). Parameters in Eq. 2.9 remain loosely constrained by theoretical limits of biochemical reactions at present.

Testing the Models

When tested, both the Roden and Ehleringer and the Barbour and Farquhar versions of the model for the oxygen isotope composition of cellulose perform well. Roden and Ehleringer found good agreement between measured and modeled $\delta^{18}O_c$ in both glasshouse-grown (Roden and Ehleringer, 1999) and field-grown (Roden and Ehleringer, 2000) riparian trees. In both cases p_{ex} was fitted from the data to be 0.42. Variation in $\delta^{18}O_c$ was driven by variation in $\delta^{18}O_s$ and humidity for the glasshouse-grown plants, and by differing $\delta^{18}O_s$, $\delta^{18}O_v$, temperature, and humidity at five sites for the field-grown trees. More recently, recalculation of these data to remove source water $\delta^{18}O$ variation (i.e., data are presented as $\Delta^{18}O$) revealed evidence of the influence of a Péclet effect on both leaf water and cellulose isotope ratios (Barbour *et al.*, 2004).

$\Delta^{18}O$ of cellulose and leaf tissue modeled by Eqs 2.7 and 2.8 were shown to predict measured variation, induced by abscisic acid application and by changes in humidity, for cotton plants grown in a glasshouse (Barbour and Farquhar, 2000). In this experiment, 95% of variation in $\Delta^{18}O_l$ was explained by Eq. 2.8, with $p_{ex}p_x$ fitted to be 0.38 and ε_{cp} to be −7.5‰.

The predictive power of Eq. 2.7 was also tested by Helliker and Ehleringer (2002a,b). $\delta^{18}O$ of cellulose from ten C_3 and C_4 grass species was well predicted by Eq. 2.7, when the model was rewritten in terms of isotope composition relative to VSMOW. As source water $\delta^{18}O$ was constant for this experiment, all variation in $\delta^{18}O_c$ was due to species variation in leaf properties (as they affect leaf water enrichment) and differences in growth humidity. Helliker and Ehleringer (2002a) found that a value for $p_{ex}p_x$ of 0.25 fit the data well. Back-calculation of p_x from $\delta^{18}O_{wm}$ measurements (Helliker and Ehleringer, 2002b) showed a range for these plants of between 0.50 and 0.62, meaning that p_{ex} ranged from 0.40 to 0.50.

Combined Measurement of Carbon and Oxygen Isotope Ratios

$\Delta^{13}C$ of plant tissue is known to be a photosynthesis-weighted integrator of carbon supply and demand (Farquhar *et al.*, 1982). Variation in $\Delta^{13}C$ may be driven by changes in g_s, or changes in photosynthetic capacity (V_1), or changes in both, because $\Delta^{13}C$ has been shown to be positively related to the ratio of intercellular to atmospheric CO_2 concentration (c_i/c_a) (Farquhar *et al.*, 1982) by:

$$\Delta^{13}C = a + (b - a)c_i/c_a \tag{2.10}$$

where a is the ^{13}C fractionation associated with diffusion through stomata and the boundary layer (4.4‰), and b is the effective biochemical fractionation by Rubisco during carbon fixation when c_i is used, rather than the CO_2 concentration at the site of carboxylation (about 27‰).

When variation in $\Delta^{13}C$ is driven by changes in g_s alone, a negative relationship between $\Delta^{13}C$ and $\Delta^{18}O$ is predicted (equivalent to a positive correlation between $\delta^{13}C$ and $\delta^{18}O$). If variation in $\Delta^{13}C$ is driven by changes in V_1 alone no relationship between $\Delta^{13}C$ and $\Delta^{18}O$ is expected, because

$\Delta^{18}O$ is not affected by V_1. If variation in $\Delta^{13}C$ is driven by increases in both g_s and V_1, then the change in $\Delta^{18}O$ per unit change in $\Delta^{13}C$ will be greater than if g_s alone had increased (Barbour *et al.*, 2002). As such, measurement of both $\Delta^{13}C$ and $\Delta^{18}O$ will allow the g_s and V_1 effects on $\Delta^{13}C$ to be teased apart (Farquhar *et al.*, 1994; Yakir and Israeli, 1995; Scheidegger *et al.*, 2000).

A review of published relationships reveals that a positive relationship between $\delta^{13}C$ and $\delta^{18}O$ (or the equivalent negative relationship between $\Delta^{13}C$ and $\Delta^{18}O$) has been found for a number of different experimental systems, including field-grown leaves (Sternberg *et al.*, 1989; Barbour *et al.*, 2000a), cellulose from field-grown trees (Saurer *et al.*, 1997; Barbour *et al.*, 2002) and cotton leaf tissue and its cellulose from plants grown in humidity-controlled glasshouses (Barbour and Farquhar, 2000). The $\Delta^{18}O{:}\Delta^{13}C$ relationship for an experimental system in which variation in $\Delta^{13}C$ is driven by V_1 alone has not been published to date. Of particular interest, Barbour *et al.* (2002) interpreted variation in the slope of the tree ring cellulose $\delta^{18}O : \delta^{13}C$ relationship between field sites as a function of average humidity at each site. As predicted by theory, a greater change in $\delta^{18}O_c$ per unit change in $\delta^{13}C$ of cellulose ($\delta^{13}C_c$) was found when humidity was lower.

Potential Applications

Breeding

As plant material has been shown to record leaf evaporative conditions, in terms of both the external evaporative environment (RH) and stomatal regulation of water loss (g_s), measurement of $\Delta^{18}O_p$ may provide a powerful tool for plant breeders. While knowledge of variation in g_s may be of interest to breeders in its own right, the link between $\Delta^{18}O_p$ and crop yield is likely to generate rather more excitement. Two important crop plants, cotton and wheat, have been shown to display strong correlations between g_s and yield when grown in non-limiting environments (e.g., Lu *et al.*, 1994; Sayre *et al.*, 1997). Stomatal conductance itself is difficult and time-consuming to measure in the field, so a simple, one-off, integrative measurement of g_s throughout the growth of a plant would be of considerable benefit. Barbour *et al.* (2000a) have shown that $\delta^{18}O$ of both whole leaf tissue and cellulose are strongly negatively related to seasonal mean g_s and to grain yield for field-grown wheat. The plants investigated were an historical series of wheat cultivars released between 1962 and 1988, and showed a yield increase of $0.88\%\ yr^{-1}$, 88% of which was explained by variation in g_s (Sayre *et al.*, 1997).

Measurement of $\Delta^{18}O_p$ may also help plant breeders interpret variation in carbon isotope discrimination. As described earlier, variation in $\Delta^{13}C$ may be driven by either changes in stomatal conductance or in photosynthetic capacity, or changes in both, while $\Delta^{18}O$ should not be affected by photosynthetic capacity. In the field-grown wheat experiment described earlier, most of the variation in $\Delta^{13}C$ was driven by changes in g_s, so a strong, positive correlation between $\delta^{18}O$ and $\delta^{13}C$ (or a negative relationship between $\Delta^{18}O$ and $\Delta^{13}C$) was expected and observed (Barbour *et al.*, 2000a).

However, before encouraging plant breeders to include measurement of $\Delta^{18}O_p$ in their existing $\Delta^{13}C$ breeding programs, or to begin new $\Delta^{18}O$ programs, we have three warnings. First, we anticipate that not all species display a strong correlation between yield and g_s, and even in species that do, in situations where the correlation does not hold (such as during significant soil water deficit), the correlation between $\Delta^{18}O_p$ and yield will break down. Secondly, we caution that further research into the extent to which leaf water ^{18}O enrichment is recorded in $\Delta^{18}O_p$ (i.e., better estimates of $p_{ex}p_x$) in a range of species and environments is required. And finally, a number of accompanying samples (including measurement of source water and water vapor $\delta^{18}O$) must be taken from the environment of interest to allow full interpretation of variation in $\Delta^{18}O_p$.

Ecophysiology

An integrative measurement of stomatal regulation of water loss will also be of relevance in studies concerning plant response to variation in the growth environment. Again, a technique that overcomes the difficulties of field measurement of g_s using gas exchange techniques would be valuable. Many studies of plant and ecosystem function use measurements of shoot g_s as a basic tool in understanding carbon and water fluxes. An integrative measure of g_s, requiring little (and unskilled) labor in the field, would be of great merit. However, the same cautions as outlined above for breeding applications apply to ecophysiological applications.

A further application of interest in ecophysiological studies is using $\Delta^{18}O_p$ to interpret studies of ecosystem fluxes and isotope ratios of CO_2 and water. The stable oxygen isotope ratio of ecosystem CO_2 and H_2O may allow partitioning of the fluxes into the components of photosynthesis and respiration (for CO_2) or transpiration and soil evaporation (for H_2O). However, to interpret variation in $C^{18}O^{16}O$ and $H^{18}O^{16}O$, the leaf water $\Delta^{18}O$ must be known. $\Delta^{18}O_p$ may provide an integrative measure of $\Delta^{18}O_L$, and would therefore dispense with time-consuming diurnal sampling of leaf water for isotope analysis.

Paleoclimate

Interest in $\Delta^{18}O_p$ was first roused, in the late 1940s, because tree ring cellulose $\Delta^{18}O$ was suggested to be an 'isotope thermometer'. Recent work (e.g., Roden and Ehleringer, 1999; Barbour and Farquhar, 2000; Helliker and Ehleringer, 2002b) has provided firm evidence that the oxygen isotope composition of plant tissue contains a record of the leaf evaporative environment as well as the $\delta^{18}O$ of source water. This means that temperature reconstruction from $\delta^{18}O$ of tree rings is not simple; the leaf evaporative environment (VPD) and plant regulation of water loss (g_s) must also be taken into account. Recognition of these important effects on $\delta^{18}O_p$ opens the door to a rather more interesting reconstruction of past climates, which includes relative humidity and water availability (via g_s), as well as temperature, as part of that climate.

Recent work in which the wide annual rings of *Pinus radiata* grown in New Zealand were divided into very small sections, representing between 3 and 30 days of growth, has shown that $\delta^{18}O_p$ may be interpreted in terms of climatic conditions (Barbour *et al.*, 2002). In this experiment specific climatic events, such as drought and high rainfall, were recorded as peaks and troughs in tree ring cellulose $\delta^{18}O$. As suggested by Barbour *et al.* (2002), multiple-isotope analysis ($\delta^{18}O$, $\delta^{13}C$, and δD) of tree ring cellulose may allow a more complete picture of past climates, and plant response to environmental variation, to be drawn.

Conclusions

Interpretation of variation in the oxygen isotope composition of plant organic material ($\Delta^{18}O_p$) has a number of exciting potential applications, from paleoclimatic reconstruction, through understanding plant and ecosystem carbon and water fluxes, to breeding for higher-yielding crop plants. Current understanding is summarized in models that predict variation in $\Delta^{18}O_p$ as a result of variation in the leaf evaporative environment, and in isotopic exchange between plant water and organic molecules. These models predict measured variation in $\Delta^{18}O$ of cellulose quite accurately, and should allow novel isotope techniques to be used to address a wide range of questions involving plant regulation of water loss.

References

Adar E. M., Gev I., Lipp J., Yakir D. and Gat J. R. (1995) Utilization of oxygen-18 and deuterium in stem flow for the identification of transpiration source: soil water versus ground water in sand dune terrain. In *Application of Tracers in Arid Zone Hydrology* (E. M. Adar and C. Leibundgut, eds) Vol 232, pp 329–338, Association of Scientific Hydrology Publications.

Barbour M. M. (1999) A physiological study of organic oxygen isotope composition. PhD thesis, Australian National University, Canberra.

Barbour M. M. and Farquhar G. D. (2000) Relative humidity- and ABA-induced variation in carbon and oxygen isotope ratios of cotton leaves. *Plant Cell Environ* **23**: 473–485.

Barbour M. M., Fischer R. A., Sayre K. D. and Farquhar G. D. (2000a) Oxygen isotope ratio of leaf and grain material correlates with stomatal conductance and grain yield in irrigated wheat. *Aust J Plant Physiol* **27**: 625–637.

Barbour M. M., Andrews A. J. and Farquhar G. D. (2001) Correlations between oxygen isotope ratios of wood constituents of *Quercus* and *Pinus* samples from around the world. *Aust J Plant Physiol* **28**: 335–348.

Barbour M. M., Schurr U., Henry B. K., Wong S. C. and Farquhar G. D. (2000b) Variation in the oxygen isotope ratio of phloem sap sucrose from castor bean: Evidence in support of the Péclet effect. *Plant Physiol* **123**: 671–679.

Barbour M. M., Walcroft A. S. and Farquhar G. D. (2002) Seasonal variation in $\delta^{13}C$ and $\delta^{18}O$ of cellulose from growth rings of *Pinus radiata*. *Plant Cell Environ* **25**: 1483–1499.

Barbour M. M., Roden J. S., Farquhar G. D. and Ehleringer J. R. (2004) Expressing leaf water and cellulose oxygen isotope ratios as enrichment above source water reveals evidence of a Péclet effect. *Oecologia* **138**: 426–435.

Bariac T., Gonzalez-Dunia J., Tardieu F., Tessier D. and Mariotti A. (1994a) Spatial variation of the isotopic composition of water (^{18}O, ^{2}H) in organs of aerophytic plants: 1. Assessment under laboratory conditions. *Chem Geol* **115**: 307–315.

Bariac T., Gonzalez-Dunia J., Tardieu F., Tessier D. and Mariotti A. (1994b) Spatial variation of the isotopic composition of water (^{18}O, ^{2}H) in the soil–plant–atmosphere system: 2. Assessment under field conditions. *Chem Geol* **115**: 317–333.

Bender M. L., Labeyrie L., Raynaud D. and Loris C. (1985) Isotopic composition of atmospheric O_2 in ice linked to deglaciation and global primary productivity. *Nature* **318**: 349–352.

Bret-Harte M. S. and Silk W. K. (1994) Nonvascular, symplastic diffusion of sucrose cannot satisfy the carbon demands of growth in the primary root tip of *Zea mays* L. *Plant Physiol* **105**: 19–33.

Cernusak L. A., Pate J. S. and Farquhar G. D. (2002) Diurnal variation in the stable isotope composition of water and dry matter in fruiting *Lupinus angustifolius* under field conditions. *Plant Cell Environ* **25**: 893–907.

Craig H. (1961) Isotopic variations in meteoric waters. *Science* **133**: 1702–1703.

Craig H. and Gordon L. I. (1965) Deuterium and oxygen-18 variations in the ocean and the marine atmosphere. In *Proceedings of a Conference on Stable Isotopes in Oceanographic Studies and Paleotemperatures* (E. Tongiorgi, ed.) pp. 9–130. Laboratory of Geology and Nuclear Science, Pisa.

Dawson T. E., Pausch R. C. and Parker H. M. (1998) The role of hydrogen and oxygen stable isotopes in understanding water movement along the soil–plant–atmospheric continuum. In *Stable Isotopes: Integration of Biological, Ecological and Geochemical Processes* (H. Griffiths, ed.) pp. 169–183. BIOS Scientific Publishers, Oxford.

DeNiro M. J. and Epstein S. (1979) Relationship between oxygen isotope ratios of terrestrial plant cellulose, carbon dioxide and water. *Science* **204**: 51–53.

DeNiro M. J. and Cooper L. W. (1989) Post-photosynthetic modification of oxygen isotope ratios of carbohydrates in the potato: Implications for paleoclimatic reconstruction based upon isotopic analysis of wood cellulose. *Geochim Cosmochim Acta* **53**: 2573–2580.

Dongmann G., Nurnberg H. E., Forstel H. and Wagener K. (1974) On the enrichment of $H_2{}^{18}O$ in the leaves of transpiring plants. *Radiat Environ Biophys* **11**: 41–52.

Edwards T. W. D. and Fritz P. (1986) Assessing meteoric water composition and relative humidity from ^{18}O and ^{2}H in wood cellulose: Paleoclimatic implications for southern Ontario. *Can J Earth Sci* **22**: 1720–1726.

Epstein S., Thompson P. and Yapp C. J. (1977) Oxygen and hydrogen isotopic ratios in plant cellulose. *Science* **198**: 1209.

Farquhar G. D., O'Leary M. H. and Berry J. A. (1982) On the relationship between carbon isotope discrimination and the intercellular carbon dioxide concentration in leaves. *Aust J Plant Physiol* **9**: 121–137.

Farquhar G. D. and Lloyd J. (1993) Carbon and oxygen isotope effects in the exchange of carbon dioxide between terrestrial plants and the atmosphere. In *Stable Isotopes and Plant Carbon–Water Relations* (J. R. Ehleringer, A. E. Hall and G. D. Farquhar, eds) pp. 47–70. Academic Press, San Diego.

Farquhar G. D., Condon A. G. and Masle J. (1994) On the use of carbon and oxygen isotope composition and mineral ash content in breeding for improved rice production under favorable, irrigated conditions. In *Breaking the Yield Barrier* (K. G. Cassman, ed.) pp. 95–101. International Rice Research Institute, Manila.

Farquhar G. D., Barbour M. M. and Henry B. K. (1998) Interpretation of oxygen isotope composition of leaf material. In *Stable Isotopes: Integration of Biological, Ecological and Geochemical Processes* (H. Griffiths, ed.) pp. 27–61. BIOS Scientific Publishers, Oxford.

Flanagan L. B., Phillips S. L., Ehleringer J. R., Lloyd J. and Farquhar G. D. (1994) Effects of changes in leaf water oxygen isotopic composition on discriminations against $C^{18}O^{16}O$ during photosynthesis. *Aust J Plant Physiol* **21**: 221–234.

Gonfiantini R. (1984) Advisory Group Meeting on Stable Isotope Reference Samples for Geochemical and Hydrological Investigations. Isotope Atomic Energy Commission, Vienna.

Gonfiantini R., Gratziu S. and Tongiorgi E. (1965) Oxygen isotopic composition of water in leaves. In *Isotopes and Radiation in Soil Plant Nutrition Studies*, Tech. Rep. Ser. No 206, pp. 405–410. Isotope Atomic Energy Commission, Vienna.

Gray J. and Thompson P. (1977) Climatic information from $^{18}O/^{16}O$ analysis of cellulose, lignin and whole wood from tree rings. *Nature* **270**: 708–709.

Guy R. D., Fogel M. F., Berry J. A. and Hoering T. C. (1987) Isotope fractionation during oxygen production and consumption by plants. In *Progress in Photosynthetic Research III* (J. Biggins, ed.) pp. 597–600. Kluwer, Dordrecht.

Helliker B. R. and Ehleringer J. R. (2000) Establishing a grassland signature in veins: ^{18}O in the leaf water of C3 and C4 grasses. *Proc. Natl Acad Sci USA* **97**: 7894–7898.

Helliker B. R. and Ehleringer J. R. (2002a) Differential ^{18}O enrichment of leaf cellulose in C_3 *versus* C_4 grasses. *Funct Plant Biol* **29**: 435–442.

Helliker B. R. and Ehleringer J. R. (2002b) Grass blades as tree-rings: environmentally induced changes in the oxygen isotope ratio of cellulose along the length of grass blades. *New Phytologist* **155**: 417–424.

Hill S. A., Waterhouse J. S., Field E. M., Switsur V. R. and apRees T. (1995) Rapid recycling of triose phosphates in oak stem tissue. *Plant Cell Environ* **18**: 931–936.

International Atomic Energy Agency (1992) *Statistical Treatment of Data on Environmental Isotopes in Precipitation.* IAEA, Austria.

Libby L. M., Pandolfi L. J., Payton P. H., Marshall J. III, Becker B. and Giertz-Sienbenlist V. (1976) Isotopic tree thermometers. *Nature* **261**: 284–288.

Lu Z. M., Radin J. W., Turcotte E. L., Percy R. and Zeiger E. (1994) High yields in advanced lines of Pima cotton are associated with higher stomatal conductance, reduced leaf area and lower leaf temperature. *Physiologia Plantarum* **92**: 266–272.

Model P., Ponticorvo L. and Rittenberg D. (1968) Catalysis of an oxygen-exchange reaction of fructose-1,6-diphosphate and fructose-1-phosphate with water by rabbit muscle aldolase. *Biochemistry* **7**: 1339–1347.

Roden J. S. and Ehleringer J. R. (1999) Hydrogen and oxygen isotope ratios of tree-ring cellulose for riparian trees grown long-term under hydroponically controlled environments. *Oecologia* **121**: 467–477.

Roden J. S. and Ehleringer J. R. (2000) Hydrogen and oxygen isotope ratios of tree ring cellulose for field-grown riparian trees. *Oecologia* **123**: 481–489.

Roden J. S., Lin G. and Ehleringer J. R. (2000) A mechanistic model for interpretation of hydrogen and oxygen isotope ratios in tree-ring cellulose. *Geochim Cosmochim Acta* **64**: 21–35.

Samuel D. and Silver B. L. (1965) Oxygen isotope exchange reactions of organic compounds. *Adv Phys Org Chem* **3**: 1885–1895.

Saurer M., Aellen K. and Siegwolf R. (1997) Correlating $\delta^{13}C$ and $\delta^{18}O$ in cellulose of trees. *Plant Cell Environ* **20**: 1543–1550.

Sayre K. D., Rajaram S. and Fischer R. A. (1997) Yield potential progress in short bread wheats in Northwest Mexico. *Crop Sci* **37**: 36–42.

Scheidegger Y., Saurer M., Bahn M. and Siegwolf R. (2000) Linking stable oxygen and carbon isotopes with stomatal conductance and photosynthetic capacity: A conceptual model. *Oecologia* **125**: 350–357.

Schmidt H.-L., Werner R. A. and Roßmann A. (2001) ^{18}O pattern and biosynthesis of natural plant products. *Phytochemistry* **58**: 9–32.

Sternberg L. and DeNiro M. (1983) Biogeochemical implications of the isotopic equilibrium fractionation factor between oxygen atoms of acetone and water. *Geochim Cosmochim Acta* **47**: 2271–2274.

Sternberg L., DeNiro M. and Savidge R. (1986) Oxygen isotope exchange between metabolites and water during biochemical reactions leading to cellulose synthesis. *Plant Physiol* **82**: 423–427.

Sternberg L., Mulkey S. S. and Wright S. J. (1989) Oxygen isotope ratio stratification in a tropical moist forest. *Oecologia* **81**: 51–56.

Walker C. D. and Lance R. C. (1991) The fractionation of ^{2}H and ^{18}O in leaf water of barley. *Aust J Plant Physiol* **18**: 411–425.

Wang X.-F. and Yakir D. (1995) Temporal and spatial variation in the oxygen-18 content of leaf water in different plant species. *Plant Cell Environ* **18**: 1377–1385.

Wang X.-F., Yakir D. and Avishai M. (1998) Non-climatic variations in the oxygen isotopic compositions of plants. *Global Change Biol* **4**: 835–849.

Yakir D., DeNiro M. J. and Rundel P. W. (1989) Isotopic inhomogeneity of leaf water: Evidence and implications for the use of isotopic signals transduced by plants. *Geochim Cosmochim Acta* **53**: 2760–2773.

Yakir D. and DeNiro M. J. (1990) Oxygen and hydrogen isotope fractionation during cellulose metabolism in *Lemna gibba* L. *Plant Physiol* **93**: 325–332.

Yakir D., DeNiro M. and Gat J. (1990) Natural deuterium and oxygen-18 enrichment in leaf water of cotton plants grown under wet and dry conditions: Evidence for water compartmentation and its dynamics. *Plant Cell Environ* **13**: 49–56.

Yakir D. and Israeli Y. (1995) Reduced solar irradiance effects on net primary productivity (NPP) and the $\delta^{13}C$ and $\delta^{18}O$ values in plantations of *Musa* sp. *Musaceae. Geochim Cosmochim Acta* **59**: 2149–2151.

Zundel G., Miekeley W., Grisi B. M. and Förstel H. (1978) The $H_2{}^{18}O$ enrichment in the leaf water of tropic trees: Comparison of species from the tropical rain forest and the semi-arid region of Brazil. *Radiat Environ Biophys* **15**: 203–212.

3

Stable Isotope Composition of Soil Organic Matter

Gerd Gleixner

Introduction

One promising alternative to help mitigate the impact of global change on world ecosystems is to promote the increased storage of atmospheric carbon dioxide in components of terrestrial ecosystems (Houghton *et al.*, 2001; McCarthy *et al.*, 2001). Photosynthesis continuously extracts CO_2 from the atmosphere and forms plant biomass, which is again mineralized by microorganisms to re-form atmospheric CO_2. Some of the plant carbon formed accumulates as plant biomass in terrestrial ecosystems, and some is transformed into microbial biomass or new molecules are synthesized from it to form soil organic matter (SOM). Excluding the ocean components, the atmosphere, aboveground biomass, and soil organic matter form three of the major pools of the global carbon cycle (Fig. 3.1). Each of these pools differs in the amount of carbon stored and its stability or lifetime. For example, the aboveground biomass and the atmospheric carbon pools store 720 GtC and 620 GtC, respectively (1 giga ton of carbon $= 10^{15}$ g of carbon). Together these two pools contain less carbon than soil organic matter, which holds 1580 GtC. Soil organic matter (SOM) is known to have high ^{14}C ages, with some carbon dating back to the last glaciation 14 000 years ago (Wang *et al.*, 1996). Both the amount and age of the carbon in the SOM pool supports the suggestion that this pool could be a target for increasing carbon sequestration from the atmosphere. Unfortunately, comparatively little is known about the mechanisms and the dynamics of carbon storage in soils (Schimel *et al.*, 2001). Even less is known about how soil carbon storage might be influenced by autotrophic and heterotrophic organisms (Catovsky *et al.*, 2002). Most of our understanding of soil carbon derives from simple input–output models that consider soil carbon only at an aggregated level of bulk carbon. Most soil carbon models developed on this aggregated carbon level make use of three pools with different longevities or time scales—yearly, decadal, and

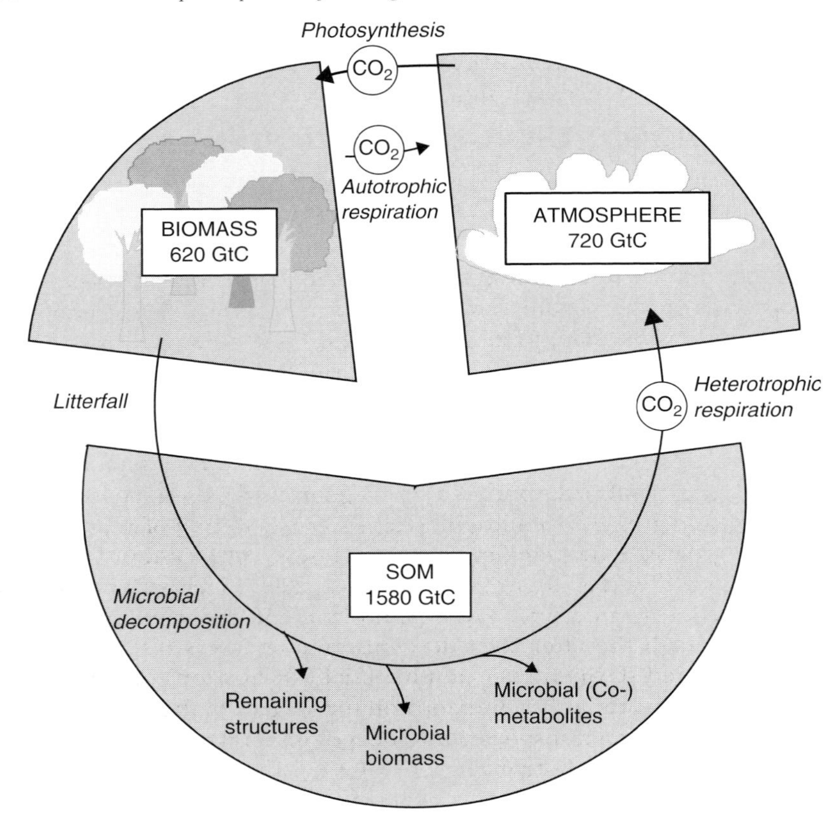

Figure 3.1 Global carbon pool sizes and carbon flow in the terrestrial carbon cycle.

millennial—to describe the dynamics of soil carbon (Jenkinson *et al.*, 1987; Parton *et al.*, 1987). However, these models neglect important knowledge derived from molecular studies of soil organic matter, such as the existence of different chemical forms of carbon, i.e., carbohydrates or lignin, having different stability against decomposition (Gleixner *et al.*, 2001a). The models also ignore the existence of chemical gradients in soil profiles (Hedges and Oades, 1997) and the role of dissolved carbon for the distribution of carbon within soil depth profiles (Neff and Asner, 2001). Moreover, the fundamental importance of soil macro- and microorganisms for the recycling of carbon in soils has not even been considered (Scheu, 2001). This chapter will summarize our current understanding of the dynamics of soil carbon, including a focus on the new insights gained from compound specific investigations. Studies of stable plant biomarkers will be described here, such as leaf waxes, to trace the flow of plant litter, and the application of labile microbial biomarkers, such as phospholipid fatty acids from microbial cell walls, to trace the carbon sources of soil microorganisms. Finally, natural stable isotope labeling experiments using C_3 and C_4 plants will give insight into the stability and turnover of soil organic matter.

Origin of Carbon in Soils

In general, the accumulation of carbon in soil is the result of ecosystem development driven by the input and decomposition of plant-derived carbon (Amundson, 2001; Jobbagy and Jackson, 2001). In the early stages of ecosystem development, after the retreat of the ice, mainly lichens and mosses add carbon to the bare surface. As a consequence, surface rocks are biologically weathered (Barker and Banfield, 1996; Banfield *et al.*, 1999) and the first soil organic matter is formed from decomposing biomass. The increase of surface substrate temperature, higher nutrient availability due to weathering, and greater water holding capacity due to input of SOM enable further progress in ecosystem development (Lucas, 2001) and the development of soil profiles (Fig. 3.2) (Tandarich *et al.*, 2002). Higher biomass and litter production form a litter layer (L horizon) of undecomposed plant litter. Underneath the litter layer develops organic layers (O horizon) of partially degraded—fermented—plant material (O_f horizon) or completely reworked humic material (O_h horizon) on the surface of the mineral layer. Organic matter is then transported into deeper mineral soil layers either by the digging action of soil organisms or by percolating rainwater (Neff and Asner, 2001; Gabet *et al.*, 2003; Wardle *et al.*, 2003). The latter process is most important for the development of the soil profiles. The transport of carbon from the O horizons in the upper mineral horizon and the export of minerals and metal oxides from this horizon through percolating soil water form a mineral-depleted A horizon in the mineral soil. Below the A horizon an often brownish or reddish mineral, the enriched B horizon, is formed due to the precipitation of leached weathering products, i.e., iron oxides/hydroxides and/or humic substances, from the percolating stream of soil water. Beneath the developed soil profile, unaltered parent substrate remains in the C horizon (Fig. 3.2).

Carbon found in soil is therefore primarily produced by plants from atmospheric CO_2 and enters the soil as root or leaf litter. Soil organisms decompose this litter, releasing most carbon back to the atmosphere as CO_2. Some of the plant-derived litter may remain untouched in soil, but most of the litter-derived carbon remaining in soil is transformed to soil organic matter by the action of soil organisms. The complex process of soil organic matter formation is consequently an achievement of the trophic networks in soil and might be influenced by its species composition (Staddon, 2004). In general, 'shredder' organisms, i.e., earthworms or woodlice, break up

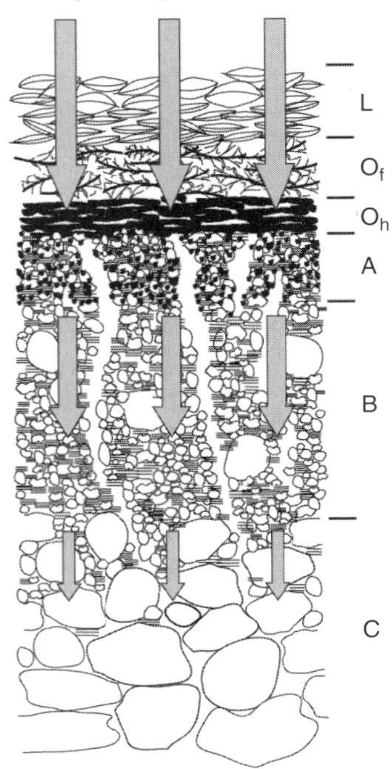

Figure 3.2 Different horizons of a soil profile. L, litter; O, organic layer; A, mineral layer with organic carbon and leached minerals; B, mineral layer with precipitation of oxides/hydroxides and/or carbon; C, unaltered parent substrate.

the nutrient poor litter and extract digestible compounds. This process increases the surface area of litter and inoculates it with decomposer microorganisms that further degrade compounds via external digestion. Soil animals like nematodes, woodlice, collembola, or mites feed on these nutrient-rich microorganisms, and predators hunt the microbe-feeding soil animals. Finally decomposers mineralize dead soil animals, closing the elemental cycle of carbon in soil. As a consequence the formation and turnover of soil carbon depends on the interaction of plants and soil organisms (Korthals *et al.*, 2001).

Chemical Structure and Stable Isotopic Ratio of Plant Carbon Forming Soil Organic Matter

The stability of soil carbon and hence the amount of stored carbon, depends mainly on two factors: (1) the chemical structure of the carbon molecules (Lichtfouse *et al.*, 1998), and (2) their interaction with mineral soil surfaces (Kaiser and Guggenberger, 2003). The latter topic will not be discussed here as it is not a limiting factor for soil carbon storage (Christensen, 1992; Baldock and Skjemstad, 2000). The main focus of this chapter will be on the chemical structure and stable isotope composition of organic matter, which enables one to distinguish plant carbon input from new carbon produced by microbial processes.

Most plant-derived carbon belongs to a small number of chemical compounds, mainly carbohydrates, lipids, lignin, and proteins. Some of these, like carbohydrates or proteins, are very good energy sources for soil organisms and are less stable in soil than lignin or lipids (Gleixner *et al.*, 2001a). As a consequence, the decomposition rate of plant litter will change with litter quality and stable plant-derived molecules may accumulate in the soil as a result of decomposition. Wood is the most abundant plant biomass component and it mainly consists of cellulose and lignin (Fig. 3.3A&B). Cellulose is less stable than lignin, and so lignin accumulates during wood decomposition, i.e., it is selectively preserved. This is well known for decomposition via brown rot fungi (Gleixner *et al.*, 1993).

Lignin itself is a complex polymer made from three different lignin monomers, coumaryl, coniferyl, and sinapyl alcohol, differing in their methoxyl substitution in the ortho position of the phenolic ring (Fig. 3.3C). The composition of the three monomers is characteristic for the origin of the lignin. Monocotyledon plants are rich in coumaryl alcohol, whereas dicotyledon plants are rich in coniferyl and sinapyl alcohol. In conifer trees coniferyl alcohol is the main lignin monomer, whereas in broad leaf trees sinapyl alcohol is the dominant molecule. Depending on the species composition of the plant community, the type of lignin biomarkers input to the soil will vary and the identification of selectively preserved lignin molecules will indicate the presence of unaltered plant remains.

The selective preservation of chemically resistant molecules is also known for lipid molecules like alkanes (Fig. 3.3D) (Lichtfouse *et al.*, 1998). Alkanes are hydrocarbons consisting of only hydrogen and carbon atoms. The lack of oxygen, nitrogen, or sulfur atoms lowers the reactivity of alkanes and so they can be found in long-lived geological samples (Yen and Moldowan, 1988). The alkanes are part of the epicuticular and root waxes, such as cutin and suberin, protecting plants against water loss and microbial attack (Nierop, 1998). The composition of alkanes is characteristic for different plant types and is widely made use of to reconstruct palaeo environments (Didyk *et al.*, 1978). Green algae synthesize mainly alkanes with chain length of 17 carbon atoms, whereas higher plants mainly synthesize alkanes with chain lengths of 27, 29, and 31 carbon atoms (Rieley *et al.*, 1991). The relative composition of these different alkanes is typical for their specific plant origin (Schwark *et al.*, 2002). The identity of different plants can be observed using a mixing diagram for the C_{27}, C_{29}, and C_{31} alkanes (Fig. 3.4). *Fagus sylvatica* synthesizes mainly the C_{27} alkane, whereas *Quercus cerris*

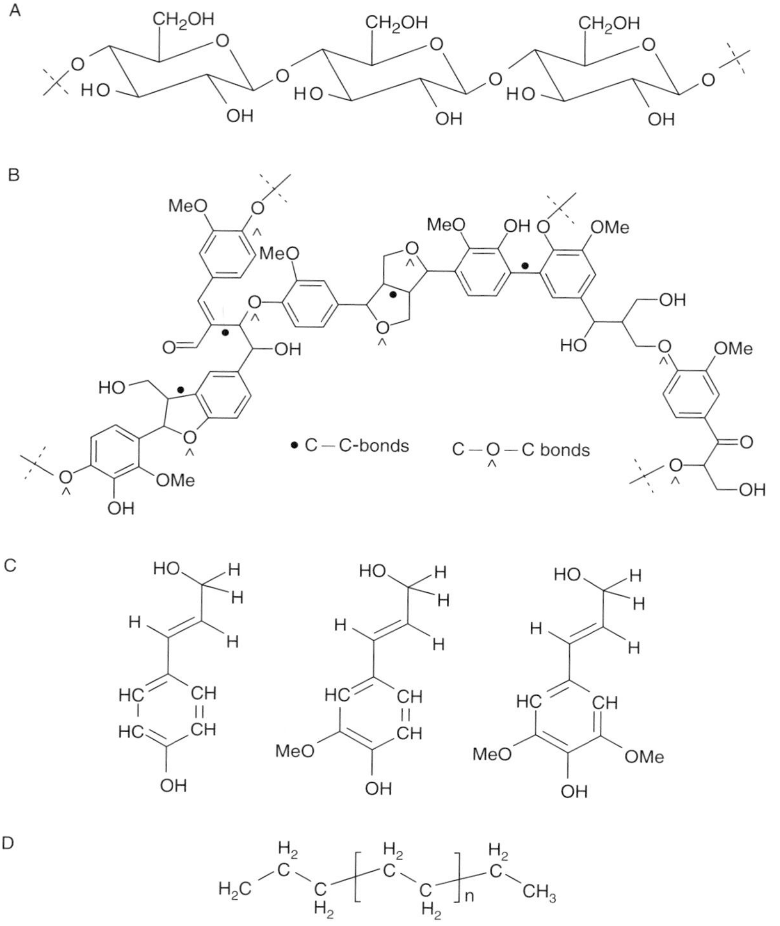

Figure 3.3 Chemical structure of main biochemical elements of plants. (A) cellulose, (B) lignin, (C) lignin monomers, (D) alkanes; 'n' indicates the number of repeated structures within the corresponding total chain length.

synthesizes mainly the C_{29} alkane. The relative abundance of these three alkanes is also reflected in lipid extracts from soil. In beech (*Fagus*) forests the O and A horizons of the soil are clearly dominated by the C_{27} alkane (Fig. 3.4). In the deeper B horizon this C_{29} dominance decreases, and, consequently, preserved plant material is less important for SOM formation at depth in this soil. In both these examples (lignin and alkanes) evidence is provided that plant-derived biomarkers can be found in soil. In contrast to results at the level of aggregated bulk soil carbon, the plant origin of soil carbon can only be identified only using a molecular approach.

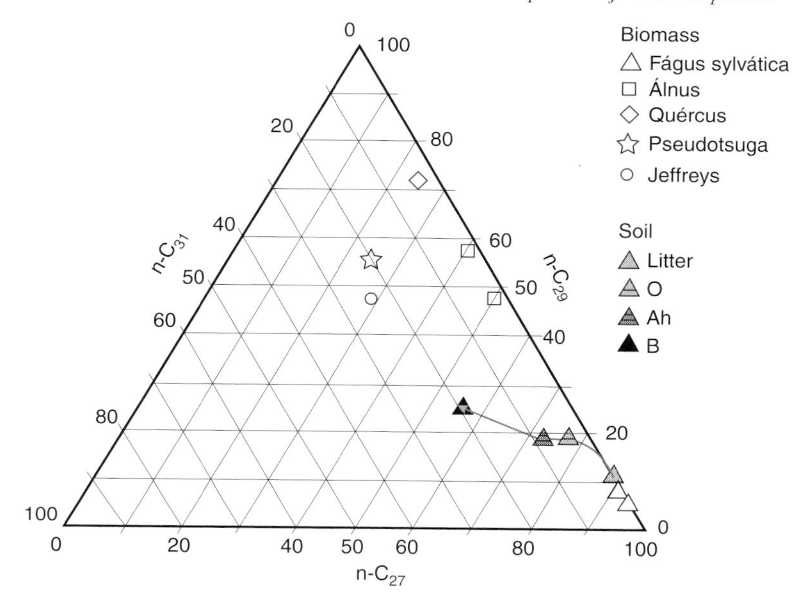

Figure 3.4 Relative composition of alkanes C_{27}, C_{29}, and C_{31} in various plant species and in depth horizons of soil under beech vegetation.

In addition to information on the chemical structure of molecules, we also can use measurements of carbon stable isotope composition to gain new insights into soil carbon input and storage in soils. Plant biomass is known to differ in its stable isotopic content (O'Leary, 1981; Schmidt and Gleixner, 1998). There are well known differences among species, like the enrichment of ^{13}C in C_4 plant vs C_3 plants of ~12–15‰, or the smaller isotopic difference between conifers and broad leaf trees of ~5‰. This natural label can be used after a change of vegetation to trace the flow of new carbon into soil organic matter (Balesdent *et al.*, 1987). Additionally intermolecular isotope differences are known. Lignin, for example, is depleted in ^{13}C relative to cellulose by up to 6‰ (Schmidt and Gleixner, 1998). Differing decomposition rates of specific chemical compounds introduce isotopic shifts to the bulk soil organic matter that mimic isotope effects or source carbon differences (Gleixner *et al.*, 1993). For example, the relative increase in the lignin content can cause an isotopic shift to more depleted $\delta^{13}C$ values in the wood remaining after decomposition. Unfortunately, most investigations on soil carbon are based on the aggregated level of bulk soil or bulk plant material (Boutton, 1996) and we have to consider possible interference from the selective preservation and isotopic shift in bulk soil carbon compounds. Using information on the isotopic composition of individual molecules overcomes this problem, and molecules isolated from the soil with the same isotopic content as plant precursors indicates their selective preservation (Kracht and Gleixner, 2000). In contrast, changing isotope ratios in individual molecules indicates other ongoing processes, such as microbial degradation, microbial synthesis, or differences in source carbon materials.

Distribution of Carbon and Nitrogen and their Stable Isotopes in Soil Profiles

The main sources for soil organic matter in natural systems are leaf litter input to the top of the soil and root litter input down to the rooting depth. We evaluated both–the relative distribution of soil carbon and root biomass with depth–from a global dataset of 2721 and 117 samples, respectively (Jobbagy and Jackson, 2001). The samples in this dataset were from all major biomes of the earth, i.e., boreal forest, crops, desert, sclerophyllous shrubs, temperate deciduous forest, temperate evergreen forest, temperate grassland, tropical deciduous forest, tropical evergreen forest, tropical grassland/savanna, and tundra. Over 60% of the root biomass (as a global average) was found in the top 20 cm of soil and it declined in a logarithmic pattern with depth (Fig. 3.5). Only 14% of root biomass was found below 40 cm. In contrast, only 40% of the soil carbon was located

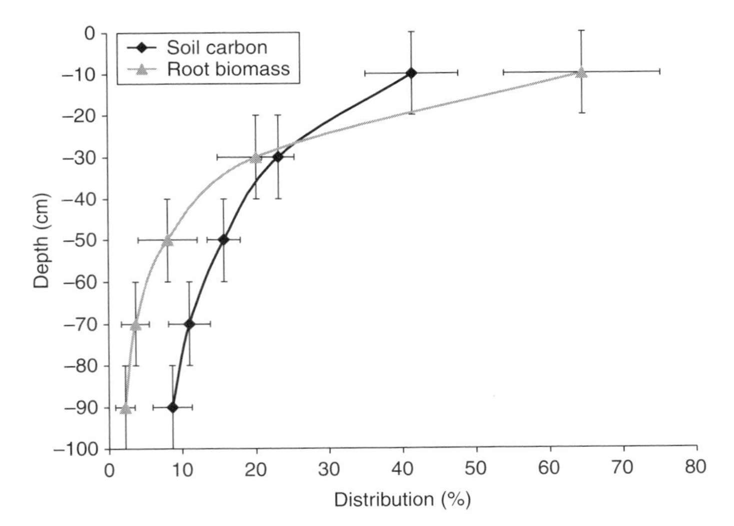

Figure 3.5 Global summary of the distribution of soil carbon and root biomass in depth profiles of the world's major ecosystems: y error bars indicate sampling interval; x error bars indicate standard deviation from 11 biomes summarizing 2721 soil samples and 117 root biomass samples. Data from Jobbagy and Jackson (2001).

in the top 20 cm of soil. Soil carbon also declined logarithmically; however, 36% of it was found at a depth below 40 cm. The strong correlation between root biomass distribution and soil carbon distribution supports the importance of root-derived carbon for the formation of soil carbon ($y = 0.0199x^{2.181}, R^2 = 0.9991$). Relative to the distribution of root biomass less carbon is found in the top 20 cm of soils and more carbon in the subsoil. This underlines the importance of (1) microbial degradation of biomass in the upper 20 cm, (2) water movement for the downward transport of dissolved organic carbon, and (3) the sorption of carbon in deeper soil horizons. These findings suggest that the distribution of root carbon to the soil, which is influenced by the plants, might be a factor for controlling carbon storage. However, in the upper 20 cm of soil profiles the decomposition (Cebrian and Duarte, 1995; Cebrian, 1999), and hence the community of soil organisms, might control carbon storage, whereas in deeper soil horizons intrinsic soil factors might be more important for carbon storage. However, a further distinction between these different processes is not possible based only on analyses of the bulk carbon content.

Coinciding with the decreasing carbon concentration with soil depth is a correlated change in the concentration of soil nitrogen. However, the decline in soil nitrogen content is less pronounced and consequently shifts occur in the C/N ratio of soil organic matter along a depth profile. The C/N ratio changes from values of above 30 ± 15, which are characteristic for plant litter, to values of 10 ± 2, which are characteristic for microbial biomass (Fig. 3.6). The effect was stronger for independent replicates from

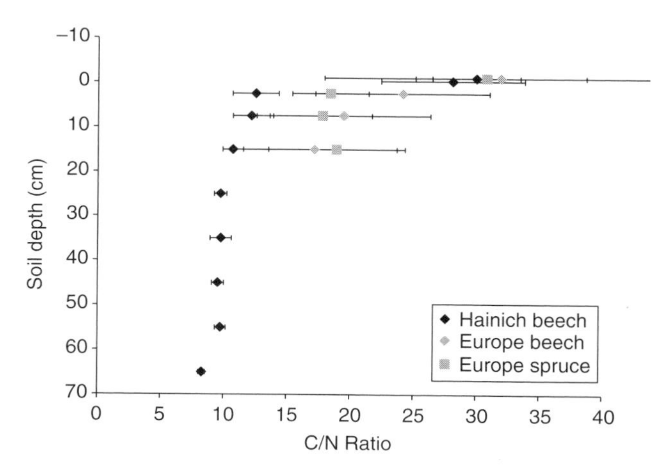

Figure 3.6 C/N ratio of soil organic matter from different depth intervals of 100 independent replicates from an old-growth beech stand in the Hainich National Park, Germany (unpublished). Similar data are also presented for 4 beech stands and 6 spruce stands from a latitudinal gradient in Europe. Data from Schulze (2000).

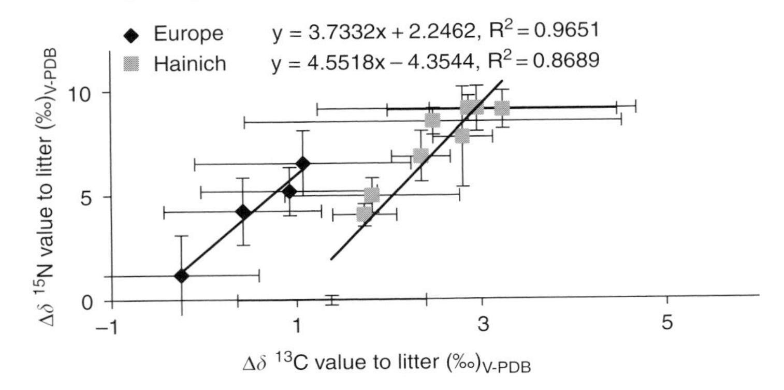

Figure 3.7 Difference between the $\delta^{13}C$ and $\delta^{15}N$ values of soil organic matter in various soil depths and the $\delta^{13}C$ and $\delta^{15}N$ values of litter from 100 independent samples from an old-growth beech stand in the Hainich National Park, Germany (unpublished). Similar data are also presented for 4 beech stands and 6 spruce stands from a latitudinal gradient in Europe. Data from Schulze (2000).

an old growth beech forest in the Hainich National Park, Germany, than for different beech and spruce stands from a European latitudinal gradient (Schulze, 2000). The decrease in C/N ratios suggests that, in the upper few centimeters of soil profiles, root and leaf litter may be a substantial part of the SOM pool, whereas in deeper horizons microbial-derived carbon structures may dominate the SOM pool.

In order to prove the microbial origin of carbon in deeper soil horizons, we compared the enrichment of ^{13}C and ^{15}N values with depth to the enrichment of both isotopes in trophic networks. This enrichment is known to be between 0 and 1‰ for C and between 3 and 4‰ for N (Rothe and Gleixner, 2000). We used $\delta^{13}C$ and $\delta^{15}N$ values of 10 different beech and spruce stands over a latitudinal gradient in Europe (Schulze, 2000) and analyzed in addition $\delta^{13}C$ and $\delta^{15}N$ values of 100 independent depth profiles from the Hainich National Park, Germany. Interestingly, for both cases the ^{13}C and ^{15}N values were highly correlated (Fig. 3.7) indicating that both the ^{13}C and the ^{15}N values of soil organic matter increased with depth at a slope of between 3.7 and 4.6. These values are in good agreement with the trophic level shift expected from food chains, and this suggests that soil carbon in deeper horizons derives mainly from soil organisms.

Dynamics of Soil Organic Matter

Storage of soil carbon depends on the carbon sources input, their chemical structure, and the decomposition rate or turnover rate of soil organic

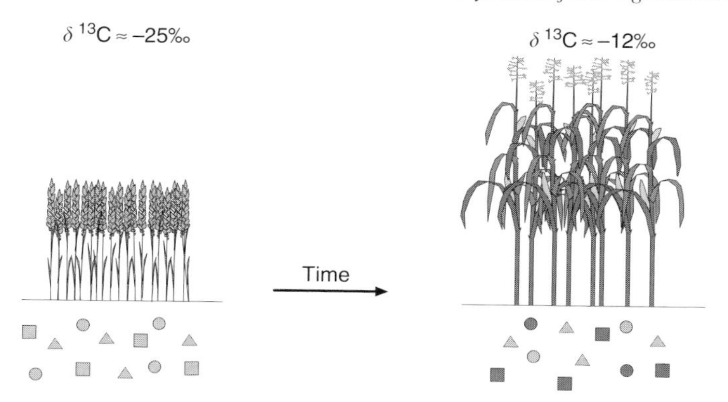

Figure 3.8 Scheme of a natural stable isotope labeling experiment.

matter. Some current state-of-the-art experimental approaches to determine the turnover rate of soil organic matter make use of natural stable isotope labeling experiments (Fig. 3.8) (Balesdent and Mariotti, 1996). In these experiments, the existing vegetation is replaced by structurally similar but isotopically different vegetation, i.e., C_3 plants like wheat or rye ($\delta^{13}C \sim -25‰$) are replaced by C_4 plants like maize ($\delta^{13}C \sim -12‰$). Initially all SOM molecules are labeled according to the isotopic signature of the C_3 crop (Fig. 3.8). Several years after the vegetation change the new vegetation differentially labels individual molecules (i.e., squares in Fig. 3.8 are already completely labeled whereas triangles are not labeled at all). The difference in $\delta^{13}C$ values of soil organic matter from a field without vegetation change and one with vegetation change can be used to calculate the fraction of remaining C_3-derived carbon (Balesdent and Mariotti, 1996). Assuming exponential decay of carbon in soils at steady state the apparent residence time of total soil carbon or of individual compounds of soil organic matter can be determined (Gleixner *et al.*, 1999). This apparent residence time indicates how much time is needed to label the complete pool of carbon with carbon from the new crop.

Corresponding turnover times for bulk soil organic matter, in the upper 25 cm of agricultural fields, are between 10 and 100 years (Balesdent and Mariotti, 1996; Collins *et al.*, 2000; Paul *et al.*, 2001). In forest systems the change from broad leaf trees to conifers or the use of labeled $^{13}CO_2$ in FACE experiments indicates that only a small amount of new plant-derived carbon enters the litter layer (Schlesinger and Lichter, 2001). Most carbon is immediately respired back to the atmosphere. For a vegetation change experiment involving a switch from a 120-year-old beech stand to a spruce stand at the Waldstein, Fichtelgebirge, Germany, the extremely low input of new carbon to soil organic matter is obvious (Fig. 3.9). The calculated mean residence time for soil organic matter increased from 60 years in the

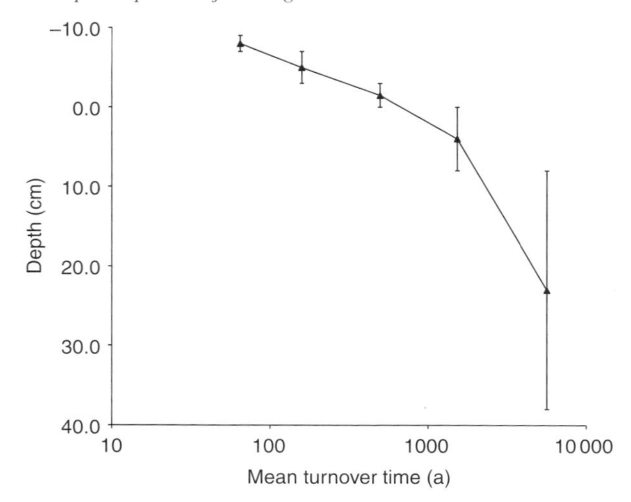

Figure 3.9 Turnover time of soil organic matter from different depths of a beech stand converted to spruce. Litter layers are indicated with negative depth.

litter layer down to more than 5000 years at 10–30 cm soil depth. However, these turnover rates contrast with age of the bulk soil organic matter in these profiles, which are calculated to be less than 500 years old using ^{14}C analyses. Similar to the bulk ^{13}C values, a mixture of old soil carbon with new highly ^{14}C enriched 'bomb' carbon ends up with a mean ^{14}C age that is difficult to interpret (Trumbore, 2000).

Physical fractionation of soil carbon partially overcomes this problem. Soil organic matter found in the sand fraction or in the light density fraction of soil has a shorter turnover time than carbon found in the silt/clay or the heavy density fraction (Balesdent and Mariotti, 1996). It is assumed that carbon derived from plant litter first enters the sand or light fraction, as this pool is labeled quickly by new vegetation. Later in the course of the degradation, mineral–organic complexes are formed and carbon is stabilized on these complexes, which belong to the heavy or silt/clay fraction, and which are labeled more slowly by the carbon input from new vegetation (Sohi *et al.*, 2001; Six *et al.*, 2002). However, so far it is not clear what source of carbon, plant or microbial carbon, enters the more stable mineral fraction. Only compound-specific isotope ratios will give insight into this process.

Molecular Insight into Soil Organic Matter Formation

To determine the compound specific isotope ratios, either individual compounds or their breakdown products are extracted from organic matter by solvents or heat, respectively (Hayes *et al.*, 1990; Gleixner and Schmidt, 1998). The isotopic content of soluble compounds is determined directly, i.e., alkanes, or after derivatization of polar groups, i.e., phospholipid fatty acids, in a gas chromatograph coupled via a combustion unit to an isotope ratio mass spectrometer (GC-C-IRMS) (Hilkert *et al.*, 1999). Alternatively, molecular fragments are produced by heat from non-soluble compounds, like proteins, carbohydrates or lignin and transferred on-line to a GC-C-IRMS system (Gleixner *et al.*, 1999). Under pyrolysis conditions intramolecular water release and intermolecular bond cleavage from specific volatile breakdown products, e.g., from carbohydrates derivatives of furane and pyrane and from lignin derivatives of phenol, are analyzed for their isotopic content. In combination with vegetation change experiments, the label of individual compounds can be evaluated (i.e., squares in Fig. 3.8 are already completely labeled with the isotopic signal of the new vegetation whereas circles are not labeled at all).

In spite of our current knowledge of SOM stability, we were able to demonstrate turnover times shorter than 1 year for the major plant-derived molecules like 'stable' lignin and cellulose (Gleixner *et al.*, 1999; Gleixner *et al.*, 2001b) indicating that plant-derived carbon skeletons are neither chemically nor physically stabilized in soil. No indication for specific, recoverable molecules with turnover times in the millenium range, as suggested by soil carbon models, could be found (Gleixner *et al.*, 2002). Moreover, pyrolysis products of carbohydrates and proteins that were only present in soil samples but not in plant samples had unexpectedly long turnover times between 20 and 100 years (Fig. 3.10). These turnover times are in agreement with turnover times of the bulk soil. However, carbohydrates and proteins are known to be unstable in soil (Trojanowski *et al.*, 1984). At the same time they are also known to be a major part of soil microorganisms. Consequently carbon turnover might be controlled by soil organisms. Moreover, the composition of the belowground food web might be of higher importance than plants for providing the major carbon source for storage in SOM.

Therefore, we focused on the role of the soil microbiota in the storage of soil carbon using labile PLFAs to trace the flow of labeled carbon into the microbial carbon pool. PFLAs were extracted from soils of a 40-year-old C_3 to C_4 vegetation change experiment in Halle, Germany. In spite of our present knowledge suggesting a preferential flow of carbon from labile plant material over labile carbon in organisms to stable carbon in soil, we detected organisms feeding on all available carbon sources in soil (Fig. 3.11). After vegetation change some organisms are completely labeled by the new vegetation, while others are only labeled according to the change in the isotopic content of the soil organic matter or less. As soil organisms have only short life cycles and PLFAs are unstable in soil, the soil organisms were obviously feeding on carbon, like soil organic matter, that was not

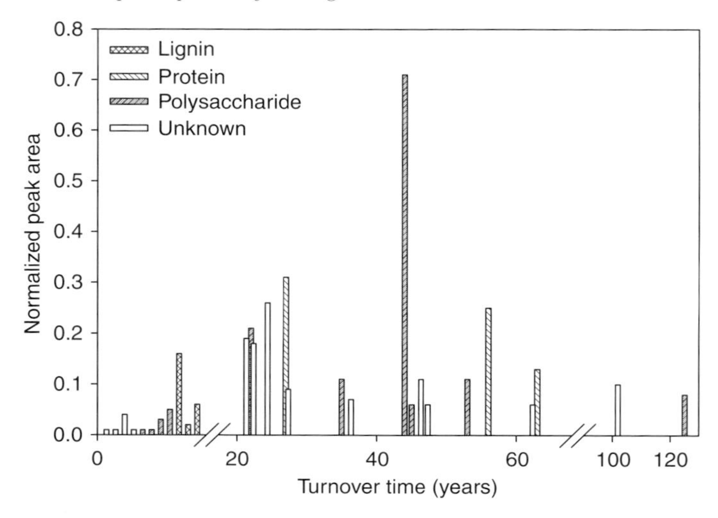

Figure 3.10 Turnover time and relative peak area of individual pyrolysis products for bulk soil submitted to vegetation change from C_3 plants to C_4 plants. From Gleixner *et al.* (2002).

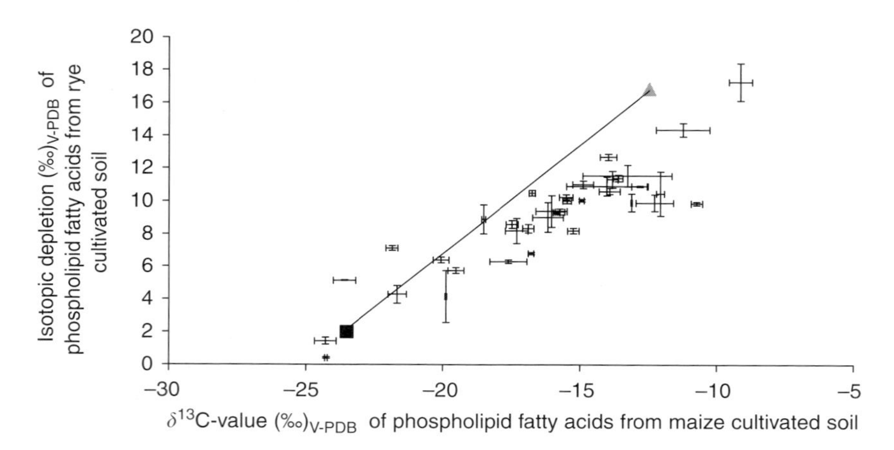

Figure 3.11 Isotopic difference between phospholipid fatty acids extracted from soil under maize (C_4) and those of soil under continuous wheat (C_3) cropping.

labeled even 40 years after the start of the experiment. This suggests that soil carbon is not 'stable' in soil but is constantly under ongoing reuse. Every process that keeps individual carbon atoms in this recycling process possibly increases the carbon storage in soils. This assumption is strongly supported by results from the Long Term Ecological Research site at the Niwot Ridge, Colorado (Neff *et al.*, 2002). The addition of nitrogen increased both the primary production and the species richness in this N-limited system.

However, neither the amount of carbon in the soil nor the ^{14}C content of soil organic matter changed significantly. Using compound-specific isotope ratios we were able to demonstrate that 'young' plant-derived carbon from cellulose and lignin was completely degraded through the addition of nitrogen. At the same time the turnover of the mineral-associated carbon was accelerated and new carbon entered this slower pool.

Summary

This chapter reviewed our current knowledge of carbon storage in soil. Based on results at the aggregated level of bulk soil and physical soil fractions, the high amount of old ^{14}C carbon stored in world soils is stabilized by physical and chemical mechanisms. This is consistent with the general concept of carbon models having a yearly, a decadal, and a millennium carbon pool. However, to date the identification of molecules from the millennium carbon pool, which would be a target for carbon sequestration, is still unresolved. Using the latest isotope-based techniques at the molecular level, we explored the fate of plant-derived carbon input to the soil carbon pool. This offered a new view about how carbon is stored in soil. Our studies suggest that carbon in soil is not stabilized, but is continuously reused by soil organisms to form their biomass. As the chemistry of plants and soil organisms is quite similar, except for the higher N contents in soil organisms, which are seldom investigated, this difference in turnover time is difficult to detect with conventional methods. However, food web (trophic level) induced isotope shifts during decomposition of plant organic matter clearly supported our view. Compound-specific isotope ratios of microbial biomarkers finally proved that soil organisms are able to use every carbon source available in soil organic matter for survival and growth. Moreover, these results suggest that the missing millennium carbon pool in soil might only be a missing process, like the use of different carbon sources, i.e., gases, or the formation of microbial resting stages. Interestingly our results highlight that carbon turnover, which might be strongly affected by climate change, is of higher importance for carbon storage than soil storage capacity, which depends on soil mineralogy.

References

Amundson R. (2001) The carbon budget in soils. *Annu Rev Earth Planetary Sci* **29**: 535–562.

Baldock J. A. and Skjemstad J. O. (2000) Role of the soil matrix and minerals in protecting natural organic materials against biological attack. *Org Geochem* **31** (7–8): 697–710.

Balesdent J., Mariotti A. and Guillet B. (1987) Natural ^{13}C abundance as a tracer for studies of soil organic matter dynamics. *Soil Biol Biochem* **19**(1): 25–30.

Balesdent J. and Mariotti A. (1996) Measurement of soil organic matter turnover using ^{13}C natural abundance. In *Mass Spectrometry of Soils* (T. W. Boutton and S. Yamasaki, eds) pp. 83–111. Marcel Dekker, New York.

Banfield J. F., Barker W. W., Welch S. A. and Taunton A. (1999) Biological impact on mineral dissolution: Application of the lichen model to understanding mineral weathering in the rhizosphere. *Proc Natl Acad Sci USA* **96**(7): 3404–3411.

Barker W. W. and Banfield J. F. (1996) Biologically versus inorganically mediated weathering reactions—relationships between minerals and extracellular microbial polymers in lithobiontic communities. *Chem Geol* **132**(1–4): 55–69.

Boutton T. W. (1996) Stable carbon isotope ratios of soil organic matter and their use as indicators of vegetation and climate change. In *Mass Spectrometry of Soils* (T. W. Boutton and S. Yamasaki, eds) pp. 47–82. Marcel Dekker, New York.

Catovsky S., Bradford M. A. and Hector A. (2002) Biodiversity and ecosystem productivity: Implications for carbon storage. *Oikos* **97**(3): 443–448.

Cebrian J. (1999) Patterns in the fate of production in plant communities. *Am Naturalist* **154**(4): 449–468.

Cebrian J. and Duarte C. M. (1995) Plant growth-rate dependence of detrital carbon storage in ecosystems. *Science* **268**(5217): 1606–1608.

Christensen B. T. (1992) Physical fractionation of soil and organic matter in primary particle size and density separates. *Adv Soil Sci* **20**: 1–90.

Collins H. P., Elliott E. T., Paustian K., Bundy L. C., Dick W. A., Huggins D. R., Smucker A. J. M. and Paul E. A. (2000) Soil carbon pools and fluxes in long-term corn belt agroecosystems. *Soil Biol Biochem* **32**(2): 157–168.

Didyk B. M., Simoneit B. R. T., Brassell S. C. and Eglinton G. (1978) Organic geochemical indicators of paleo-environmental conditions of sedimentation. *Nature* **272**(5650): 216–222.

Gabet E. J., Reichman O. J. and Seabloom E. W. (2003) The effects of bioturbation on soil processes and sediment transport. *Annu Rev Earth Planetary Sci* **31**: 249–273.

Gleixner G., Danier H. J., Werner R. A. and Schmidt H.-L. (1993) Correlations between the ^{13}C content of primary and secondary plant products in different cell compartments and that in decomposing basidiomycetes. *Plant Physiol* **102**: 1287–1290.

Gleixner G. and Schmidt H. L. (1998) On-line determination of group-specific isotope ratios in model compounds and aquatic humic substances by coupling pyrolysis to GC-C-IRMS. In *Nitrogen-Containing Macromolecules in the Bio- and Geosphere, ACS Symposium Series 707* (A. B. Stankiewicz and P. F. van Bergen, eds) pp. 34–46. American Chemical Society, Washington.

Gleixner G., Bol R. and Balesdent J. (1999) Molecular insight into soil carbon turnover. *Rapid Commun Mass Spect* **13**(13): 1278–1283.

Gleixner G., Czimczik C. J., Kramer C., Lühker B. and Schmidt M. W. I. (2001a) Plant compounds and their turnover and stabilization as soil organic matter. In *Global Biogeochemical Cycles in the Climate System* (E. D. Schulze, M. Heimann, S. Harrison, E. A. Holland, J. Lloyd, I. C. Prentice and D. S. Schimel, eds) pp. 201–215. Academic Press, San Diego.

Gleixner G., Kracht O. and Schmidt H. L. (2001b) Group specific isotope ratios of humic substances. In *Understanding and Managing Organic Matter in Soils, Sediments and Waters* (S. S. Swift and K. M. Spark, eds) pp. 195–201. International Humic Substance Society, St Paul.

Gleixner G., Poirier N., Bol R. and Balesdent J. (2002) Molecular dynamics of organic matter in a cultivated soil. *Org Geochem* **33**(3): 357–366.

Hayes J. M., Freeman K. H., Popp B. N. and Hoham C. H. (1990) Compound-specific isotopic analyses—a novel tool for reconstruction of ancient biogeochemical processes. *Org Geochem* **16**(4–6): 1115–1128.

Hedges J. I. and Oades J. M. (1997) Comparative organic geochemistries of soils and marine sediments. *Org. Geochem.* **27**(7–8): 319–361.

Hilkert A. W., Douthitt C. B., Schluter H. J. and Brand W. A. (1999) Isotope ratio monitoring gas chromatography mass spectrometry of D H by high temperature conversion isotope ratio mass spectrometry. *Rapid Commun Mass Spectrometry* **13**(13): 1226–1230.

Houghton J. T., Ding Y., Griggs D. J., Noguer M., van der Linden P. J., Dai X., Maskell K. and Johnson C. A. (2001) *Climate Change 2001: The Scientific Basis. Contribution of Working Group I to the Third Assessment Report of the Intergovernmental Panel on Climate Change.* Cambridge University Press, Cambridge.

Jenkinson D. S., Hart P. B. S., Rayner J. H. and Parry L. C. (1987) Modelling the turnover of organic matter in long-term experiments at Rothamsted. *INTECOL Bull* **15**: 1–8.

Jobbagy E. G. and Jackson R. B. (2001) The distribution of soil nutrients with depth: Global patterns and the imprint of plants. *Biogeochemistry* **53**(1): 51–77.

Kaiser K. and Guggenberger G. (2003) Mineral surfaces and soil organic matter. *Eur J Soil Sci* **54**(2): 219–236.

Korthals G. W., Smilauer P., van Dijk C. and van der Putten W. H. (2001) Linking above- and below-ground biodiversity: Abundance and trophic complexity in soil as a response to experimental plant communities on abandoned arable land. *Funct Ecol* **15**(4): 506–514.

Kracht O. and Gleixner G. (2000) Isotope analysis of pyrolysis products from Sphagnum peat and dissolved organic matter from bog water. *Org Geochem* **31**(7–8): 645–654.

Lichtfouse E., Chenu C., Baudin F., Leblond C., Da Silva M., Behar F., Derenne S., Largeau C., Wehrung P. and Albrecht P. (1998) A novel pathway of soil organic matter formation by selective preservation of resistant straight-chain biopolymers—chemical and isotope evidence. *Org Geochem* **28**(6): 411–415.

Lucas Y. (2001) The role of plants in controlling rates and products of weathering: Importance of biological pumping. *Annu Rev Earth Planetary Sci* **29**: 135–163.

McCarthy J. J., Canziani O. F., Leary N. A., Dokken D. J. and White K. S. (2001) *Climate Change 2001: Impacts, Adaptation, and Vulnerability. Contribution of Working Group II to the Third Assessment Report of the Intergovernmental Panel on Climate Change.* Cambridge University Press, Cambridge.

Neff J. C. and Asner G. P. (2001) Dissolved organic carbon in terrestrial ecosystems: Synthesis and a model. *Ecosystems* **4**(1): 29–48.

Neff J. C., Townsend A. R., Gleixner G., Lehman S. J., Turnbull J. and Bowman W. D. (2002) Variable effects of nitrogen additions on the stability and turnover of soil carbon. *Nature* **419**(6910): 915–917.

Nierop K. G. J. (1998) Origin of aliphatic compounds in a forest soil. *Org Geochem* **29**(4): 1009–1016.

O'Leary M. H. (1981) Carbon isotope fractionation in plants. *Phytochemistry* **20**(4): 553–567.

Parton W. J., Schimel D. S., Cole C. V. and Ojima D. S. (1987) Analysis of factors controlling soil organic matter levels in great plains grassland. *Soil Sci Soc Am J* **51**: 1173–1179.

Paul E. A., Collins H. P. and Leavitt S. W. (2001) Dynamics of resistant soil carbon of midwestern agricultural soils measured by naturally occurring C-14 abundance. *Geoderma* **104**(3–4): 239–256.

Rieley G., Collier R. J., Jones D. M., Eglinton G., Eakin P. A. and Fallick A. E. (1991) Sources of sedimentary lipids deduced from stable carbon isotope analyses of individual compounds. *Nature* **352**(6334): 425–427.

Rothe J. and Gleixner G. (2000) Do stable isotopes reflect the food web development in regenerating ecosystems? *Isotopes Environ Health Stud* **36**(3): 285–301.

Scheu S. (2001) Plants and generalist predators as links between the below-ground and above-ground system. *Basic Appl Ecol* **2**(1): 3–13.

Schimel D. S., House J. I., Hibbard K. A., Bousquet P., Ciais P., Peylin P., Braswell B. H. *et al.* (2001) Recent patterns and mechanisms of carbon exchange by terrestrial ecosystems. *Nature* **414**(6860): 169–172.

Schlesinger W. H. and Lichter J. (2001) Limited carbon storage in soil and litter of experimental forest plots under increased atmospheric CO_2. *Nature* **411**(6836): 466–469.

Schmidt H. -L. and Gleixner G. (1998) Carbon isotope effects on key reactions in plant metabolism and ^{13}C-patterns in natural compounds. In *Stable Isotopes: Integration in Biological, Ecological and Geochemical Processes* (H. Griffiths, ed.) pp. 13–26. BIOS, Oxford.

Schulze E. D. (2000) *Carbon and Nitrogen Cycling in European Forest Ecosystems.* Springer, Berlin.

Schwark L., Zink K. and Lechterbeck J. (2002) Reconstruction of postglacial to early Holocene vegetation history in terrestrial Central Europe via cuticular lipid biomarkers and pollen records from lake sediments. *Geology* **30**(5): 463–466.

Six J., Conant R. T., Paul E. A. and Paustian K. (2002) Stabilization mechanisms of soil organic matter: Implications for C-saturation of soils. *Plant Soil* **241**(2): 155–176.

Sohi S. P., Mahieu N., Arah J. R. M., Powlson D. S., Madari B. and Gaunt J. L. (2001) A procedure for isolating soil organic matter fractions suitable for modeling. *Soil Sci Soc Am J* **65**(4): 1121–1128.

Staddon P. L. (2004) Carbon isotopes in functional soil ecology. *Trends Ecol Evol* **19**(3): 148–154.

Tandarich J. P., Darmody R. G., Follmer L. R. and Johnson D. L. (2002) History of soil science—Historical development of soil and weathering profile concepts from Europe to the United States of America. *Soil Sci Soc Am J* **66**(2): 335–346.

Trojanowski J., Haider K. and Huettermann A. (1984) Decomposition of carbon-14-labeled lignin holocellulose and lignocellulose by mycorrhizal fungi. *Arch Microbiol* **139**(2–3): 202–206.

Trumbore S. (2000) Age of soil organic matter and soil respiration: Radiocarbon constraints on belowground C dynamics. *Ecol Appl* **10**(2): 399–411.

Wang Y., Amundson R. and Trumbore S. (1996) Radiocarbon dating of soil organic matter. *Quat Res* **45**(3): 282–288.

Wardle D. A., Yeates G. W., Williamson W. and Bonner K. I. (2003) The response of a three trophic level soil food web to the identity and diversity of plant species and functional groups. *Oikos* **102**(1): 45–56.

Yen T. F. and Moldowan J. M. (1988) *Geochemical Biomarkers.* Harwood Academic Publishers, Chur.

4

Factors Determining the ^{13}C Abundance of Soil-Respired CO_2 in Boreal Forests

Peter Högberg, Aif Ekblad, Anders Nordgren, Agnetu H. Plamboeck, Anders Ohlsson, Bhupinderpal-Singh, Mona N. Högberg

Introduction

Variations in carbon isotope composition ($\delta^{13}C$) enable tracing sources of and sinks for atmospheric CO_2 (e.g., Tans *et al.*, 1993; Ciais *et al.*, 1995a,b; Fung *et al.*, 1997). A pioneer of this field, Keeling (1958), used the low $\delta^{13}C$ of urban air to demonstrate that the elevated atmospheric CO_2 in such areas originated from the burning of fossil fuels. Others have used the difference in $\delta^{13}C$ between C_3 and C_4 plants (Smith and Epstein, 1971) of about 14‰ to determine the contributions from these groups of plants to soil respiration. When, for example, a C_4 plant is planted in a soil previously covered by C_3 plants, the new input, first of C_4 plant root respiration and later also from decomposition of C_4 soil organic matter (SOM), can be distinguished from the respiration derived from decomposition of old C_3 SOM (e.g., Rochette *et al.*, 1999). Common to such approaches is the use of isotopic mixing-models, in which the total amount of CO_2 has an average isotopic composition but is made up of various fractions with different isotopic compositions.

Analysis of the isotopic composition and rate of CO_2 evolution from soil has thus increasingly been used in studies of C dynamics in the soil–plant–atmosphere system. The C isotope fractionation in plants has been described in great detail starting with the seminal work by Farquhar and others (Farquhar *et al.*, 1982; Farquhar *et al.*, 1989). However, the causation of the isotopic composition of the CO_2 efflux from soils is less well understood, with the exception of differences relating to C_3 and C_4 plants (e.g., Cerling and Wang, 1996), as mentioned above.

Even in pure C_3 plant ecosystems, like the boreal forests, there is significant temporal and spatial variability in the $\delta^{13}C$ of the soil CO_2 efflux. Analysis of the isotopic composition of the CO_2 respired from soils may thus reveal information about this important component of the ecosystem C balance (Flanagan and Ehleringer, 1998; Ehleringer *et al.*, 2000). This is crucial, since a large terrestrial sink for atmospheric CO_2 has been located in the northern hemisphere (e.g., Ciais *et al.*, 1995b; Keeling *et al.*, 1996), and the vast boreal forests may be largely responsible (e.g., Myneni *et al.*, 1997). At the same time, boreal and arctic ecosystems have large amounts of C stored in the soil (Post *et al.*, 1982), and could potentially become a source of CO_2 in a warmer climate promoting more rapid decomposition of SOM. Furthermore, the northern hemisphere has complex dynamics in terms of annual fluctuations in both the concentration of CO_2 in the atmosphere and its $\delta^{13}C$ (e.g., Ciais *et al.*, 1995a,b; Trolier *et al.*, 1996). It is of utmost importance to understand the causes of this variability, since it interferes with the partitioning between the ocean and the terrestrial contributions in global models.

The purpose of this review is thus to provide an update on the reviews by Flanagan and Ehleringer (1998) and Ehleringer *et al.* (2000) on the causation of the $\delta^{13}C$ of the soil CO_2 efflux and, in doing this, focus on the boreal forests.

CO_2 Production in Soils

Who is Respiring?

There are two major sources contributing to the efflux of CO_2 from soils: root (autotrophic) respiration and heterotrophic respiration. The C sources used are sugars, derived more or less directly from plant photosynthesis, and a range of more or less complex plant macromolecules. In the soil food-web, of course, secondary products such as dead bacteria, fungi, soil animals, etc. are also ultimately decomposed by the heterotrophic community. However, although the above divisions may seem simple and straightforward, the belowground world is in reality more complex (see Gleixner, Chapter 3).

Plant roots are commonly mycorrhizal, i.e., forming symbiotic associations with fungi, which by definition are heterotrophs. Yet these fungi receive, like the roots of the autotrophic plants, sugars more or less directly from the plant foliage (Smith and Read, 1997). In boreal forests, the dominant coniferous trees engage in ectomycorrhizal symbiosis, in which up to several tens of micrometer-thick fungal sheaths commonly cover close to 100% of the fine root tips (Taylor *et al.*, 2000; Fig. 4.1). From these sheaths an extramatrical mycelium, with a biomass of roughly the same size (Wallander *et al.*, 2001), ramifies further out into the soil. Although the growth and activity of this fungal tissue is supported by photosynthates from

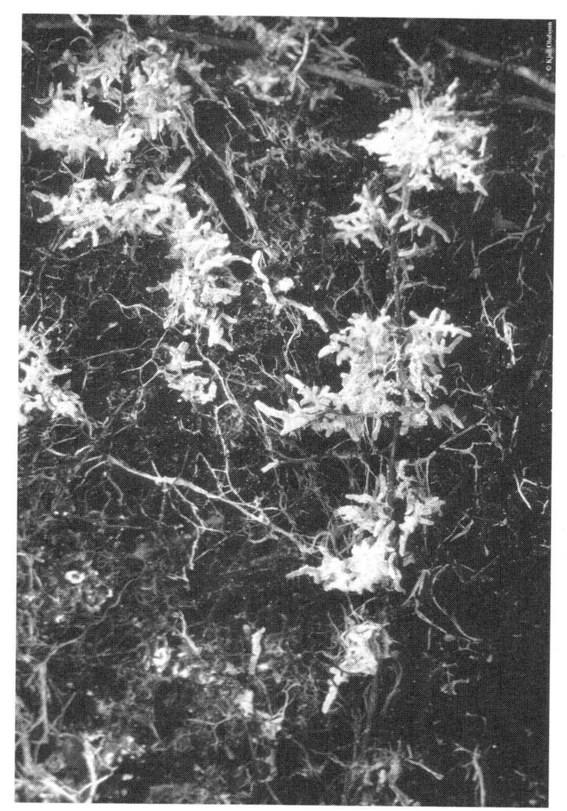

Figure 4.1 The organic mor-layer of a boreal forest soil. Tree roots, with the fine roots colonized by a white ectomycorrhizal fungus, are clearly visible, while bacteria, and many other heterotrophic microorganisms decomposing the organic matter constituting the dark background matrix cannot be seen by the naked eye. Note the rhizomorphs and fungal strands extending outwards from clusters of ectomycorrhizal fine roots; these structures are damaged if the system is disturbed by soil or root sampling. Photo courtesy of Kjell Olofsson.

the trees, there also seems to be a significant leakage of C compounds from the fungal cells to the wider community of soil organisms (Garbaye, 1994; Timonen *et al.*, 1998; Högberg and Högberg, 2002). Heterotrophic activity, on the other hand, encompasses anything from decomposition of simple sugars leaked from plant roots to the degradation of secondary products like the large complex macromolecules collectively termed humus. Hence, the division between autotrophic and heterotrophic is not clear-cut.

One of the simpler and technically feasible ways (see later) of classifying components of soil respiration is to group organisms into those that receive photosynthates more or less directly from the plant canopy and those that receive their C mainly through decomposition of dead or dying organic matter. This means that the former group would include the classic autotrophic component plant roots, but also heterotrophic organisms like mycorrhizal fungi and a wide range of other organisms in the rhizosphere and mycorrhizosphere, which largely depend on the flux of recent photosynthates. We will call this the autotrophic component or root respiration, although in a classic sense it also involves heterotrophic organisms. Thus, the heterotrophs will, in our view, be those organisms that degrade more complex molecules, and do not depend on recent plant photosynthates.

With the above division in major components, the C mass balance and the C isotope mass balance of the CO_2 efflux from the soil become:

$$CO_{2\,\mathrm{total}} = CO_{2\,\mathrm{hetero}} + CO_{2\,\mathrm{auto}} \tag{4.1}$$

and

$$CO_{2\,\mathrm{total}} \times \delta^{13}C_{\mathrm{total}} = (CO_{2\,\mathrm{hetero}} \times \delta^{13}C_{\mathrm{hetero}}) + (CO_{2\,\mathrm{auto}} \times \delta^{13}C_{\mathrm{auto}}) \tag{4.2}$$

in which CO_2 is the respiration rate, $\delta^{13}C$ is the ^{13}C abundance of the CO_2, and subscripts total, hetero, and auto stand for total, heterotrophic, and autotrophic respiration, respectively. We will examine the different factors controlling the variables in this equation. It follows from the above that:

$$CO_{2\,\mathrm{auto}} = CO_{2\,\mathrm{total}} - CO_{2\,\mathrm{hetero}} \tag{4.3}$$

which means that it is sufficient to determine total respiration and one of its major components in order to calculate the other component.

How Large is the Fraction of Root Respiration?

Root respiration is highly variable among plant species, and in space and time. It is, furthermore, technically difficult to separate from heterotrophic respiration because mycorrhizal roots and true heterotrophs occur in a complex and delicate mix in the soil (Fig. 4.1). As a result, estimates of the contribution by root respiration to total soil respiration vary from 10 to 90% (Hanson *et al.*, 2000). Part of this variability clearly relates to methodological problems.

In an attempt to obtain a robust division of the major components of soil respiration, we recently girdled (ring-barked) a boreal Scots pine forest (Högberg *et al.*, 2001). This involved girdling of all trees on 30 m × 30 m plots, with an average of 120 trees per plot. The treatment terminates the flux of recent photosynthates to roots and their associated microorganisms, but does not, in a shorter time perspective, affect the uptake of water by the trees or the soil temperature. Nor does the treatment involve any physical disturbance of the soil biota. During the first year we found that up to 56%

of the soil respiratory activity was lost from the girdled plots when compared to the untreated control plots. Apparently, this was a conservative estimate of the contribution by root respiration, as we observed that starch reserves were more rapidly depleted in roots of the girdled trees. It was thus no surprise that the respiration on girdled plots during the second vegetation period after girdling, when the stores of starch were depleted in roots of girdled trees, was up to 65% lower than in control plots (Bhupinderpal-Singh *et al.*, 2003).

The decline in soil respiration after girdling could be remarkably rapid (Högberg *et al.*, 2001). When girdling was carried out in August, the time of the year when the C allocation to roots is supposed to be at maximum in northern temperate conifers (Hansen *et al.*, 1997), 40% of the soil respiratory activity was lost in 5 days and 56% in 14 days. These data clearly show that root respiration can account for 50%, or more, of total soil respiration in boreal forests. Ongoing girdling experiments in two other boreal forests support this estimate (Högberg *et al.*, unpublished; Nordgren *et al.*, unpublished). Planned experiments of the same kind, but outside the boreal zone, will determine if the large contributions from root respiration found in boreal forests are exceptional or not.

The amount of plant biomass produced, which ultimately in the longer term feeds the heterotrophic community, is constrained by the supply of nutrients. Likewise, the fraction of plant C allocated to roots and mycorrhizal fungi is strongly affected by the nutrient supply, but decreases as the nutrient supply increases. This means that the nutrient supply should have opposing effects on the C allocation to roots and heterotrophs, and that the ratio root/heterotrophic respiration should decrease with increasing nutrient supply (Högberg *et al.*, 2003). The sum of the two components should increase in response to greater nutrient supply when measured per unit area, but there are as yet not many data from boreal forest ecosystems to test this assumption. An exception is the detailed study of Scots pine by Linder and Axelsson (cited in Cannell, 1989) demonstrating that the fraction of photosynthate C allocated belowground was 32% in fertilized trees while it was 59% in N-limited control trees, at the same time as fertilization doubled photosynthetic C fixation. The partitioning between root growth and respiration was not measured, but it seems likely that root respiration accounts for 75% of the C allocated belowground in a strongly nutrient-limited boreal forest (Högberg *et al.*, 2002).

Disturbances may create large variations in the ratio root/heterotrophic respiration. For example, clear-felling should cause a decrease in soil respiration, and hence a shift in the ratio autotrophic/heterotrophic respiration. Another example is that of a forest with large stores of soil C in SOM, and which acted as a source of CO_2 (Valentini *et al.*, 2000), which was attributed to increased heterotrophic respiration caused by forest drainage several decades earlier of the previously anaerobic soil system.

What Determines the $\delta^{13}C$ of Root Respiration?

The $\delta^{13}C$ of photosynthates is determined by the isotopic signature of the source air and fractionation against ^{13}C during photosynthesis. The latter, Δ, is described for C_3 plants in a simplified equation provided by Farquhar (1991):

$$\Delta = 4.4 + 22.6(C_i/C_a) \quad (4.4)$$

in which 4.4 is the fractionation against ^{13}C during diffusion of CO_2 in air, and C_i and C_a are the partial pressures of CO_2 inside leaves and in ambient air, respectively. To calculate the $\delta^{13}C$ of the photosynthates, one has to add the signature of the source, atmospheric CO_2, which today is c. −8‰ at the level of the upper canopy (and change the signs in Eq. 4.4 from plus to minus). For example, when C_i/C_a varies between 0.4 and 0.7, the $\delta^{13}C$ of the plant varies between −21.4 and −28.2‰. It also follows from the above that understorey plants obtain low $\delta^{13}C$, because they refix some isotopically depleted CO_2 respired from soils (which frequently is from −30 to −20‰) and also as shade plants operate at high C_i/C_a (Brooks *et al.*, 1997). The ratio C_i/C_a decreases as stomata close during dry conditions and increases when they open more fully. It decreases at higher rates of photosynthesis, caused by, for example, higher levels of irradiation and improved plant nutrient status. Hence, drought-stressed overstorey trees obtain high $\delta^{13}C$, as they are operating at low C_i/C_a (Brooks *et al.*, 1997).

We are not aware of any study showing that root respiration by understorey plants has lower $\delta^{13}C$ than that of overstorey plants, although this can be inferred from the difference in the signature of foliage (Brooks *et al.*, 1997). However, a difference of 2‰ between over- and understorey trees was shown to occur in both foliage and fine roots in a mixed boreal forest, although the $\delta^{13}C$ of roots of spruce, the co-dominant, showed greater temporal variability than roots of the dominant pine and the understorey birch (Table 4.1; Plamboeck, 1999). We also found that sporocarps of ectomycorrhizal fungi specific to the different host tree species displayed the same 2‰ difference between over- and understorey trees. There was also an additional 2‰ enrichment of the ectomycorrhizal fungi relative to their respective plant hosts (Högberg *et al.*, 1999; Fig. 4.2), which will be discussed later.

Managed boreal forests commonly have a reduced stratification of the canopy (i.e., few understorey trees), but there is mostly a significant understorey of ericaceous dwarf shrubs, which usually is 0.1–0.3 m tall. The latter component is even more depleted in ^{13}C than are understorey trees.

Table 4.1 The δ^{13}C of Roots of Scots Pine (*Pinus sylvestris* L.), Norway Spruce (*Picea abies* (L.) Karst.) and Birch (*Betula pendula* L.) in a Boreal Forest; Pine is the Dominant, Spruce the Co-dominant, and Birch the Understorey Tree Species (Plamboeck, 1999). Data are Means ±1 S.E.

	Year of sampling	
	1994	1995
Species	δ^{13}C (‰)	
Pine		
Roots diam. <1 mm	−27.8 ± 0.2	−28.0 ± 0.2
Roots diam. 1–2 mm	−27.2 ± 0.3	−26.8 ± 0.2
Spruce		
Roots diam. <1 mm	−29.4 ± 0.3	−27.4 ± 0.4
Roots diam. 1–2 mm	−29.8	−27.1
Birch		
Roots diam. <1 mm	−29.8 ± 0.2	−30.0 ± 0.1
Roots diam. 1–2 mm	−29.4 ± 0.4	−30.0 ± 0.2

As already mentioned this is partly because they work at high C_i/C_a, partly because they refix some CO_2 respired from the soil. The latter is most important in closed forests with little turbulent air movement (Buchmann *et al.*, 1996). Brooks *et al.* (1997) reported that the former effect, the physiological effect, explained c. 80% of variations of c. 3‰ in δ^{13}C with plant height in boreal forests. In a survey of a wider range of forests globally, Buchmann *et al.* (2002) ascribed 70% of the variations in leaf δ^{13}C to the physiological effect. It seems that the dominant trees contribute more to root respiration than understorey trees and other plants (Flanagan *et al.*, 1999). In accordance with this assumption, Högberg *et al.* (1999) demonstrated, based on their data on δ^{13}C of fungal sporocarps, that the bulk of the C in the population of ectomycorrhizal fungi came from the dominant trees.

Several factors related to tree age, e.g., decreased hydraulic conductance in tree stems, increase the δ^{13}C of photosynthates (Fessenden and Ehleringer, 2002). This should lead to an increase in δ^{13}C of the ecosystem-respired CO_2 efflux as tree stands age, as was supported by a study of a chronosequence in Washington, USA (Fessenden and Ehleringer, 2002).

Differences in δ^{13}C among tree-rings (e.g., Leavitt and Long, 1986; Saurer *et al.*, 1997) occur in response to differences in weather conditions between years. The response in δ^{13}C signature of the sugars produced during photosynthesis to environmentally induced shifts in C_i/C_a or CO_2

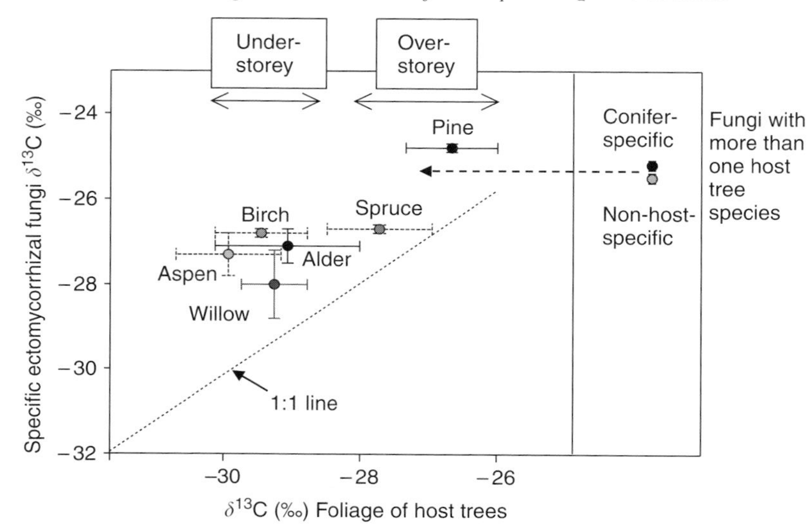

Figure 4.2 The ^{13}C abundance of trees and their symbiotic ectomycorrhizal fungi in a temperate forest with two conifer species in the overstorey and a mix of broadleaf species in the understorey (data from Högberg *et al.*, 1999). Some fungal species are host-specific; there is a correlation between the $\delta^{13}C$ of these fungi and that of their tree hosts. Other fungi are less specific; their $\delta^{13}C$ indicates that they receive most of their C from the overstorey Scots pine trees. Bars show ±1 S.E.

source is, however, immediate (Brugnoli *et al.*, 1988; Brugnoli *et al.*, 1998). This can cause substantial variability in the shorter term. Pate and Arthur (1998) found variations in $\delta^{13}C$ of phloem sap in Australian *Eucalyptus*, which correlated with seasonal variations in plant water deficit. In line with this, Ekblad and Högberg (2001) reported substantial variation in the $\delta^{13}C$ of the soil CO_2 efflux in a boreal forest. They found a strong negative correlation between $\delta^{13}C$ and the air relative humidity (ARH) days before the sampling of the soil efflux (Fig. 4.3). The dominant trees in that boreal forest were 20–25 m tall, and the data indicated a time delay of 1–4 days between canopy photosynthesis and root respiration. This delay should reflect the distance of transport, i.e., tree height, and is likely variable also depending on tree species, temperature, etc. Bowling *et al.* (2002) found a delay of 5–10 days in the response in $\delta^{13}C$ of ecosystem-respired CO_2 to changes in air vapor pressure deficit (VPD) at a number of forest sites in Oregon.

We conclude that the $\delta^{13}C$ of the root respiration is largely determined by a number of influences on C isotope fractionation during photosynthesis in the dominant trees in a forest. Air relative humidity and irradiation play important roles and can affect the isotopic signature of the soil respiration

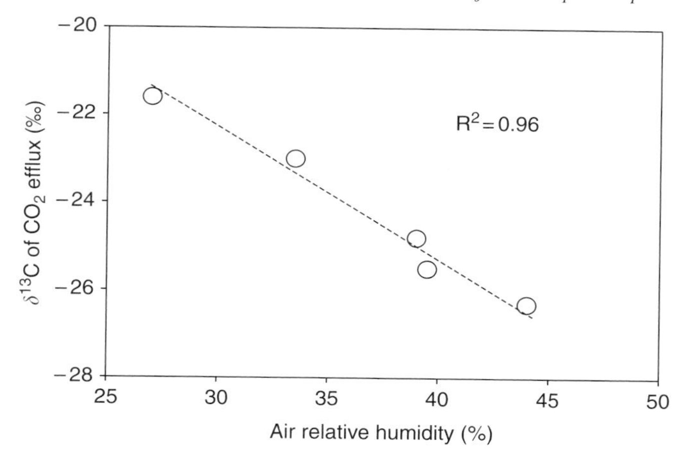

Figure 4.3 Correlation between $\delta^{13}C$ of the soil CO_2 efflux in a boreal coniferous forest and air relative humidity 3–4 days before soil respiration was sampled (Ekblad and Högberg, 2001, by permission of Springer-Verlag).

dynamically within days. Other factors, such as stand age and nutritional status, are also important, but cause much less dynamic effects.

What Determines the $\delta^{13}C$ of Heterotrophic Soil Respiration?

The ^{13}C of soil organic matter (SOM) is originally dependent on the average ^{13}C of plant photosynthates at the site. Balesdent *et al.* (1993) proposed, therefore, that the SOM at dry sites should have a higher ^{13}C than at wet sites. Data on the $\delta^{13}C$ of the soil C from a range of sites in Oregon were consistent with this hypothesis (Bowling *et al.*, 2002). However, there are circumstances when other factors influencing C_i/C_a may be more important, for example N supply, as discussed below. It is also most important to note that, with regard to heterotrophic respiration, there may be considerable time-lags between C fixation through photosynthesis and the decomposition in the soil of products of plant photosynthesis. For example, in boreal spruce trees, needles may be retained for 5 years or more before they are dropped as litter, and although much of the needle litter C is lost from the soil within a few years, fractions of this C may reside in the soil for hundreds of years (Harrison *et al.*, 2000). Because of this, and the large and dynamic component of root respiration, the $\delta^{13}C$ of soil-respired CO_2 may substantially deviate from that of SOM in field measurements (Fig. 4.4; see also Fig. 4.6).

Photosynthates are used for respiration, growth, or storage of C in the plants. Chemical compounds in plant cells have a non-random (non-statistical) distribution of ^{13}C as a result of isotope fractionations at

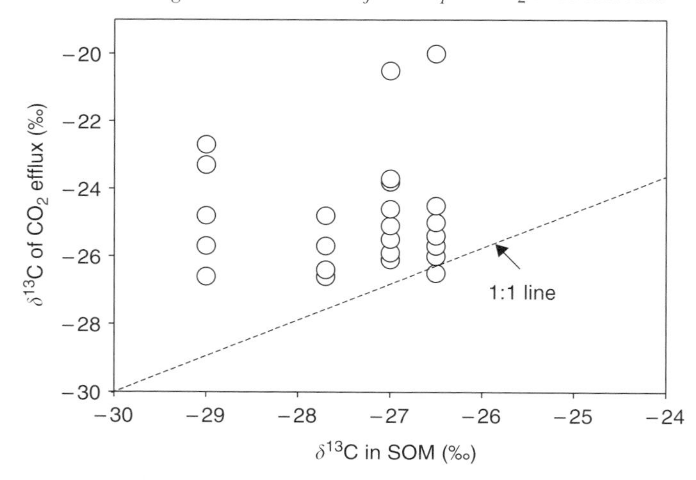

Figure 4.4 The ^{13}C abundance of the soil CO_2 efflux in the field, and the average $\delta^{13}C$ of root-free SOM from the mor-layer at the same four boreal forest sites (A. Ekblad, unpublished).

metabolic branch-points during their synthesis (Schmidt and Gleixner, 1998; Hobbie and Werner, 2004). Lignin and lipids are usually depleted relative to cellulose, while starch and some, but not all, sugars are enriched, depending on their positions along metabolic pathways (Schmidt and Gleixner, 1998; Hobbie and Werner, 2004). Hence, the $\delta^{13}C$ of litter deposited on the soil (above-ground litter) or in the soil (root litter) has been determined by dynamic effects of abiotic conditions on C isotope fractionation during photosynthesis as well as by subsequent isotopic fractionations during the synthesis of various compounds in the plants.

Once the organic matter reaches the soil system, it will be colonized and degraded by microbes. These microbes will alter its isotopic composition by being selective among components of the bulk substrate and potentially also because of fractionations during decomposition. With regard to the latter, several authors suggest that there may be a considerable isotopic fractionation during microbial decomposition and respiration (e.g., Blair *et al.*, 1985; Mary *et al.*, 1992). However, a problem in these studies has been that the C isotope budgets are not complete and that the fractionation reported may relate to shifts between C sources of different origins. For example, when microbes are grown on C_3 C and then shifted to a C_4 C source, they will continue to respire C_3 C for some time (Ekblad and Högberg, 1996; Ekblad and Högberg, 2000), which may be interpreted as evidence of isotopic fractionation.

In our experiments, we have not found evidence of significant isotopic fractionation during microbial respiration. When C_3 sucrose (−25.7‰) was added to root-free mor-layer soil from a boreal forest (Ekblad and

Högberg, 2000), the respiration increased threefold, but the C isotopic composition ($-26.5 \pm 0.6‰$) was not different from that of the control ($-26.4 \pm 1.9‰$) to which only water was added. When C_4 sucrose was added, respiration also increased threefold, but the isotopic composition changed and indicated a contribution from C_4 C as well as increased respiration of endogenous microbial C_3 C. In another study, we (Ekblad *et al.*, 2002) studied the possible importance of variations in fractionation related to the intramolecular distribution of C isotopes in the sugars applied (Henn and Chapela, 2000). We tested C_4-derived glucose against a mix of C_3-derived glucose and universally ^{13}C-enriched glucose. The calculated fractional contributions of the added C sources to soil respiration were very similar, suggesting that ^{13}C-discrimination is minor during microbial respiration.

Studies of plants also report conflicting evidence in this context (O'Leary, 1988; Ghashghaie *et al.*, 2003). A detailed study of dark respiration in protoplasts using substrates with known $\delta^{13}C$ provided strong evidence that fractionation during respiration is negligible (Lin and Ehleringer, 1997). However, as reviewed by Ghashghaie *et al.* (2003), there are reports of substantial enrichment in ^{13}C of dark-respired CO_2, although these studies could not define the exact $\delta^{13}C$ of the substrates. Metabolic fractionations occur, and the latter reports may, for example, reflect time lags in the studied system, and problems of identifying the substrates used and their $\delta^{13}C$. We suggest, therefore, that the idea of isotopic fractionation during respiration in plants and microbes should be subject to more critical testing.

There is, nevertheless, commonly a 3–4‰ enrichment in $\delta^{13}C$ of SOM down soil profiles in mature boreal forests (Fig. 4.5). According to biogeochemical theory this profile is caused by isotopic fractionations during decomposition and respiration (e.g., Nadelhoffer and Fry, 1988; Ågren *et al.*, 1996). As a consequence, the $\delta^{13}C$ of the respired CO_2 by heterotrophs decomposing the SOM should be isotopically lighter than the SOM. In our further discussion of the causation of variations in isotopic composition of SOM in the soil profile, we will rather closely follow that of Ehleringer *et al.* (2000), but add a few additional comments. Ehleringer *et al.* (2000) discussed four hypotheses as follows.

Hypothesis 1: Influence of Change in the $\delta^{13}C$ of the Source, Atmospheric CO_2

Since the onset of the Industrial Revolution the $\delta^{13}C$ of atmospheric CO_2 has decreased along with the increase in the concentration of CO_2 because of combustion of ^{13}C- (and ^{14}C-) depleted fossil fuels. This is sometimes referred to as the 'Suess' effect. The effect on the $\delta^{13}C$ of atmospheric CO_2 has been a decrease of about 1.3‰ (Friedli *et al.*, 1986; Trolier *et al.*, 1996), which has been transmitted through the plants via litter to the SOM. This

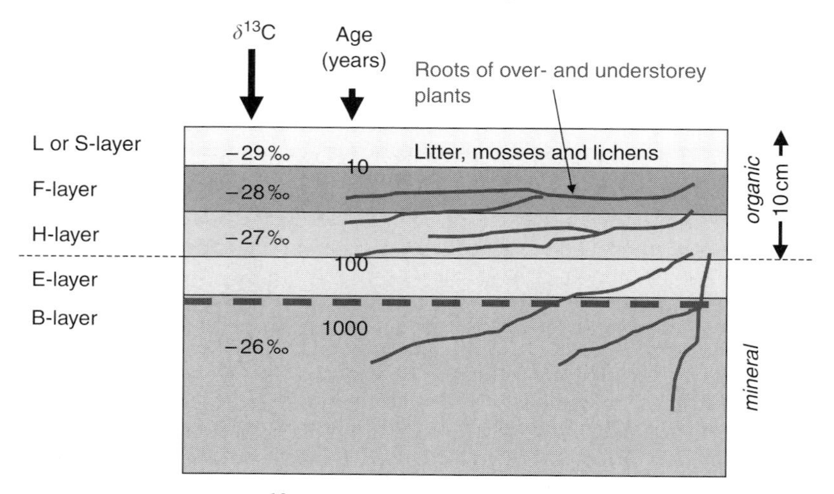

Figure 4.5 Variations in $\delta^{13}C$ (Bauer *et al.*, 2000; Högberg and Högberg, unpublished) and 'bomb' ^{14}C age of root-free SOM in the soil profile of boreal forest (Harrison *et al.*, 2000; Franklin *et al.*, 2003).

means that young organic C in the superficial layer is 1.3‰ lighter than the old organic C found deeper down, and can thus explain up to 50% of the enrichment of SOM with increasing soil depth.

Hypothesis 2: Microbial Fractionation During Litter Decomposition This is the classic explanation of the ^{13}C enrichment of SOM with increasing soil depth (e.g., Nadelhoffer and Fry, 1988). As was discussed earlier, there is no unequivocal evidence that this fractionation is significant. However, even a minimal fractionation during metabolism and respiration could cause a substantial enrichment if the fraction of C remaining in the soil becomes very small. To test this, one can make calculations using a modified Rayleigh equation (Mariotti *et al.*, 1981):

$$\delta_s = \delta_{s0} + \varepsilon \log_e f \tag{4.5}$$

where s stands for substrate, s0 stands for the substrate at time 0, f for the fraction of the original substrate that remains, and ε for the isotopic discrimination. We can thus test, as an example, the isotopic fractionation needed to produce a 1.5‰ enrichment of the remaining substrate (i.e., what is needed in addition to the Suess effect to produce a 3‰ gradient in the soil profile) when 1% of the C in a litter cohort remains in the soil (after several hundred years, or so, cf. Harrison *et al.*, 2000; Fig. 4.5). Equation 4.5 then shows that the fractionation needed is −0.3‰. Such a small instantaneous fractionation is hard to detect, but the above calculation tells us that we

in both the dominant trees and the SOM (Fig. 4.6). We agree, however, with Balesdent *et al.* (1993) that the $\delta^{13}C$ of the SOM is strongly correlated with that of the long-term C inputs from plants. Note that the isotopic signature of the surface of the organic mor-layer rather closely followed that of needle litter of the dominant trees (Fig. 4.6), suggesting that dominant trees contribute most C to SOM. We suspect that the unexpected increase in $\delta^{13}C$ of dominant plants and SOM in the direction of the discharge area is caused by an effect of an improved nutrient (N) supply on the ratio C_i/C_a in the dominant plants, and that this is more important than effects of variations in soil moisture under these generally humid conditions. Plant-available N increases several times along the gradient, and this should have a large effect on the rate of photosynthesis (Field and Mooney, 1986) and, hence, on the ratio C_i/C_a. Understorey plants had very low $\delta^{13}C$ values (down to $-33‰$) in the dense nutrient-rich forest (Fig. 4.6), but the ^{13}C data on SOM indicated that they did not contribute substantially to SOM.

Relevant to the discussion above, the $\delta^{13}C$ of CO_2 respired from root-free soil samples did not differ from that of the SOM (Ekblad, unpublished), while the CO_2 efflux in the field, as measured in 2002, was 2–5‰ enriched in ^{13}C relative to the SOM (Fig. 4.6). We suggest that this enrichment of the efflux in the field indicates a large influence of the root respiratory component (cf. Ekblad and Högberg, 2001). Note that the opposing trends in soil CO_2 efflux and SOM most likely reflect the fact that 2002 was a dry year; the high average $\delta^{13}C$ of the soil CO_2 efflux at 0 m is probably due to the high value for May, which was unusually dry. Note also that the difference in $\delta^{13}C$ between soil-respired CO_2 and SOM decreases in the direction of higher nutrient supply, which probably reflects a decrease in root respiration (Högberg *et al.*, 2003).

Concluding Remarks

There is a significant spatial and temporal variability in $\delta^{13}C$ in the soil respiration in boreal ecosystems. This encompasses variation in the ratio between the two main components, root and heterotrophic respiration, as well as dynamic effects of environmental conditions on the root respiration component, in particular. This compounded variability may seem discouraging to modelers. However, as it does not appear to be random, it seems to offer unique possibilities to clarify aspects of the C dynamics in boreal ecosystems and their contribution to the atmospheric CO_2 balance. We are convinced that approaches combining measurements of ^{13}C with analyses of ^{14}C and ^{18}O in the soil respiratory efflux will significantly increase our ability to understand the dynamics of these systems. Further detailed process studies in the laboratory are needed to clarify the complex isotopic fractionations occurring in soil microorganisms and their influence on the $\delta^{13}C$ of the soil CO_2 efflux. There is also a need for a detailed isotopic mass balance at the scale of a forest stand, especially in view of our incomplete understanding of the ^{13}C enrichment with soil depth in settings where the soil CO_2 efflux commonly has a higher $\delta^{13}C$ than the SOM at the surface. This discrepancy probably in part reflects the large contribution made by root respiration and temporal disequilibria among ecosystem C pools, but we would like to see further and more conclusive evidence that this is the case.

References

Ågren G. I., Bosatta E. and Balesdent J. (1996) Isotope discrimination during decomposition of organic matter—a theoretical analysis. *Soil Sci Soc Am J* **60**: 1121–1126.

Amundson R., Stern L., Baisden T. and Wang Y. (1998) The isotopic composition of soil and soil-respired CO_2. *Geoderma* **82**: 83–114.

Balesdent J., Girardin A. and Mariotti A. (1993) Site-related $\delta^{13}C$ of tree leaves and soil organic matter in a temperate forest. *Ecology* **74**: 1713–1721.

Bauer G., Gebauer G., Harrison T., Högberg P., Högbom L., Schinkel H., Taylor A. F. S., Novak M., Harkness D. D., Persson T. and Schulze E.-D. (2000) Biotic and abiotic controls over ecosystem cycling of stable natural nitrogen, carbon and sulphur isotopes. *Ecol Stud* **142**: 189–214.

Berg B. (1986) Nutrient release from litter and humus in coniferous forest soils—a mini review. *Scand J For Res* **1**: 359–369.

Bhupinderpal-Singh, Nordgren A., Ottosson-Löfvenius M., Högberg M. N., Mellander P.-E. and Högberg P. (2003) Mycorrhizal root and soil heterotrophic respiration as revealed by girdling of boreal Scots pine forest: extending observations beyond the first year. *Plant Cell Environ* **26**: 1287–1296.

Blair N., Leu A., Munos E., Olsen J., Kwong E. and DesMarais D. (1985) Carbon isotope fractionation in heterotrophic microbial metabolism. *Appl Environ Microbiol* **50**: 996–1001.

Bowling D. R., McDowell N. G., Bond B. J., Law B. E. and Ehleringer J. R. (2002) ^{13}C content of ecosystem respiration is linked to precipitation and vapor pressure deficit. *Oecologia* **131**: 113–124.

Brooks J. R., Flanagan L. B., Buchmann N. and Ehleringer J. R. (1997) Carbon isotope composition of boreal plants: functional grouping of life forms. *Oecologia* **110**: 301–311.

Brugnoli E., Hubick K. T., von Caemmerer S., Wong S. C. and Farquhar G. D. (1988) Correlation between the carbon isotope discrimination in leaf starch and sugars of C3 plants and the ratio of intercellular and atmospheric partial pressures of carbon dioxide. *Plant Physiol* **88**: 1418–1424.

Brugnoli E., Scartazza A., Lauteri M., Monteverdi M. C. and Maguas C. (1998) Carbon isotope discrimination in structural and non-structural carbohydrates in relation to productivity and adaptation to unfavourable conditions. In *Stable Isotopes* (H. Griffiths, ed.) pp. 133–146. BIOS Scientific Publishers, Oxford.

Buchmann N., Brooks J. R. and Ehleringer J. R. (2002) Predicting daytime carbon isotope ratios of atmospheric CO_2 within forest canopies. *Funct Ecol* **16**: 49–57.

Buchmann N., Kao W.-Y. and Ehleringer J. R. (1996) Carbon dioxide concentrations within forest canopies—variation with time, stand structure, and vegetation type. *Global Change Biol* **2**: 421–432.

Cannell M. G. R. (1989) Physiological basis of wood production: review. *Scand J For Res* **4**: 459–490.

Cerling T. E., Solomon D. K., Quade J. and Bowman J. R. (1991) On the isotopic composition of carbon in soil carbon dioxide. *Geochim Cosmochim Acta* **55**: 3404–3405.

Cerling T. E. and Wang Y. (1996) Stable carbon and oxygen isotopes in soil CO_2 and soil carbonate: Theory, practice, and application to some prairie soils of upper Midwestern North America. In *Mass Spectrometry of Soils* (T. W. Boutton and S. Yamasaki, eds) pp. 113–131. Marcel Dekker, New York.

Ciais P., Tans P. P., White J. C. W., Trolier M., Francey R. J., Berry J. A., Randall D. R., Sellers P. J., Collatz J. G. and Schimel D. S. (1995a) Partitioning of ocean and land uptake of CO_2 as inferred by $\delta^{13}C$ measurements from the NOAA climate monitoring and diagnostics laboratory global air sampling network. *J Geophys Res* **100**: 5051–5070.

Ciais P., Tans P. P., Trolier M., White J. W. C. and Francey R. J. (1995b) A large northern hemisphere terrestrial CO_2 sink indicated by the $^{13}C/^{12}C$ ratio of atmospheric CO_2. *Science* **269**: 1098–1102.

Ehleringer J. R., Buchmann N. and Flanagan L. B. (2000) Carbon isotope ratios in belowground carbon cycle processes. *Ecol Appl* **10**: 412–422.

Ekblad A. and Högberg P. (2000) Analysis of $\delta^{13}C$ of CO_2 distinguishes between microbial respiration of added C4-sucrose and other soil respiration in a C3-ecosystem. *Plant Soil* **219**: 197–209.

Ekblad A. and Högberg P. (2001) Natural abundance of ^{13}C in CO_2 respired from forest soils reveals speed of link between tree photosynthesis and root respiration. *Oecologia* **127**: 305–308.

Ekblad A., Nyberg G. and Högberg P. (2002) ^{13}C discrimination during microbial respiration of added C3-, C4- and ^{13}C-labelled sugars to a C3-forest soil. *Oecologia* **131**: 245–249.

Farquhar G. D. (1991) Use of stable isotopes in evaluating plant water use efficiency. In *Stable Isotopes in Plant Nutrition, Soil Fertility and Environmental Studies*, pp. 475–488. IAEA, Vienna.

Farquhar G. D., O'Leary M. H. and Berry J. A. (1982) On the relationship between carbon isotope discrimination and photosynthesis. *Aust J Plant Physiol* **9**: 121–137.

Farquhar G. D., Ehleringer J. R. and Hubick K. T. (1989) Carbon isotope discrimination and photosynthesis. *Annu Rev Plant Physiol Mol Biol* **40**: 503–537.

Fessenden J. E. and Ehleringer J. R. (2002) Age-related variations in $\delta^{13}C$ of ecosystem respiration across a coniferous forest chronosequence in the Pacific Northwest. *Tree Physiol* **22**: 159–167.

Field C. B. and Mooney H. A. (1986) The photosynthesis–nitrogen relationships in wild plants. In *On the Economy of Plant Form and Function* (T. Givnish, ed.) pp. 22–55. Cambridge University Press, Cambridge.

Flanagan L. B. and Ehleringer J. R. (1998) Ecosystem–atmosphere CO_2 exchange: interpreting signals of change using stable isotope ratios. *Tree* **13**: 10–14.

Flanagan L. B., Kubien D. S. and Ehleringer J. R. (1999) Spatial and temporal variability in the carbon and oxygen stable isotope ratio of respired CO_2 in a boreal forest ecosystem. *Tellus* **51**B: 367–384.

Franklin O., Högberg P., Ekblad A. and Ågren G. I. (2003) Pine forest floor carbon accumulation in response to N and PK additions—bomb ^{14}C modelling and respiration studies. *Ecosystems* **6**: 644–658.

Friedli H., Lötscher H., Oeschger H., Siegenthaler U. and Stauffer B. (1986) Ice core records of the $^{13}C/^{12}C$ ratio of atmospheric CO_2 in the past two centuries. *Nature* **324**: 237–238.

Fung I. Y., Field C. B., Berry J. A., Thompson M. V., Randerson J. T., Malmstrom C. M. and Vitousek P. M. (1997) Carbon-13 exchanges between the atmosphere and biosphere. *Global Biogeochem Cycles* **11**: 507–533.

Garbaye J. (1994) Helper bacteria: a new dimension to the mycorrhizal symbiosis. *Tansley Rev* 76. *New Phytol* **128**: 197–210.

Gaudinski J., Trumbore S. E., Davidson E. A. and Zheng S. (2000) Soil carbon cycling in temperate forest: radiocarbon-based estimates of residence times, sequestration rates and partitioning of fluxes. *Biogeochemistry* **51**: 33–69.

Gaudinski J., Trumbore S. E., Davidson E. A., Cook A. C., Markewitz D. and Richter D. D. (2001) The age of fine-root carbon in three forests of the United States measured by radiocarbon. *Oecologia* **129**: 420–429.

Ghashghaie J., Badeck F.-W., Lanigan G., Nogues S., Tcherkez G., Deléens E., Cornic G. and Griffiths H. (2003) Carbon isotope fractionation during dark respiration and photorespiration in C3 plants. *Phytochem Rev* **2**: 145–161.

Giesler R., Högberg M. and Högberg P. (1998) Soil chemistry and plants in Fennoscandian boreal forest as exemplified by a local gradient. *Ecology* **79**: 119–137.

Gleixner G., Danier H.-J., Werner R. A. and Schmidt H.-L. (1993) Correlations between the ^{13}C content of primary and secondary plant products in different cell compartments and that in decomposing basidiomycetes. *Plant Physiol* **102**: 1287–1290.

Hansen J., Türk R., Vogg G., Heim R. and Beck E. (1997) Conifer carbohydrate physiology: updating classical views. In *Trees—Contributions to Modern Tree Physiology* (H. Rennenberg, W. Eschrich and H. Ziegler, eds), pp. 97–108. Backhuys, Leiden.

Hanson P. J., Edwards N. T., Garten C. T. and Andrews J. A. (2000) Separating root and soil microbial contributions to soil respiration: a review of methods and observations. *Biogeochem* **48**: 115–146.

Harrison A. F., Harkness D. D., Rowland A. P., Garnett J. S. and Bacon P. J. (2000) Annual carbon and nitrogen fluxes in soils along the European forest transect, determined using ^{14}C-bomb. *Ecol Stud* **142**: 237–256.

Henn M. R. and Chapela I. H. (2000) Differential C isotope discrimination by fungi during decomposition of C3- and C4-derived sucrose. *Appl Environ Microbiol* **66**: 4180–4186.

Henn M. R. and Chapela I. H. (2001) Ecophysiology of ^{13}C and ^{15}N isotopic fractionation in forest fungi and the roots of the saprotrophic-mycorrhizal divide. *Oecologia* **128**: 480–487.

Henn M. R., Gleixner G. and Chapela I. H. (2002) Growth-dependent stable carbon isotope fractionation by basidiomycete fungi: $\delta^{13}C$ pattern and physiological process. *Appl Environ Microbiol* **68**: 4956–4964.

Hobbie E. A., Macko S. A. and Shugart H. H. (1999) Insights into nitrogen and carbon dynamics of ectomycorrhizal and saprotrophic fungi from isotopic evidence. *Oecologia* **118**: 353–360.

Hobbie E. A., Weber N. S., Trappe J. M. and van Klinken G. J. (2002) Using radiocarbon to determine the mycorrhizal status of fungi. *New Phytol* **156**: 129–136.

Hobbie E. A. and Werner R. A. (2004) Intramolecular, compound-specific, and bulk carbon isotope patterns in C3 and C4 plants: a review and synthesis. *New Phytol* **161**: 371–385.

Högberg P. and Ekblad A. (1996) Substrate-induced respiration measured *in situ* in a C3-plant ecosystem using additions of C4-sucrose. *Soil Biol Biochem* **28**: 1131–1138.

Högberg P., Högbom L., Schinkel H., Högberg M., Johannisson C. and Wallmark H. (1996) ^{15}N abundance of surface soils, roots and mycorrhizas in profiles of European forest soils. *Oecologia* **108**: 207–214.

Högberg P., Plamboeck A. H., Taylor A. F. S. and Fransson P. M. A. (1999) Natural ^{13}C abundance reveals trophic status of fungi and host-origin of carbon in mycorrhizal fungi in mixed forests. *Proc Natl Acad Sci USA* **96**: 8534–8539.

Högberg P., Nordgren A., Buchmann N., Taylor A. F. S., Ekblad A., Högberg M. N., Nyberg G., Ottosson-Löfvenius M. and Read D. J. (2001) Large-scale forest girdling shows that current photosynthesis drives soil respiration. *Nature* **411**: 789–792.

Högberg M. N. and Högberg P. (2002) Extramatrical ectomycorrhizal mycelium contributes one-third of microbial biomass and produces, together with associated roots, half the dissolved organic carbon in a forest soil. *New Phytol* **154**: 791–795.

Högberg P., Nordgren A. and Ågren G. I. (2002) Carbon allocation between tree root growth and root respiration in boreal pine forest. *Oecologia* **132**: 579–581.

Högberg M. N., Bååth E., Nordgren A., Arnebrant K. and Högberg P. (2003) Contrasting effects of nitrogen availability on plant carbon supply to mycorrhizal fungi and saprotrophs—a hypothesis based on field observations in boreal forests. *New Phytol* **160**: 225–238.

Jenkinson D. S. and Rayner J. H. (1977) The turnover of soil organic matter in some of the Rothamstead classical experiments. *Soil Sci* **123**: 298–305.

Keeling C. D. (1958) The concentration and isotopic abundances of carbon dioxide in rural areas. *Geochim Cosmochim Acta* **13**: 322–334.

Keeling R. F., Piper S. C. and Heimann M. (1996) Global and hemispheric CO_2 sinks deduced from changes in atmospheric O_2 concentration. *Nature* **381**: 218–221.

Leavitt S. W. and Long A. (1986) Stable-carbon isotope variability in tree foliage and wood. *Ecology* **67**: 1002–1010.

Lin G. and Ehleringer J. R. (1997) Carbon isotopic fractionation does not occur during dark respiration in C3 and C4 plants. *Plant Physiol* **114**: 391–394.

Mariotti A., Germon G. C., Hubert P., Kaiser P., Létolle R., Tardieux A. and Tardieux P. (1981) Experimental determination of nitrogen kinetic isotope fractionation: some principles; illustrations for the denitrification and nitrification processes. *Plant Soil* **62**: 413–430.

Mary B., Mariotti A. and Morel J. L. (1992) Use of ^{13}C variations at natural abundance for studying the biodegradation of root mucilage, roots and glucose in soil. *Soil Biol Biochem* **24**: 1065–1072.

Melillo J. M., Aber J. D. and Muratore J. F. (1982) Nitrogen and lignin control of hardwood leaf litter decomposition dynamics. *Ecology* **63**: 621–626.

Myneni R. B., Keeling C. D., Tucker C. J., Asrar G. and Nemani R. R. (1997) Increased plant growth in the northern high latitudes from 1981 to 1991. *Nature* **386**: 698–702.

Nadelhoffer K. J. and Fry B. (1988) Controls on natural nitrogen-15 and carbon-13 abundances in forest soil organic matter. *Soil Sci Soc Am J* **52**: 1633–1640.

O'Leary M. H. (1988) Carbon isotopes in photosynthesis. *BioScience* **38**: 328–336.

Pate J. and Arthur D. (1998) $\delta^{13}C$ analysis of phloem sap as a novel means of evaluating seasonal water stress and interpreting carbon isotope signatures of foliage and trunk wood of *Eucalyptus globulus*. *Oecologia* **117**: 301–311.

Plamboeck A. H. (1999) Root activity in Scots pine dominated stands assessed by isotopic methods. PhD thesis, Series Silvestria, No. 112. Swedish University of Agricultural Sciences, Umeå.

Post A. M., Emanuel W. R., Zinke P. J. and Strangenberger A. G. (1982) Soil carbon pools and world life zones. *Nature* **298**: 156–159.

Rayment M. B. (2000) Closed chamber systems underestimate soil CO_2 efflux. *Eur J Soil Sci* **51**: 107–110.

Rochette P., Flanagan L. B. and Gregorich E. G. (1999) Separating soil respiration into plant and soil components using analyses of the natural abundance of carbon-13. *Soil Sci Soc Am J* **63**: 1207–1213.

Saurer M., Borella S., Schweingruber F. and Siegwolf R. (1997) Stable carbon isotopes in tree rings of beech: climatic versus site-related influences. *Trees* **11**: 291–297.

Schmidt H.-L. and Gleixner G. (1998) Carbon isotope effects on key reactions in plant metabolism and ^{13}C-patterns in natural compounds. In *Stable Isotopes* (H. Griffiths, ed.) pp. 13–25. BIOS Scientific Publishers, Oxford.

Smith B. N. and Epstein S. (1971) Two categories of $^{13}C/^{12}C$ ratios for higher plants. *Plant Physiol* **47**: 380–384.

Smith S. E. and Read D. J. (1997) *Mycorrhizal Symbiosis.* Academic Press, London, UK.

Tans P. P., Berry J. A. and Keeling R. F. (1993) Oceanic $^{13}C/^{12}C$ observations: a new window on ocean CO_2 uptake. *Global Biogeochem Cycles* **7**: 353–368.

Taylor A. F. S., Högbom L., Högberg M., Lyon T. E. J., Näsholm T. and Högberg P. (1997) Natural ^{15}N abundance in fruit bodies of ectomycorrhizal fungi from boreal forest. *New Phytol* **136**: 713–720.

Taylor A. F. S., Martin F. and Read D. J. (2000) Fungal diversity in ectomycorrhizal communities of Norway spruce [*Picea abies* (L.) Karst.] and beech (*Fagus sylvatica* L.) along north–south transects in Europe. *Ecol Stud* **142**: 343–365.

Taylor A. F. S., Fransson P. M. A., Högberg P., Högberg M. N. and Plamboeck A. H. (2003) Species level patterns in ^{15}N and ^{13}C abundance of ectomycorrhizal and saprotrophic fungal sporocarps. *New Phytol* **159**: 757–774.

Timonen S., Jörgensen, K. S., Haatela K. and Sen R. (1998) Bacterial community structure at defined locations of *Pinus sylvestris-Suillus bovinus* and *Pinus sylvestris-Paxillus involutus* mycorrhizosphere in dry pine forest humus and nursery peat. *Can J Microbiol* **44**: 499–513.

Trolier M., White J. W. C., Tans P. P., Masarie K. A. and Gemery P. A. (1996) Monitoring the isotopic composition of atmospheric CO_2: measurements from NOAA global air sampling network. *J Geophys Res* **101**: 25897–25916.

Valentini R., Matteuci G., Dolman A. J., Schulze E.-D., Rebmann C., Moors E. J., Granier A., *et al.* (2000) Respiration as the main determinant of carbon balance in European forests. *Nature* **404**: 861–865.

Wallander H., Nilsson, L.-O., Hagerberg D. and Bååth E. (2001) Estimation of the biomass and seasonal growth of ectomycorrhizal fungi in the field. *New Phytol* **151**: 753–760.

Walton A., Ergin M. and Harkness D. D. (1970) Carbon-14 concentrations in the atmosphere and carbon exchange rates. *J Geophys Res* **75**: 3089–3098.

5

Factors that Control the Isotopic Composition of N_2O from Soil Emissions

Tibisay Pérez

Introduction

Nitrous oxide (N_2O) is an important greenhouse gas. On a per-molecule basis its radiative forcing potential is 296 times that of CO_2 averaged over a 100-year period (Yung *et al.*, 1976; Ehalt *et al.*, 2001). In addition, nitrous oxide plays other important roles in global environmental change. For example, destruction of N_2O in the stratosphere via photolysis and photo-oxidation provides the main source of NO_X, a family of chemical species that contribute to the catalytic destruction of stratospheric ozone. Because of these important roles, there is strong interest in understanding the mechanisms contributing to nitrous oxide production and consumption and how these sources and sinks are changing over time.

The concentration of N_2O has varied significantly during geologic times with very low values during glacial periods (~200 ppbv) and higher concentrations during interglacial periods (~270 ppbv) (Fluckiger *et al.*, 1999). Currently the tropospheric mixing ratio of nitrous oxide is about 316 ppbv and it is increasing at a rate of ~0.25% yr^{-1} (Weiss, 1981). The known sources of N_2O do not balance the known tropospheric accumulation and stratospheric sink (Cicerone, 1989; Prather *et al.*, 1995).

The primary source of N_2O is that produced by bacterial activity during nitrification and denitrification in soil and in the ocean (Khalil and Rasmussen, 1992). Nitrous oxide is also produced during biomass burning and during combustion of fossil fuel in automobiles and industrial processes (Thiemens and Trogler, 1991). The main sink for N_2O is stratospheric destruction (Khalil and Rasmussen, 1992). Nitrous oxide emissions from soil and the ocean have high spatial and temporal variability. Due to the uncertainty in the magnitude of all the sources, it has been difficult to develop a global budget for N_2O.

Stable isotope studies may contribute significantly to understanding the global budget for nitrous oxide. The nitrogen and oxygen isotope composition of N_2O differs among sources and varies with the mechanism of production (nitrification and denitrification). A preliminary simplified global budget for tropospheric N_2O showed that it represents a balance between the input of sources from soil and the ocean that are depleted in heavier isotopes (^{15}N and ^{18}O) and a return flux of nitrous oxide from the stratosphere that is enriched in the heavier isotopes (Kim and Craig, 1993; Dore *et al.*, 1998; Rahn and Wahlen, 2000; Pérez *et al.*, 2001).

The objective of this chapter is to describe in further detail the mechanisms that cause variation in the stable isotope composition of N_2O produced and emitted from soil in terrestrial ecosystems. This information will illustrate how stable isotope techniques can be used to differentiate between nitrous oxide produced during nitrification and denitrification, and how stable isotope measurements can help to improve our understanding of the relative contributions of different sources to the global tropospheric budget for N_2O.

Factors that Control the Rate of Production and Emission of Soil N_2O

Nitrous oxide is produced and consumed by bacterial activity associated with the soil nitrogen cycle. Production of N_2O can occur in association with both nitrification and denitrification. During nitrification NH_4^+ is oxidized to NO_3^-, and in the process NO and N_2O can be produced as intermediate by-products. Under anaerobic conditions (or in anaerobic microsites within an otherwise aerobic soil) nitrate is reduced to N_2, again with the intermediate by-products NO and N_2O. The NO and N_2O that are dissolved in the soil water phase will diffuse out of the soil into the atmosphere if they are not chemically transformed (oxidized to form nitrate or reduced to form N_2). There are three general factors that affect the rate of N_2O production and its emission from soils: (1) availability of substrate (ammonium and nitrate), which affects the total flux rate of material through the nitrogen cycle and therefore the amount of nitrous oxide formed (Keller *et al.*, 1988); (2) the efficiency of N_2O production and consumption during nitrogen cycle processes. This is in turn controlled by the relative rates of nitrification and denitrification, which change with environmental conditions; (3) efficiency of N_2O diffusion out of the soil, which is controlled by the amount of water in soils (water-filled pore space) and by soil texture (Matson and Vitousek, 1990; Matson *et al.*, 1990; Davidson, 1992; Keller and Reiners, 1994; Davidson and Schimel, 1995). A number of recent studies have examined the importance of these factors, as is briefly reviewed below.

A major reason for variation in N_2O and NO emissions between different soil systems may be changes in the relative contributions of nitrification and denitrification. Several parameters that would allow prediction of the relative contribution of nitrification and denitrification to the emitted NO, N_2O, and N_2 gases have been proposed. First, it has been proposed that at an optimum water-filled pore space (WFPS) of 60%, nitrification is the dominant process, and NO is the predominant nitrogen gas emitted. At higher values of WFPS (between 60 and 90%), soil aeration becomes limited, denitrification activity approaches a maximum, and N_2O becomes the most abundant gas emitted. However, when WFPS is greater than 90%, N_2O produced by denitrification is reduced to N_2, which then becomes the most abundant gas emitted (Davidson, 1991, 1993; Keller and Reiners, 1994; Dendooven *et al.*, 1996). Second, a widely accepted assumption is that when the N_2O/NO ratio is larger than one, denitrification is the main process producing N_2O emissions (Davidson, 1993). Third, some studies found that N_2O emissions were higher after NO_3^- rather than NH_4^+ fertilization (Keller *et al.*, 1988; Livingston *et al.*, 1988; Bakwin *et al.*, 1990). This suggested that the greater part of N_2O is produced by denitrification in soils.

The generalizations presented above have controversial aspects, however, and recent experimental results have shown that the production of N_2O by nitrification or denitrification in a particular soil is not easy to predict. For instance, it has been found that nitrifier bacteria can produce significant amounts of NO and N_2O under very anaerobic soil conditions (Bollmann and Conrad, 1998). De Klein and Van Logtestijn (1996) found no correlation between denitrification rates and the rate of NO_3^- addition. The available techniques (non-isotope techniques) for estimating the relative contribution of nitrification and denitrification to the emitted N_2O are invasive methods that might not adequately represent the soil under natural field conditions. Since most of the techniques used to partition N_2O sources are not done *in situ* but in laboratory incubations, the possibility of creating experimental artefacts is high.

Measurement of the stable isotope composition of N_2O is an important tool that can be used to determine the relative contribution of production, consumption, and diffusion to N_2O emissions and test some of the ideas discussed above. The advantage of using stable isotopes is that they have the potential of differentiating processes *in situ*. The isotope effects for nitrification and denitrification are different and vary in accordance with the same factors that control the rate of production and consumption of nitrous oxide.

Processes Controlling the Stable Isotope Composition of N_2O

Stable isotope compositions are presented in this chapter using δ notation, where $\delta = ((R_{sample}/R_{standard}) - 1)$, and $R = {}^{15}N/{}^{14}N$ or ${}^{18}O/{}^{16}O$. The standards used are atmospheric N_2 for ${}^{15}N/{}^{14}N$, and standard mean ocean water (SMOW) for ${}^{18}O/{}^{16}O$. The δ values are conveniently expressed in parts per thousand or per mil (‰).

Effect of Substrate Availability on the $\delta^{15}N$ of N_2O

Mass-dependent isotope effects occur during chemical transformations, with molecules containing the lighter isotope (${}^{14}N$) reacting faster than the molecules containing the heavier isotope (${}^{15}N$). As a consequence the product of a reaction is depleted in ${}^{15}N$ relative to the isotope composition of the substrate pool. Assuming complete removal of products and no further inputs of substrate, the substrate pool will become enriched in the heavy isotope as the reaction proceeds over time. The change in the isotope composition of the substrate pool will in turn result in a progressive increase in the heavy isotope composition of the product pool as the substrate is consumed. Changes in the availability of substrates (ammonium and nitrate) for N_2O production should cause, therefore, temporal shifts in the $\delta^{15}N$ of N_2O produced in soils. This idea has been tested under field conditions by adding pulses of substrate (urine, fertilizer) to soil and monitoring temporal shifts in the isotope composition of N_2O emitted from soils, as will be discussed below.

After urea fertilization in agricultural fields it has been observed that the final product of nitrification (NO_3^-) has a lower $\delta^{15}N$ value than the substrate (NH_4^+), and nitrate becomes more enriched in ${}^{15}N$ as time progresses when substrate availability is limiting (Nadelhoffer and Fry, 1994). The $\delta^{15}N$ of N_2O from fertilized soils and agricultural fields has also been shown to increase with time after soil amendments (Pérez *et al.*, 2001; Yamulki *et al.*, 2001). For example, in a subtropical area of Mexico the isotope composition of N_2O emitted from an agricultural soil shifted dramatically over a 2-week period after urea fertilization and irrigation, with $\delta^{15}N$ values ranging from highly depleted values (−46‰) during the first week when N_2O emissions were high, to enriched $\delta^{15}N$ values (+5‰) at the end of the second week when emissions were low (Pérez *et al.*, 2001). It was also observed that the $\delta^{15}N$ values of substrate NH_4^+ changed from 2 to 25‰ during the 2-week period. This suggests that the N_2O isotopic composition changes are related to the substrate availability, if nitrification was the major pathway of N_2O production. Although it is difficult to deduce that all the N_2O emitted was derived from nitrification in this study, there was remarkable similarity

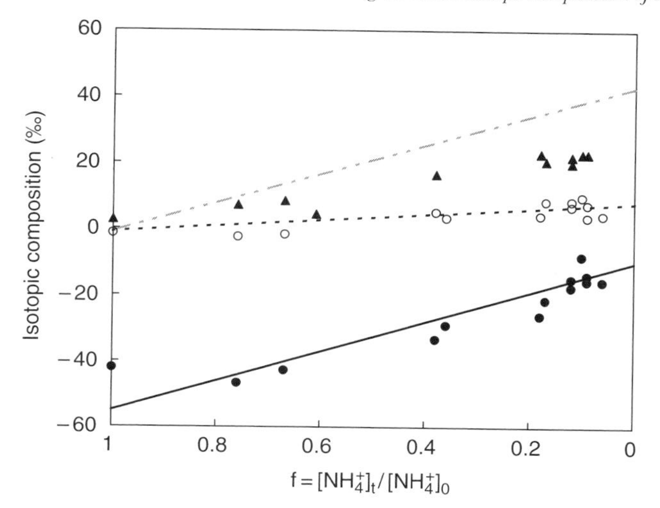

Figure 5.1 Comparison of measurements of the stable isotope composition of nitrous oxide and ammonium made in a field study with urea-fertilized soil (Pérez *et al.*, 2001) and a lab study using a pure culture of ammonium-oxidizing bacteria *Nitrosomonas* sp. (Ueda *et al.*, 1999). The stable isotope data (‰) are plotted as a function of changes in the concentration of ammonium during the experiment (NH_4^+ t), expressed relative to initial starting concentration of ammonium (NH_4^+ 0). For the field study the filled and open circles are the δ^{15}N-N_2O and δ^{18}O-N_2O values, respectively; and the filled triangles are the δ^{15}N-NH_4^+ values. The data for the incubation study of Ueda *et al.* (1999) are shown as regression lines; the black and dashed lines represent the δ^{15}N-N_2O and δ^{18}O-N_2O values, respectively; and the dot-dash line represents the δ^{15}N-NH_4^+ values.

between these results and those of a study conducted with a pure culture of *Nitrosomonas* sp. (Ueda *et al.*, 1999; Fig. 5.1), which strongly suggests that the N_2O isotopic changes in the agricultural field were mostly a product of nitrification. *Nitrosomonas* bacteria are exclusively involved with nitrification. In another agricultural study the δ^{15}N value of N_2O emitted from a soil fertilized with urine showed a change over a 24-hour period from −12 to 0‰ (Yamulki *et al.*, 2001). The magnitude of the emission rate did not correlate with the isotopic shift, but the enrichment was progressively higher with time. This suggests that substrate consumption led to a progressive enrichment of the produced N_2O. Unfortunately, in this study no measurements were made of either the concentration or δ^{15}N of NH_4^+ in order to support this conclusion.

Isotope Effects (δ^{15}N) During Diffusion of N_2O in Soil

The effect of diffusion on the isotopic composition of N_2O has been studied in Brazilian tropical rain forest soils (Pérez *et al.*, 2000). Soil air was collected at different depths and the $\delta^{15}N$ values of N_2O in the soil air were measured. The authors found an enrichment of ^{15}N in N_2O at a depth interval of 75–100 cm and the $\delta^{15}N$ values of N_2O decreased in soil layers above and below this depth. The high $\delta^{15}N$ values in the 75–100 cm depth interval suggested that this was a zone of N_2O production via denitrification and that N_2O diffused both upward towards the surface and downward to greater soil depths from this zone. Molecular diffusion of N_2O would be expected to have an isotope effect of 4.35‰ for $\delta^{15}N$. The soil showed a difference in $\delta^{15}N$ values of 4.6 ± 0.6‰ between 75 cm and deeper soil layers, which was consistent with that expected from fractionation caused by molecular diffusion. Therefore the isotopic composition of the N_2O emitted from these soils should reflect the $\delta^{15}N$ of N_2O production and the isotopic shift due to molecular diffusion.

Isotope Effects ($\delta^{15}N$) During Nitrification and Denitrification

As discussed above, normal mass-dependent effects result in products of a chemical reaction being depleted in heavy isotopes relative to the initial substrate pool. Consistent with this, it has been observed that N_2O formed during nitrification is depleted in ^{15}N relative to NH_4^+ (Létolle 1980; Mariotti *et al.* 1981; Yoshida, 1988), and that N_2O produced during denitrification is depleted in ^{15}N relative to NO_3^- (Table 5.1). Since different bacterial populations are involved in nitrification and denitrification, the magnitudes of the isotope effects are different for these two production pathways. There is a much larger isotope effect for nitrification than for denitrification, when the isotopic composition of N_2O is compared to that of the substrates ammonium and nitrate, respectively (Table 5.1; Fig. 5.2).

If it was assumed that the nitrogen cycle processes started with ammonium and proceeded in a linear sequence, $NH_4^+ \rightarrow N_2O \rightarrow NO_3^- \rightarrow N_2O$, the $\delta^{15}N$ of N_2O produced by denitrification should be lower than that of nitrous oxide produced by nitrification, because of mass-dependent isotope fractionation. However, a number of studies have reported that N_2O produced via denitrification has higher $\delta^{15}N$ values than that produced via nitrification (Wahlen and Yoshinari, 1985; Yoshida, 1988; Yoshinari and Koike, 1994; Webster and Hopkins, 1996; Barford, 1997; Yoshinari *et al.*, 1997; Barford *et al.*, 1999). The apparent contradiction can be resolved by considering two other isotope effects that operate simultaneously. First, as N_2O is formed some may be lost from the soil via diffusion to the atmosphere. Fractionation during diffusion results in the remaining soil pool of N_2O becoming enriched in ^{15}N by approximately 4.35‰. The isotopic composition of the soil N_2O pool is also influenced by the extent to which it is consumed to form nitrate (during nitrification) and N_2 (during denitrification).

Table 5.1 Nitrogen and Oxygen Isotope Effects for Several Reactions Involved in N_2O Production and Consumption During Nitrogen Cycle Reactions

Stable isotope	Reaction	Organism	Isotope effect (‰)	Reference
^{15}N	$NH_4^+ \rightarrow N_2O$	*Nitrosomonas europeae*	-66.5 ± 2.3	Yoshida *et al.* (1989)
		Nitrosomonas sp.	-45.3 to -46.6 (range)	Ueda *et al.* (1999)
	$NH_2OH \rightarrow N_2O$	*Nitrosomonas europeae*	-26.0	Sutka *et al.* (2003)
		Methylococcus capsulatus (methane oxidizer)	2.3	
	$NO_2^- \rightarrow N_2O$	*Nitrosomonas europeae*	-35.9	Sutka *et al.* (2003)
	$NO_3^- \rightarrow N_2O$	*Paracoccus denitrificans*	-28.6 ± 1.9	Barford *et al.* (1999)
		Soil denitrifier	-27 to -16 (range)	Wada and Ueda (1996) (and references therein)
	$N_2O \rightarrow N_2$	*Paracoccus denitrificans*	-13 ± 2.6	Barford *et al.* (1999)
		Soil denitrifier	-27	Wada and Ueda (1996)
^{18}O	$NO_3^- \rightarrow N_2O$	*Paracoccus denitrificans*	-105	Barford (1997)
	$N_2O \rightarrow N_2$	*Pseudomonas aeruginosa*	-37 to -42 (range)	Wahlen and Yoshinari (1985)

The isotope effects are expressed as:

$$\varepsilon(‰) = (\alpha - 1) * 1000$$

where α is the ratio $R_{product}/R_{substate}$, and $R = {}^{15}N/{}^{14}N$ (or ${}^{18}O/{}^{16}O$) for the reaction product and the reaction substrate. The negative isotope effects indicate that the reaction product is depleted in the heavy isotope relative to the reaction substrate.

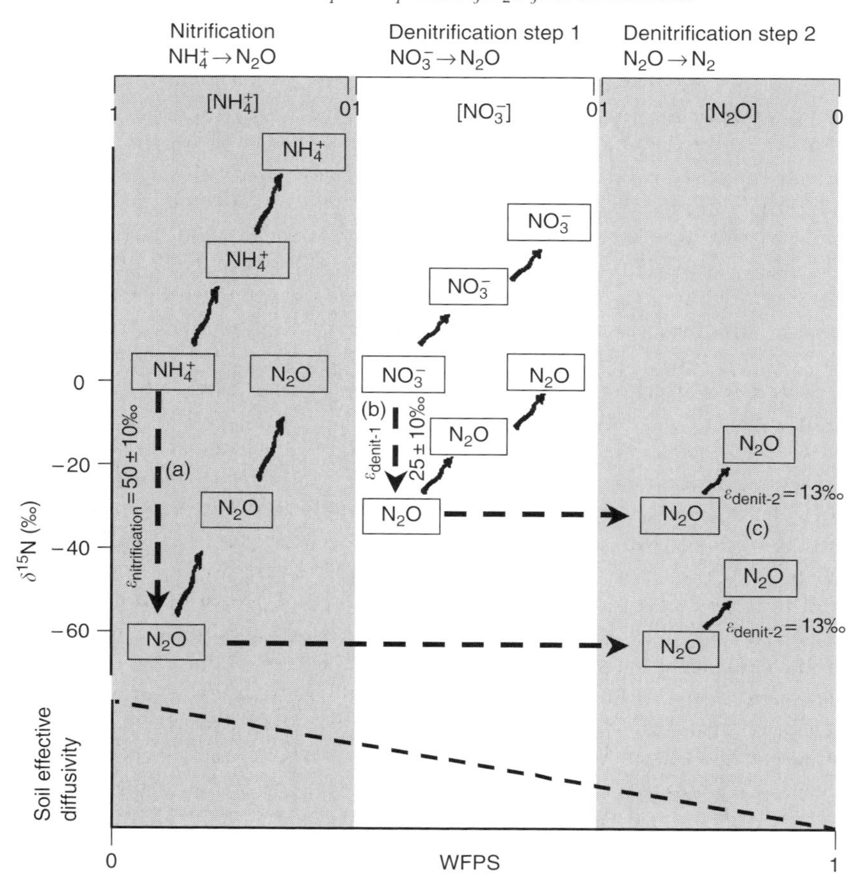

Figure 5.2 Illustration of the multiple factors influencing the stable isotope composition of nitrous oxide in soil. Separate isotope effects occur during the production of N_2O via nitrification and denitrification according to the enrichment factors (a) and (b and c) for each process, respectively. The isotopic composition of emitted N_2O can shift to more enriched δ values when: (1) substrate availability (NH_4^+ and NO_3^-) becomes limited, and (2) the effective soil diffusivity decreases, so more anaerobic microsites are present and the reduction of N_2O to N_2 via the second step of denitrification is favored. WFPS is water free pore space.

Wada and Ueda (1996) and Barford *et al.* (1999) have shown that N_2O becomes enriched in ^{15}N as it is consumed in the denitrification process to form N_2 (Table 5.1). The final value for the isotopic composition of N_2O formed by nitrification and denitrification will depend on the outcome of these interacting processes as is illustrated in Fig. 5.2. The only way of differentiating the relative contribution of each process at a particular time is to: (1) measure the isotopic composition of the substrates (NH_4^+ and NO_3^-);

(2) determine the effective soil diffusivity at different depths; (3) measure the N_2O isotopic composition at the surface and in soil air at different depths; (4) determine the soil-enrichment factors for nitrification and denitrification for each soil; and (5) model the isotopic shifts due to production, consumption, and diffusion at each soil depth. Studies that provide partial support for these concepts are briefly reviewed below.

Measurements of isotopic composition of N_2O in sewage treatment facilities (Wahlen and Yoshinari, 1985; Ueda *et al.*, 1999) and in ocean profiles (Yoshinari and Koike, 1994; Yoshinari *et al.*, 1997; Dore *et al.*, 1998; Naqvi *et al.*, 1998) demonstrated that 'light' and 'heavy' N_2O were produced from nitrification and denitrification, respectively. A soil incubation study showed the production of heavy N_2O due to denitrification: wet soils that were treated with sodium succinate (to stimulate N_2O reduction to N_2 during denitrification) and sodium ethanoate (to suppress N_2O reduction to N_2) had $\delta^{15}N$ values for N_2O of −36.8‰ and −55.3‰, respectively (Webster and Hopkins, 1996). These data support the idea that the $\delta^{15}N$ values of N_2O released via denitrification will depend on the amount of N_2 produced (heavier N_2O will be emitted if N_2 production is substantial). In other incubation studies, Webster and Hopkins (1996) also found that wetter soils produced N_2O more enriched in ^{15}N than did drier soils (i.e., anaerobic sites are more likely to have greater fraction of N_2O reduced to N_2). Pérez *et al.* (2000) measured the isotopic composition of N_2O in tropical rain forest soils in Costa Rica and in the Brazilian Amazon. They found that the $\delta^{15}N$ values of N_2O from the Brazilian soil were enriched in ^{15}N compared to the Costa Rican soil, despite the fact that previous studies concluded that denitrification was the dominant process to generate N_2O in both sites. It was proposed that reduction of N_2O to N_2 was more favorable in the Brazilian site because the clay texture (oxisols) and the higher bulk density of the Brazilian soil promoted the formation of anaerobic microsites compared to the sandy texture of the soil in Costa Rica (ultisols and inceptisols). In a study of grassland soil after urine amendments, Yamulki *et al.* (2000) observed that the $\delta^{15}N$ values of N_2O were enriched in ^{15}N after a heavy precipitation event. This was interpreted to be a result of higher reduction of N_2O to N_2 in anaerobic soil microsites produced by the heavy precipitation.

A recent study carried out by Sutka *et al.* (2003) showed that two pure cultures of oxidizing bacteria (*Nitrosomonas europeae* and *Methylococcus capsulatus*) during hydroxylamine oxidation have significantly different enrichment factors (relative to the ^{15}N of hydroxylamine) (Table 5.1). This study suggests that, additional to the factors mentioned above, the N_2O isotopic differences found in soils could be also attributed to variations in the microbial community present in the soil.

Oxygen Isotope Composition ($\delta^{18}O$) of N_2O

Similar to the processes described earlier for ^{15}N, the $\delta^{18}O$ of soil N_2O will depend on the isotope composition of the precursor substrates, substrate availability and consumption, fractionation during diffusion of N_2O ($\delta^{18}O = 8.56‰$ for molecular diffusion), and the mechanism of N_2O production. The $\delta^{18}O$ of soil N_2O is also complicated by additional factors that affect the isotopic composition of oxygen substrates [molecular oxygen (O_2) and water], processes that are not directly involved in the microbial pathways for N_2O production. For example, the $\delta^{18}O$ of soil oxygen can become enriched in ^{18}O because of fractionation (isotope effect of 15‰) during bacterial oxidative respiration (O_2 consumption) (Kendall and McDonnell, 1998). The $\delta^{18}O$ of soil water can vary with seasonal changes in precipitation inputs, and the remaining soil water pool can also become enriched in ^{18}O during evaporation of water from the soil surface (Merlivat and Jouzel, 1979; Mathieu and Bariac, 1996). In the discussion below primary focus is placed on the isotope effects that occur during production and consumption of N_2O in soil.

The $\delta^{18}O$ value of nitrification-derived N_2O should reflect the oxygen isotope composition of hydroxylamine (NH_2OH), soil air molecular oxygen, and soil water as is suggested by the following reaction sequences (Kendall and McDonnell, 1998):

Nitrosomonas oxidation mechanism:

$$NH_4^+ + H_2O \rightarrow NH_2OH + 2[H] + H^+$$

$$NH_2OH + O_2 \rightarrow NO_2^- + 2[H] + H^+$$

$$\uparrow\downarrow$$

$$H_2O$$

Nitrobacter oxidation mechanism:

$$NO_2^- + H_2O \rightarrow NO_3^- + 2[H]$$

To date there is no controlled study in the literature documenting the oxygen isotope effects associated with N_2O formation via nitrification. However, indirect measurements in field studies by Pérez *et al.* (2001) have provided some insight into the isotope effects involved. These studies showed that N_2O derived from nitrification had $\delta^{18}O$ values a few per mil higher than those of atmospheric O_2 ($\delta^{18}O$ of O_2 = 23.5‰), while the N_2O was enriched in ^{18}O by 22 to 30‰ compared to soil water. This suggested that incorporation of oxygen from molecular O_2 during N_2O formation via nitrification was greater than that of oxygen from water. The observation that $\delta^{18}O$ values for N_2O were slightly higher than those for atmospheric O_2 may indicate that the molecular oxygen in the soil airspace became enriched by microbial consumption during aerobic respiration.

The $\delta^{18}O$ of denitrification-derived N_2O should reflect the isotopic composition of the substrate (NO_3^-) and intermediate by-products, and the oxygen isotope effects associated with each denitrification step. The $\delta^{18}O$ of the NO_3^- pool may become enriched in ^{18}O as denitrification progresses, but since the nitrification–denitrification processes are variable and dependent on soil conditions, a constant enrichment of ^{18}O in NO_3^- is not expected. NO emissions from the soil does not seem to have a strong influence on the $\delta^{18}O$ of N_2O, since no drastic isotopic changes were observed in the latter under conditions where NO emissions were high (Pérez *et al.*, 2001). Enrichments of ^{18}O have been observed in the N_2O remaining after reduction to N_2 has occurred (Barford, 1997; Yoshinari *et al.*, 1997; Dore *et al.*, 1998; Naqvi *et al.*, 1998). Barford (1997) suggested that the isotope effect for N_2O reduction to N_2 was 131‰. However, such a large enrichment of ^{18}O in the emitted N_2O has not been observed in field studies. Instead the range reported in the literature for the $\delta^{18}O$ values of soil-emitted N_2O is 19.6 to 57.8‰ (Table 5.2). These results are in agreement with the ^{18}O isotope effect of 18.5‰ for N_2O reduction to N_2 calculated by Mandernack *et al.* (2000) for landfill soils.

Table 5.2 Comparison of Field Measurements of the Stable Isotope Composition of Nitrous Oxide Emitted from Soils

Site	$\delta^{15}N$, ‰ (range)	$\delta^{18}O$, ‰ (range)	Reference
Tropical rain forest, Costa Rica	−22.5 to −71.0	27.2 to 36.1	Kim and Craig (1993)
Sugar cane plantation, Maui, Hawaii	−25 to 0.1	34.6 to 43.1	Kim and Craig (1993)
Various soils	−24 to 1	35 to 45	Wada and Ueda (1996) (and references therein)
Temperate grassland, England	−34 to −15	Not measured	Yamulki *et al.* (2000)
Temperate landfill	−5.1 to 19.0	43.0 to 57.8	Mandernack *et al.* (2000)
Tropical rain forest:			Pérez *et al.* (2000)
Costa Rica	−33.9 to −23.0	21.3 to 34.6	
Brazil	−34.1 to 1.9	19.6 to 42.3	
Agricultural soil, Mexico	−46.7 to 5.9	21.0 to 33.5	Pérez *et al.* (2001)
Temperate grassland, England	−12 to 0	34 to 47	Yamulki *et al.* (2001)

It is difficult to model the $\delta^{18}O$ data for N_2O currently available in the literature because of the lack of information on ^{18}O isotope effects. Unknown fractionation factors could be estimated or inferred if the $\delta^{18}O$ values of H_2O, O_2, and NO_3^- in the soil were routinely measured and reported. Only a complete study of the oxygen isotope pool in soils will help in improving our understanding of the microbiological pathways for production and consumption of soil N_2O. The data available and the analytical difficulties relating to the characterization of ^{18}O in the N_2O precursors, suggest that ^{18}O might not be an appropriate proxy to study the N_2O microbial soil-related mechanisms.

^{15}N Intramolecular Position in N_2O Emitted from Soils

Nitrous oxide is a linear molecule (NNO). Recent analytical techniques have allowed the determination of the intramolecular ^{15}N position in the N_2O molecule (Brenninkmeijer and Rockmann, 1999; Toyoda and Yoshida, 1999; Esler *et al.*, 2000; Kaiser *et al.*, 2004). The ^{15}N position in the N_2O molecule can be located in the central N atom, referred to as the α site (or position 2), and the terminal N atom, referred to as the β site (or position 1) (Toyoda and Yoshida, 1999; Rockmann *et al.*, 2003). Variation in the intramolecular distribution of ^{15}N results in two major isotopologues, $^{15}N^{14}NO$ and $^{14}N^{15}NO$, molecules with the same mass but different chemical structures [more detailed information is given in this volume by Rahn (Chapter 15)]. Park *et al.* (2004) compared all the available data for the stratospheric site-specific $\delta^{15}N$ isotopic composition of N_2O and found that the $\delta^{15}N^{\alpha}$ values (relative to the $\delta^{15}N^{\alpha}$ of tropospheric N_2O) are more enriched in the stratosphere than those found in the troposphere (Rockmann *et al.*, 2001; Toyoda *et al.*, 2001; Park *et al.*, 2004). The difference was attributed to the N_2O photolysis and photooxidation processes in the stratosphere that preferentially fractionate the central nitrogen in the N_2O molecule (Rockmann *et al.*, 1998; Brenninkmeijer and Rockmann, 1999; Toyoda and Yoshida, 1999; Rockmann *et al.*, 2001; Kaiser *et al.*, 2004; Park *et al.*, 2004). To date, very few site-specific N_2O isotopic composition data sets for the soil N_2O sources are available. It is also difficult to make intercomparisons of the N_2O isotopologue data available due to the fact that each author reported their data relative to their own working standard. Yoshida and Toyoda (2000), using a mass balance approach, suggested that the sources (biological- and combustion-derived) would have low site preference ($\delta^{15}N^{\alpha}$–$\delta^{15}N^{\beta}$) values (-0.5 to $+15$) in comparison with the tropospheric and stratospheric site preference values (details explained in this volume by Rahn, Chapter 15). It was also predicted by these authors that N_2O produced by nitrification and denitrification would have variation in the α site ^{15}N enrichment. A preliminary study done by Toyoda *et al.* (2003) found very distinct site preference values from two denitrifying bacterial species evaluated. Sutka *et al.* (2003) showed that two pure cultures of oxidizing bacteria subject to hydroxylamine oxidation, *Nitrosomonas europeae* and *Methylococcus capsulatus*, have site preference values of $-2.3 \pm 1.9‰$ and $5.5 \pm 3.5‰$, respectively. On the other hand, nitrite reduction by *Nitrosomonas europeae* showed significantly depleted site preference values ($-8.3 \pm 3.6‰$). This study corroborates the usefulness of these measurements to differentiate N_2O microbial processes in soils. Pérez *et al.* (2001) presented some limited data in support of this hypothesis in a urea-fertilized agricultural field where the site preference of N_2O emitted from soil during a 4-day period shifted from 4.8‰, when nitrification seemed to be the process regulating the emissions, to 14.2‰ when denitrification seemed to be predominant. Yamulki *et al.* (2001) found the same site preference shift (−14 to 1‰) one week after fertilization in a dung and urine fertilized temperate grassland soil. This implies that nitrification-derived N_2O has smaller site preference values than denitrification-derived N_2O. Future studies of the intramolecular distribution of ^{15}N from pure culture and field studies will provide new insights to differentiate mechanisms of N_2O production in soil.

Conclusions

Natural abundance isotopic characterization is a useful non-invasive tool to determine the relative contribution of nitrification and denitrification to N_2O production in soils. Several factors affect the isotopic composition of N_2O emitted from soils and based on these processes a procedure for differentiating N_2O sources was proposed. This procedure includes measurements of the isotopic composition of the substrates used to form N_2O (NH_4^+ and NO_3^-); determination of the effective soil diffusivity at different depths; measurement of the isotopic composition of N_2O at the surface and in soil air at different depths; determination of the soil-enrichment factors for nitrification and denitrification; and finally modeling of the isotopic shifts due to production, consumption, and diffusion at each soil depth to infer the relative contribution of each process to the emitted N_2O at the soil surface. In addition, further measurements of the intramolecular distribution of ^{15}N in N_2O will help to differentiate the operation of specific microbial processes important for the formation of N_2O in soils.

References

Bakwin P. S., Wofsy S. C., Fan S.-M., Keller M., Trumbore S. and da Costa J. M. (1990) Emission of nitric oxide (NO) from tropical forest soils and exchange of NO between the forest canopy and atmospheric boundary layers. *J Geophys Res* **95**: 16745–16755.

Barford C. C. (1997) Stable isotope dynamics of denitrification. Ph.D. Harvard University.

Barford C. C., Montoya J. P., Altabet M. A. and Mitchell R. (1999) Steady-state nitrogen isotope effect of N_2 and N_2O production in *Paracoccus denitrificans*. *Appl Environ Microbiol* **65**: 989–994.

Bollmann A. and Conrad R. (1998) Influence of O_2 availability on NO and N_2O release by nitrification and denitrification in soils. *Global Change Biol* **4**: 387–396.

Brenninkmeijer C. A. M. and Rockmann T. (1999) Mass spectrometry of the intramolecular nitrogen isotope distribution of environmental nitrous oxide using fragment-ion analysis. *Rapid Commun Mass Spectrometry* **13**: 2028–2033.

Cicerone R. J. (1989) Analysis of sources and sinks of atmospheric Nitrous Oxide (N_2O). *J Geophys Res* **94**: 18265–18271.

Davidson E. A. (1991) Fluxes of nitrous oxide and nitric oxide from terrestrial ecosystems. In *Microbial Production and Consumption of Greenhouse Gases: Methane, Nitrogen Oxides, and Halomethanes* (J. E. Rogers and W. B. Whitman, eds) pp. 219–235. American Society of Microbiology, Washington, D.C.

Davidson E. A. (1992) Sources of nitric oxide and nitrous oxide following wetting of dry soil. *Soil Sci Soc Am J* **56**: 95–102.

Davidson E. A. (1993) Soil water content and the ratio of nitrous oxide and nitric oxide emitted from soil. In *Biogeochemistry of Global Change: Radiatively Active Trace Gases*: selected papers from the Tenth International Symposium on Environmental Biogeochemistry, San Francisco, August 19–24, 1991 (R. S. Oremland, ed.) pp. 369–386. Chapman & Hall, New York.

Davidson E. A. and Schimel J. P. (1995) Microbiological processes of production and consumption of nitric oxide, nitrous oxide and methane. In *Biogenic Trace Gases: Measuring Emissions from Soil and Water* (P. A. Matson and R. C. Harriss eds). University Press, Cambridge.

De Klein C. A. M. and Van Logtestijn R. S. P. (1996) Denitrification in grassland soils in The Netherlands in relation to irrigation, N-application rate, soil water content and soil temperature. *Soil Biol Biogeochem* **28**: 231–237.

Dendooven L., Duchateau L. and Anderson J. M. (1996) Gaseous products of denitrification process affected by the antecedent water regime of the soil. *Soil Biol Biogeochem* **28**: 239–245.

Dore J. E., Popp B. N., Karl D. M. and Sansone F. J. (1998) A large source of atmospheric nitrous oxide from subtropical North Pacific surface waters. *Nature* **396**: 63–66.

Ehhalt D., Dentener P. M. F., Derwent R., Dlugokencky E., Holland E., Isaksen I., Katima J., Kirchhoff V., Matson P., Midgley P. and Wang M. (2001) Atmospheric chemistry and greenhouse gases. In *Climate Change 2001: The Scientific Basis.* Contribution of Working Group 1 to the Third Assessment Report of the Intergovernmental Panel on Climate Change. Cambridge University Press, Cambridge, UK and New York, NY.

Esler M. B., Griffith D. W. T., Turatti F., Wilson S. R. and Rahn T. (2000) N_2O concentration and flux measurements and complete isotope characterization using FTIR spectroscopy. *Chemosphere Global Change Sci* **2**: 445–454.

Fluckiger J., Dallenbach A., Blunier T., Stauffer B., Stocker T. F., Raynaud D. and Barnola J. M. (1999) Variations in atmospheric N_2O concentration during abrupt climatic changes. *Science* **285**: 227–230.

Kaiser J., Park S., Boering K., Brenninkmeijer C. A. M., Hilkert A. and Rockmann T. (2004) Mass spectrometric method for the absolute calibration of the intramolecular nitrogen isotope distribution in nitrous oxide. *Anal Bioanal Chem* **378**: 256–269.

Keller M., Kaplan W. A., Wofsy S. C. and deCosta J. M. (1988) Emissions of N_2O from tropical forest soils: Response to fertilization of NH_4^+, NO_3^- and PO_4^{3-}. *J Geophys Res* **93**: 1600–1604.

Keller M. and Reiners W. A. (1994) Soil-atmosphere exchange of nitrous oxide, nitric oxide and methane under secondary succession of pasture to forest in the Atlantic lowlands of Costa Rica. *Global Biogeochem Cycles* **8**: 399–409.

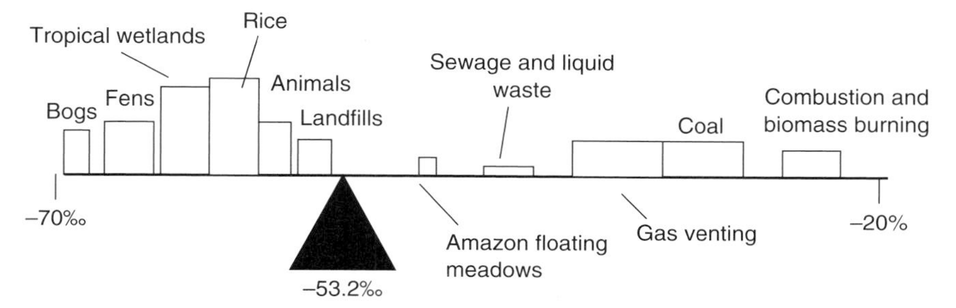

Figure 6.2 A representation of the mass-weighted isotope mass balance approach for atmospheric methane inputs (redrawn from Chanton *et al.*, 2000). The representation is similar to a 'see-saw' in that both the size of the bar and distance from the fulcrum point influence the importance of the source. The x-axis (the solid line resting on the fulcrum) represents the $\delta^{13}C$ of the methane source, and the total size of the bars is roughly related to the source strength in this representation. Bar thickness represents the $\delta^{13}C$ range of the source. The position of the fulcrum represents the isotopic composition of the global methane input, which is determined from the values of atmospheric methane corrected for isotopic fractionation associated with removal processes. The sizes of the bars in this representation are for illustrative purposes for the most part and only roughly depict the mass of the fluxes.

variations are typically considered to indicate the activity of methane-oxidizing bacteria, methanotrophs (Fig. 6.3). These bacteria preferentially consume CH_4 containing the lighter isotopes, leaving residual CH_4 that is enriched in the heavier isotopes, ^{13}C and D. The fraction of CH_4 that is oxidized can be calculated from changes in CH_4 isotopic composition across a spatial gradient from anoxic to oxidizing conditions (Happell *et al.*, 1994; Liptay *et al.*, 1998; Chanton *et al.*, 1999; Chanton and Liptay, 2000).

Sympathetic shifts in the ^{13}C and D isotopic composition of methane associated with oxidation were first reported by Coleman *et al.* (1981) in incubation studies, and subsequently have been verified in a number of field studies. The amount of the D shift is always greater than for the C shift, because the mass change of 1 is proportionally greater for H than it is for C. This means that the fractionation factor, αH for hydrogen–deuterium is larger than the fractionation factor for carbon, αC. Isotopic fractionation in this case is a kinetic process so the term α is defined as

$$\alpha = k_L/k_H \tag{6.5}$$

where k_L refers to the first-order rate constants for the reaction of $^{12}CH_4$ or CH_4, and k_H refers to the rate constants of $^{13}CH_4$ and CH_3D. When α is expressed in this manner, it is greater than 1 since the molecules containing the lighter isotopes react faster than those containing the enriched or heavier elements; estimates of αC for aerobic methane oxidation vary from 1.008 to 1.031 (Reeburgh, 1996; Liptay *et al.*, 1998; Chanton and Liptay, 2000).

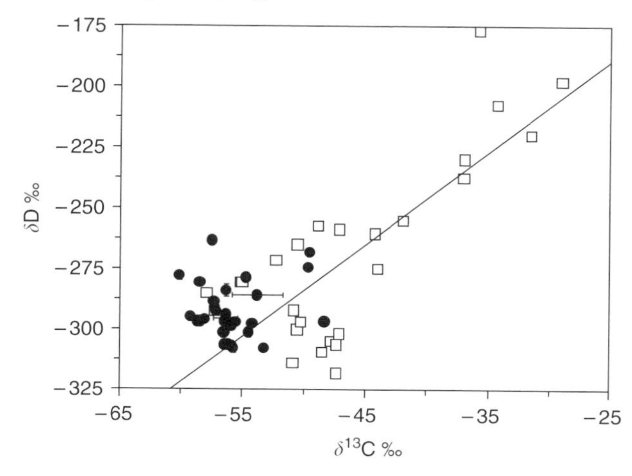

Figure 6.3 Effect of methane oxidation on $\delta^{13}C$ and δD of CH_4. Samples from several landfills in New England (Liptay *et al.*, 1998). Filled circles represent anoxic zone methane which has not been subject to oxidation. Chamber samples (open squares) captured emitted methane which has experienced the activity of methane-oxidizing bacteria. The slope of the line fit to the data is 3.8.

Some measurements of the size of the $\Delta\delta D/\Delta\delta^{13}C$ shift for CH_4 oxidation are tabulated in Table 6.1. Interestingly, observations from the field are smaller than initial laboratory results. Further investigation comparing $\delta^{13}C$ and δD fractionation driven by methanotrophy in lab and field studies is needed.

Isotopic Effects Associated with Methane Production Mechanisms

Variation in the two predominant mechanisms of methane production, CO_2 reduction and acetate fermentation, has been hypothesized to be associated with inverse or antipathetic shifts in δD and $\delta^{13}C$ of CH_4 (Fig. 6.1;

Table 6.1 Examples of Relative Changes in δD of CH_4 Relative to $\delta^{13}C$ of CH_4 Caused by Methanotrophic Activity

Environment	$\Delta\delta D/\Delta\delta^{13}C$	References
Landfill soil	2.5–3.7	Liptay *et al.* (1998)
Landfill soil	4	Lubina *et al.* (1996a,b)
Laboratory incubations	8.5–13.5	Coleman *et al.* (1981)
Wetlands	5.5	Happell *et al.* (1994)
Wetlands	2.5	Burke *et al.* (1988b)

Whiticar *et al.*, 1986; Schoell, 1988; Burke *et al.*, 1988a,b; Whiticar, 1993, 1999; Hornibrook *et al.*, 1997). Carbon isotope separation between CO_2 and CH_4 ranges from 60 to 90‰ for CO_2 reduction and 40 to 60‰ for acetate fermentation (Whiticar, 1999; Conrad *et al.*, 2002). Hydrogen isotope effects lead to CH_4–H_2O separation of 170 to 250‰ for CO_2 reduction and 250 to 400‰ for acetate fermentation. The hypothesis is that if all things are relatively equal, fermentation of acetate will result in CH_4 which is ^{13}C-enriched and D-depleted relative to CH_4 produced via the CO_2 reduction pathway. There are two key words in the pronouncement above, 'relatively equal.'

Above we referred to understanding the changes in CH_4 isotopic composition along a gradient. The scale of this gradient must be sufficiently small that the isotopic composition of methane precursors does not change significantly in such a way that interpretation will be compromised. For example, Waldron *et al.* (1999) demonstrated that on a grand spatial gradient from the tropics to the poles, methane δD composition is controlled by the δD of water present. However, at smaller spatial scales, where the δD of water is fairly constant, δD variation of CH_4 is most likely due to variation in the oxidation or production mechanism, as discussed below. Similarly, the $\delta^{13}C$ of methane can be affected by the isotopic composition of the parent organic matter. Methane produced from the decay of C_3 vegetation has been observed to be ^{13}C-depleted relative to methane produced from the decay of C_4 vegetation (Chanton and Smith, 1993).

Across what kind of spatial and temporal gradients can one expect to find transitions in the methane production mechanism? Methane is produced primarily in flooded basins and sedimentary environments; in these environments, the production of acetate and its subsequent utilization are generally greater in the upper portion of the peat or sediment column than in deeper layers (Whiticar, 1999 and references therein). This acetate production in the surface layer of peat or sediment may be associated with the breakdown of more labile organic matter, which is derived from plant root exudates in the rhizosphere (Whiting and Chanton, 1993; Shannon and White, 1994; Chanton *et al.*, 1993, 1995; Shannon *et al.*, 1996; Chasar *et al.*, 2000a,b).

Because marine systems contain abundant sulfate, which is an energy-yielding electron acceptor, bacterial respiration in marine and salt-marsh surface sediments is dominated by sulfate reduction. In these systems sulfate-reducing bacteria are able to out-compete methanogens for acetate because sulfate respiration is a more energy-yielding process than methane production. Consequently, in marine systems sulfate and acetate are consumed in surficial layers and methanogenesis is confined to deeper layers where sulfate is depleted and where there is much less acetate. Methane production in marine systems is thus formed primarily by CO_2 reduction (see Whiticar, 1999).

In contrast, freshwater systems have much lower concentrations of sulfate and it is possible for methane production to occur in surface layers of sediment or peat, i.e., more acetate is available to methanogens. This supports the idea that in marine systems methane production proceeds primarily via CO_2 reduction while in freshwater systems acetate fermentation is the dominant processes (Whiticar *et al.*, 1986; Whiticar, 1999). However, isotopic studies in freshwater wetlands indicate that with increasing depth in the peat there is a transition from acetate fermentation to CO_2 reduction (Hornibrook *et al.*, 1997, 2000a,b; Chasar *et al.*, 2000a,b). There is also evidence that vegetation affects the relative importance of methane production pathways in freshwater wetlands, although how this occurs is not clear. For example, CH_4 in *Sphagnum*-dominated bogs is produced by CO_2 reduction predominantly, while acetate fermentation is of more relative importance in *Carex*-dominated fens (Kelley *et al.*, 1992; Lansdown *et al.*, 1992; Shannon and White, 1994; Chasar *et al.*, 2000a,b).

Blair (1998) described the relative importance of methane production mechanisms in marine sediments as being controlled by the balance between the rate of deposition of organic matter and the relative importance of electron acceptors other than methane produced in the respiration of that organic matter. As more organic matter passes through the gauntlet of the higher-energy respiration modes, more methane production results and a greater percentage of that methane is produced from acetate fermentation. Measurements of fractionation factors for acetate production during anaerobic fermentation are scarce (Conrad *et al.*, 2002). Blair *et al.* (1985) reported that the carboxyl group of acetate produced in an anaerobic culture was ^{13}C enriched relative to the parent organic matter (glucose) by about 12‰. However, the carbon of acetate transferred to methane comes from the methyl group. Values for the methyl carbon are related to the ^{13}C of the parent organic matter but are ^{13}C-depleted relative to the carboxyl carbon (Blair *et al.*, 1987; Blair and Carter, 1992; Sugimoto and Wada, 1993; Hornibrook *et al.*, 2000a). Methyl carbon $\delta^{13}C$ values vary from -40 to -15‰ for anaerobic incubation of C-3 plant material, and -14 to $+4$‰ for acetate extracted from a marine sediment (references above). Kruger *et al.* (2002) measured the $\delta^{13}C$ of acetate extracted from the soil pore water of an Italian rice field and obtained values ranging from -16 to -21‰ for the average of both carbons.

An isotopic fractionation of -21‰ has been reported for the transformation of acetate methyl carbon into methane from microbial culture experiments (Gelwicks *et al.*, 1994). Depending upon the fraction of acetate that is funneled into methane production in natural systems, that fractionation may not be fully expressed due to closed system isotope effects.

Conrad *et al.* (2002), in an elegant paper calculating the relative amounts of methane derived from the two production mechanisms, assumed that the methyl group was enriched similar to the carboxyl group and then applied the fractionation factor of −21‰ to calculate the $\delta^{13}C$ of methane produced from acetate. Clearly more research needs to be performed in this area.

The validity of considering δD variation to be a result of changes in production mechanism has recently been questioned (Waldron *et al.*, 1998a,b, 1999; Sugimoto and Wada, 1995). In incubation studies where substrate lability or changes in acetate production have been manipulated to vary production mechanism, ^{13}C shifts have been observed but concurrent shifts in δD have not (Waldron *et al.*, 1998a; Sugimoto and Wada, 1995). While the H of CH_4 produced by CO_2 reduction clearly comes from environmental water (Daniels *et al.*, 1980), it has long been thought that in methane produced from acetate, one H atom comes from water and the others come from the acetate methyl group (Pine and Barker, 1956). Recent evidence has suggested that there is an exchange of hydrogen between the acetate methyl group and water in the final stages of methanogenic acetate metabolism (de Graaf *et al.*, 1996). Presumably this isotopic exchange is accompanied by an isotopic fractionation but this has not been measured.

The scientific community also needs to better describe the source of hydrogen and the isotopic fractionations involved during the production of methane from acetate. It is difficult to explain the very depleted values of δD for acetate-derived methane. An excellent example is given by Whiticar (1999). If one assumes a δD organic value of about −120‰ for the three methyl hydrogens donated, no isotopic fractionation associated with this donation, and a methane produced with a δD of −400‰, mass balance would require the strange value of −1240‰ for the final hydrogen added. Therefore it seems likely that some sort of isotopic fractionation is involved in the transfer of methyl H to methane.

Burke (1993) proposed that hydrogen concentration could affect the fractionation of H between water and methane, resulting in greater fractionation at higher H_2 concentrations but not affecting the $\delta^{13}C$ of CH_4. Burke (1993) observed that in incubation experiments and in the rumen of cows, where H_2 concentrations are greater, the δD of CH_4 is consistently depleted relative to observations made in wetlands or sediments. If the H_2 concentration varies in soils and sediments in a consistent manner along with the production mechanism, it may be that H_2 concentration alone determines variations in the δD of methane when environmental water δD is constant.

In the next section we present a number of field observations that demonstrate antipathetic shifts in methane C and D isotopic composition across seasonal, depth, and vegetation gradients. These data are consistent with the hypothesis that shifts in dominant mechanisms in methane production, i.e., CO_2 reduction and acetate fermentation, are associated with antipathetic changes in these isotopes. Further, in the same manner that sympathetic change in C and D isotopes driven by methane oxidation are described (Table 6.1), we want to develop consistent field estimates of the relative difference in the changes in C and D driven by changes in production mechanisms. The mechanisms controlling the observed variations certainly need considerable elucidation; future research should be focused in this direction.

Supporting Field Data

Field data are presented from research conducted in natural wetlands in boreal, temperate, and tropical regions. Data collected in the boreal peatlands of northern Minnesota include profiles that span depths ranging from 0.3 to 3.0 m and were taken from Chasar *et al.* (2000a,b; Fig. 6.4). The D isotopic composition of CH_4 is plotted vs ^{13}C for porewaters from both a fen and a bog. The two sites were within 3 km of each other. Shallow fen samples (<1 m) are separated from deeper fen samples (>1 m), with shallow samples falling in the range of the graph suggesting a relatively greater degree of acetate fermentation. Bog samples from all depths (from 0.3 m to 3.0 m), along with the deeper fen samples, fall within the range of values suggesting CO_2 reduction. The complete data set falls along a regression line with a slope of -4.2. The fen site is populated by vascular plants, primarily sedges and grasses (e.g., *Carex lasiocarpa, Rhyncospora fusca* and *R. alba*) that may stimulate methane production through the release of labile root exudates (Whiting and Chanton, 1993; Shannon and White, 1994; Chanton *et al.*, 1995; Shannon *et al.*, 1996; Chasar *et al.*, 2000a,b). In contrast, sedges are not present in the bog. Vegetation in the bog consists of an overstorey of Black Spruce with some Tamarack (*Picea mariana* and *Larix laricinia*) and woody shrubs, and a thick ground cover of *Sphagnum* spp.

Root exudates of *Carex* and other non-woody vascular plants may produce more labile organic matter in the surface of the fen than at depth in the fen or at the *Sphagnum*-vegetated bog. This expectation is consistent with natural abundance ^{14}C (CH_4) data that were enriched (i.e., younger) in surficial fen porewaters relative to the bog (Chasar *et al.*, 2000a).Other studies have observed similar patterns. For example, Lansdown *et al.* (1992) reported that King's Lake Bog was dominated by CO_2 reduction with little acetate fermentation supporting methane production in either surficial or deeper peat. Shannon and White (1994) and Shannon *et al.* (1996) report greater acetate accumulation in *Carex* fens relative to *Sphagnum* bogs.

Alternatively Hines *et al.* (2001) and Duddleston *et al.* (2002) have shown that acetate production can occur in some acidic environments, but this

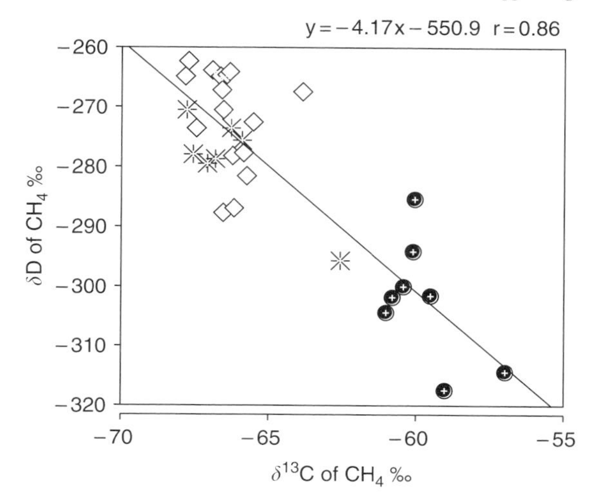

Figure 6.4 Data from the Glacial Lake Agassiz peatland in northern Minnesota (Chasar *et al.*, 2000a,b). The δD isotopic composition of CH_4 is plotted vs $\delta^{13}C$. Bog samples are represented by open diamonds, fens samples above 1 m by filled circles, and fen samples deeper than 1 m by stars. The upper left corner of the graph has relatively depleted ^{13}C and relatively enriched δD values and characterizes a zone of relatively greater CO_2 reduction while the lower right corner of the graph indicates greater acetate fermentation. Shallow fen samples (<1 m) are distinct from deeper fen samples (>1 m) and bog samples. Shallow fen sites are populated by rooted vascular plants that are thought to stimulate methane production through the release of labile root exudates.

acetate is not utilized in the anoxic porewaters. Instead the acetate accumulates until it eventually diffuses into oxic environments where it is then consumed. It may be that the difference in methane production mechanisms between bogs and fens is driven by pH. Low pH in bogs appears to inhibit acetate utilization. At the fen site pH varied from 5.5 to 6.5 while in the bog site pH varied from 3.5 to 4.5 (Chasar *et al.*, 2000a).

Data from an Alberta fen were taken from Popp *et al.* (1999) while bog data are from Popp (1998) (Fig. 6.5). Depth intervals span only 60 cm in both systems, which were about 20 km apart. Again, the bog and surficial fen samples fall in separate areas of the δD vs $\delta^{13}C$ plot. Bog samples are consistently ^{13}C-depleted and δD enriched relative to surficial fen samples, as was observed in the Minnesota data. Two lines are fit to the data: one includes all samples and has a slope of -3.1; while a line fit only to the bog data and the 60 cm fen samples has a slope of -4.9. The 20 cm fen data fall above the line fit to the 60 cm fen data, possibly due to rhizospheric methane oxidation within this zone (Popp *et al.*, 2000). Overall, we consider the line with the slope of -4.9 to better represent the effect of production mechanism variation.

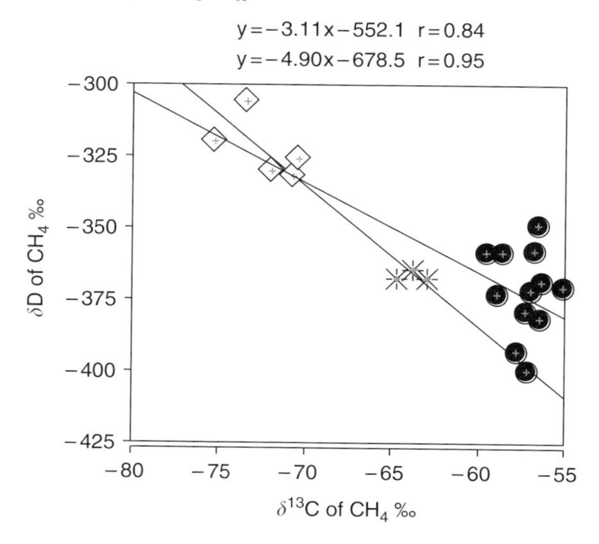

Figure 6.5 Data from Alberta boreal wetlands, Suzanne's Fen, and Bleak Lake Bog, which are within 20 km of each other (Popp *et al.*, 1999; Popp, 1998). The δD isotopic composition of CH_4 is plotted vs $\delta^{13}C$. Bog samples are represented by open diamonds, fen samples at 20 cm by filled circles, and fen samples from 60 cm by stars. The upper left corner of the graph has relatively depleted ^{13}C and relatively enriched δD values and characterizes a zone of relatively greater CO_2 reduction while the lower right corner of the graph indicates greater acetate fermentation. Bog samples are consistently ^{13}C-depleted and δD-enriched relative to fen samples, as was observed in Minnesota. Two lines are fit to the data; one includes all of the samples and has a slope of −3.1, while a line fit only to the bog data and the 60 cm fen samples has a slope of −4.9. The 20 cm fen data fall above the line fit to the 60 cm fen data, possibly due to rhizospheric methane oxidation.

The pattern produced by data collected in an Ontario *Typha* marsh, Point Pellee (Fig. 6.6; Hornibrook *et al.*, 1997) is similar to that for the fen samples described above. As with the fen, surficial samples in the marsh fall in the lower right corner of the graph with enriched C and depleted D values while deeper samples are ^{13}C depleted and more D-enriched. These data were collected from the surface to 200 cm depth, but samples from the surficial 30 cm were not considered because they were influenced by oxidation effects (Hornibrook *et al.*, 1997).

Data collected at monthly intervals over several years from bubbles in an organic-rich marine basin, Cape Lookout Bight, NC, USA also exhibit antipathetic changes in $\delta^{13}C$ and δD of methane (open symbols; Fig. 6.7; Burke *et al.*, 1988a). These isotopic variations were interpreted by the authors to represent shifts in production mechanism with acetate fermentation becoming relatively more important in summer, and CO_2 reduction becoming more important in winter. The seasonal variation in

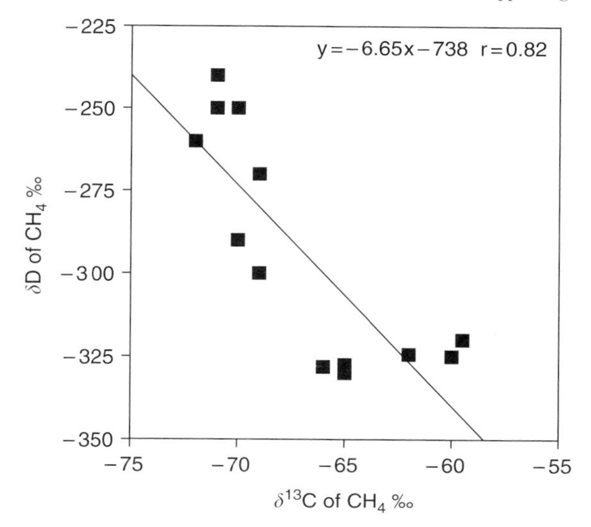

Figure 6.6 Data collected in an Ontario *Typha* marsh, Point Pellee (Hornibrook *et al.*, 1997). The δD isotopic composition of CH_4 is plotted vs δ^{13}C. The pattern produced is similar to the pattern observed from fen samples. As with the fen, surficial samples fall in the lower right corner of the graph with enriched C and depleted D values while greater depths are ^{13}C-depleted and more D-enriched. The data represent samples collected over a depth gradient of 30 to 200 cm.

production mechanism was driven by the higher summer temperatures, which resulted in sulfate depletion at shallow depths and subsequently made more acetate available to methanogens. Summer samples fall in the lower right corner of the graph indicating increased methane production from acetate while winter samples are segregated in the upper left (CO_2 reduction) corner.

Orinoco floodplain (Venezuela) data are represented by closed symbols (Fig. 6.7; Smith *et al.*, 2000). Values in the lower right hand side of the graph were obtained from aquatic macrophyte beds during the flooded season while symbols in the upper left portion of the graph represent methane collected from sediments underlying open water and flooded forest areas. As with the preceding examples, the methane produced in flooded soils populated with rooted aquatic macrophytes is consistent with a greater amount of methane produced via acetate fermentation than CO_2 reduction.

A spatial transect across a wetland in the Florida Everglades also shows antipathetic trends in ^{13}C and δD (Fig. 6.8; Burke *et al.*, 1988b).

The mean value of all of these observations for changes in $\Delta\delta D/\Delta^{13}C$ is $-5.1 \pm 1.0, n = 6$ (Table 6.2). Our examination of the covariance of these two isotopes of methane across these wetlands suggests that the variation is associated with the transition in CH_4 production mechanisms, i.e.,

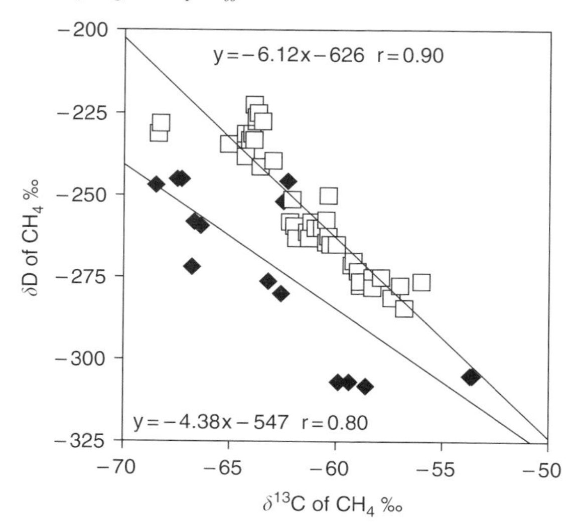

Figure 6.7 Data collected at monthly intervals over several years from CH_4 bubbles in an organic-rich marine basin, Cape Lookout Bight, NC (open squares, Burke *et al.*, 1988a). Summer samples, which represent a period of greater acetate fermentation, fall in the lower right corner of the graph while winter samples are segregated in the upper left corner. Orinoco floodplain (Venezuela) data are represented by filled diamonds (Smith *et al.*, 2000). Values in the lower right hand side of the graph were obtained from macrophyte beds during the flooded season while symbols in the upper left portion of the graph represent methane collected from sediments underlying open water areas.

a shift from acetate fermentation to CO_2 reduction. The strong build-up of acetate in summer in Cape Lookout Bight sediments (Sansone and Martens, 1982; Sansone, 1986) and its subsequent depletion via methanogenesis (Crill and Martens, 1986), concurrent with seasonal shifts in the isotopic composition of CH_4 (Fig. 6.7), is certainly evidence that increases in the importance of acetate fermentation can be associated with ^{13}C enriched and δD depleted isotopic values relative to CH_4 produced from CO_2 reduction (Burke *et al.*, 1988a). Seasonal variations in CH_4 isotopic composition have also been observed in northern wetlands by Kelley *et al.* (1992) and in freshwater wetlands by Chanton and Martens (1988), and attributed to seasonally forced variations in the production mechanism.

We have also observed relatively ^{13}C-enriched and δD-depleted values associated with the presence of rooted vascular aquatic macrophytes in Minnesota, Alberta, and the Orinoco floodplain. This observation may be extended to the Ontario marsh as well if one considers the absence of rooted macrophytes at depth in the peat column and their presence in the peat surface. Consistent with our interpretation, Shannon *et al.* (1996)reported high acetate accumulations in sites dominated by grasses and sedges,

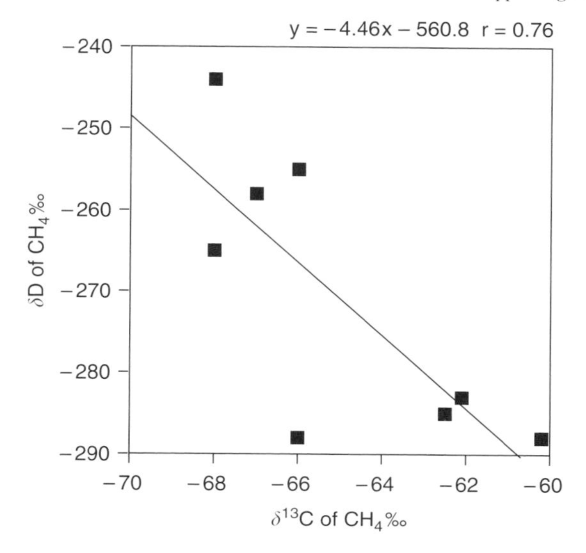

Figure 6.8 A spatial transect across a wetland in the Florida Everglades (Burke *et al.*, 1988b). Trends in the data reflect a shift in production mechanism across the wetland.

and methane emissions were greater at these sites (Shannon and White, 1994) relative to bog sites. Subsequent work by this group then linked ^{13}C-enriched methane to acetate fermentation, and ^{13}C-depleted methane to CO_2 reduction (Avery *et al.*, 1999).

Variations in the δD value of water do not have an apparent effect on the robustness of these Ontario or Minnesota results (Fig. 6.9; Hornibrook *et al.*, 1997; Siegel *et al.*, 2001). It seems to us highly unlikely that the δD of water would vary seasonally and spatially across both horizontal and depth scales to result in the variations that we have observed at sites

Table 6.2 Examples of Relative Changes in δD of CH_4 Relative to $\delta^{13}C$ of CH_4 Caused by Production Effects

Environment	$\Delta\delta D/\Delta\delta^{13}C$	Reference scales
Minnesota boreal wetlands	−4.2	Spatial depth
Alberta boreal wetlands	−4.9	Spatial depth
Ontario *Typha* wetland	−6.6	Depth
Cape Lookout Bight	−6.1	1 site, seasonal variation
Orinoco floodplain	−4.4	Spatial
Everglades	−4.5	Spatial

The mean value for $\Delta\delta D/\Delta^{13}C$ is -5.1 ± 1.0.

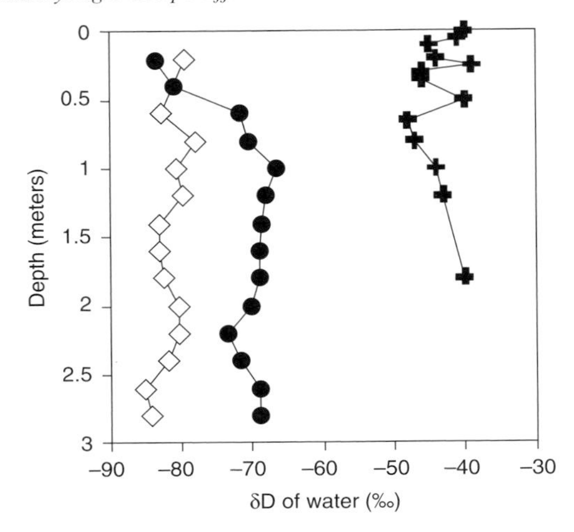

Figure 6.9 Spatial and depth variations in the δD of pore water from Minnesota and Ontario sites are not sufficient to explain the observed trends in δD of CH_4. Minnesota bog samples are represented by open diamonds, Minnesota fen samples by filled circles, and Ontario marsh samples by the plus signs.

where the δD of environmental water was not measured coincidentally with the methane measurements.

We do agree with Waldron *et al.* (1999) that strict interpretation of methane isotopic data is not warranted. For example, it may not be possible to say that acetate-produced methane falls within certain specific values whereas CO_2-produced methane has certain other specific values. Certainly, the δD of environmental water (Waldron *et al.*, 1999), the ^{13}C of DIC (Boehme *et al.*, 1996), or the acetate present (Blair and Carter, 1992) exert an enormous influence on the ultimate isotopic composition of methane. However, we think that a secondary factor influencing methane isotopic composition is the production mechanism by which the methane is formed. For δD, the mechanisms underlying these observations remain unclear, but, nonetheless, variations in production mechanism, e.g., a shift towards more acetate fermentation, result in the production of methane with relatively greater ^{13}C enrichment and greater δD depletion, if other factors are more or less constant.

Isotope Cross Plots

The relationship of $\Delta\delta D/\Delta\delta^{13}C$ described in Table 6.2 is empirical, and while it reflects the relatively larger value of αH relative to αC it does not describe $\alpha H/\alpha C$ in an exact manner. Variations in the value of the methane precursors (H_2O, acetate, and CO_2) influence CH_4 isotopic variation in addition to the αs. In most sediments and peatlands the $\delta^{13}C$ of CO_2 becomes more positive with depth (Boehme *et al.*, 1996; Hornibrook *et al.*, 1997; Chasar *et al.*, 2000a), due to preferential reduction of $^{12}CO_2$ to form CH_4, while the $\delta^{13}C$ of CH_4 becomes more negative with depth due to increasing values of α resulting from shifts in methane production mechanism. Thus changes in the apparent fractionation factor, αC (Whiticar *et al.*, 1986; Whiticar, 1999; Hornibrook *et al.*, 1997, 2000a,b; Chasar *et al.*, 2000a) as

$$\alpha C = [(\delta^{13}CO_2 + 1000)/(\delta^{13}CH_4 + 1000)] \qquad (6.6)$$

can be larger than would be expected from the decrease in methane $\delta^{13}C$ with depth. The apparent αC values illustrated in these plots represent changes in α driven by changes in production mechanism (see discussion in Whiticar, 1999); larger αC is typical of CO_2 reduction while smaller values are typical of acetate fermentation (Whiticar *et al.*, 1986; Sugimoto and Wada, 1993; Whiticar, 1999). Similarly, αH can be described as

$$\alpha H = [(\delta D\text{-}H_2O + 1000)/(\delta D\text{-}CH_4 + 1000)] \qquad (6.7)$$

Typical cross plots drawn from the Minnesota data (Fig. 6.10; Chasar *et al.*, 2000a,b) indicate increasing values of αC and decreasing αH with depth in the fen system and in the bog system relative to the surficial fen samples, consistent with a trend towards more CO_2 reduction at bog and deep fen sites relative to surface fen sites. Cross plots such as these represent a more rigorous approach to defining trends in CH_4 isotopic data. However, these cross plots involve measurements ($\delta D\text{-}H_2O$, $\delta^{13}C\text{-}CO_2$) that may not always be available; and as we have demonstrated, the δD and $\delta^{13}C$ data on CH_4 alone contain useful information.

Summary, Speculation, and Recommendation

In summary, the field studies presented here provide strong evidence that methane produced from acetate fermentation is relatively ^{13}C-enriched and δD-depleted relative to methane produced from reduction of CO_2, if precursor isotopic values are relatively constant. Laboratory incubation studies have produced similar results for $\delta^{13}C$; methane produced from acetate is ^{13}C-enriched relative to methane produced from CO_2 reduction (Sugimoto and Wada 1993; Waldron *et al.*, 1998a). However, the lab studies have not produced the apparent δD variation with production mechanism observed in the field studies. Burke (1993) noted that CH_4 produced in laboratory incubations and in the rumen with elevated H_2 concentrations exhibits

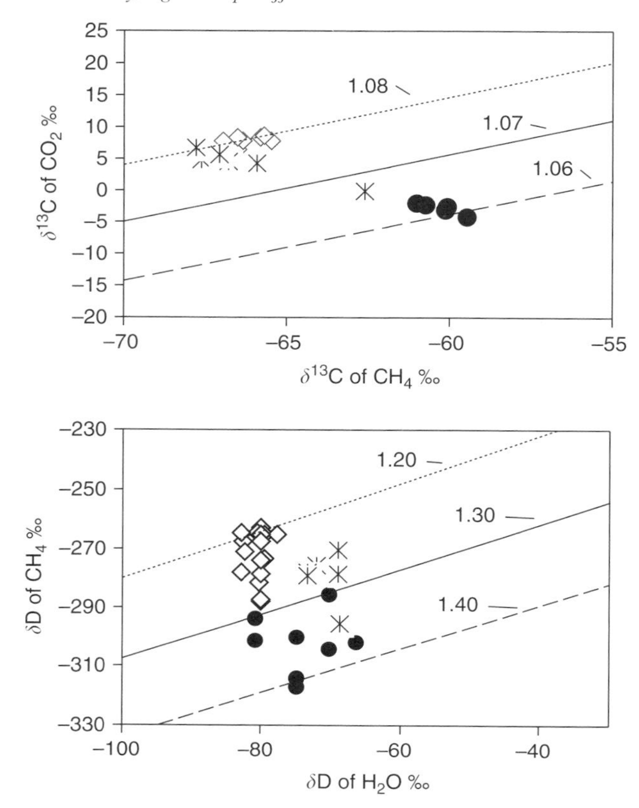

Figure 6.10 Cross plots of CH_4, ^{13}C and δD and precursors, CO_2 (upper panel), and H_2O (lower panel). In the upper panel, the upper dotted line represents $\alpha = 1.080$, the solid middle line $\alpha = 1.070$, and the lower dashed line $\alpha = 1.060$. In the lower panel, the upper dotted line represents $\alpha = 1.200$, the solid middle line $\alpha = 1.300$, and the lower dashed line $\alpha = 1.400$. Bog samples are represented by open diamonds, fen samples above 1 m by filled symbols, and fen samples deeper than 1 m by stars.

depleted δD values relative to values found in wetlands and sediments. He hypothesized that fractionation between the δD of water and the δD of methane increased with increasing partial pressure of H_2. Possibly, variations of CH_4-δD associated with variations in H_2 concentration may be an additional factor complicating comparisons of field samples with those of laboratory incubations. While H_2 was not added in the studies of Sugimoto and Wada (1995) or Waldron *et al.* (1998a) it seem possible that H_2 becomes elevated in these highly reactive closed systems. Both studies resulted in CH_4 with relatively depleted δD values relative to isotopic values observed in wetlands and sediments. Carrying this idea further, it is interesting to speculate that the variation in δD that we observed in these field studies and attributed to production mechanism variations may have been due to covariation of H_2 concentration with production mechanism. Perhaps H_2 concentration is greater in soils and sediments with more labile organic matter where acetate fermentation is of more relative importance (Burke, 1993). Overall, there is a clear need for additional investigation of H-isotope fractionation between organic matter, acetate, methane, hydrogen, and water. Measurements of H_2 concentrations and methane isotopic values across a variety of wetlands and sediments would be useful. The development of on-line continuous-flow H isotopic capability on new isotope ratio mass spectrometers makes this a very promising and exciting area for research.

Acknowledgments

This work was supported by the National Science Foundation, DEB #9615429 to Glaser, Chanton, and Siegel, and OPP 093677 to Hines, Chanton, and Juliette Rooney-Varga. We thank Larry Flanagan for a thorough and thought-provoking review.

References

Avery G. B. Jr., Shannon R. D., White J. R., Martens C. S. and Alperin M. J. (1999) Effect of seasonal changes in the pathways of methanogenesis on the $\delta^{13}C$ values of pore water methane in a Michigan peatland. *Global Biogeochem Cycles* **13**(2): 475–484.

Barker J. F. and Fritz P. (1981) Carbon isotope fractionation during microbial methane oxidation. *Nature* **293**: 289–291.

Bellisario L. M., Moore T. R., Bubier J. L. and Chanton J. P. (1999) Controls on methane emission from a northern peatland. *Global Geochem Cycles* **13**: 81–91.

Blair N. E. (1998) The $\delta^{13}C$ of biogenic methane in marine sediments: The influence of C-org deposition rate. *Chem Geol* **152**: 139–150.

Blair N. E. and Carter W. D. (1992) The carbon isotope biogeochemistry of acetate from a methanogenic marine sediment. *Geochim Cosmochim Acta* **56**: 1247–1258.

Blair N. E., Leu A., Munoz E., Olsen J., Kwong E. and DesMarais D. (1985) Carbon isotopic fractionation in heterotrophic microbial metabolism. *Appl Environ Microbiol* **50**: 996–1001.

Blair N. E., Martens C. S. and DesMarias D. (1987) Natural abundances of carbon isotopes in acetate from a coastal marine sediment. *Nature* **236**: 66–68.

Boehme S. E., Blair N. E., Chanton J. P. and Martens C. S. (1996) A mass balance ^{13}C and ^{12}C in an organic-rich methane-producing marine sediment. *Geochim Cosmochim Acta* **60**: 3835–3848.

Brune A., Frenzel P. and Cypionka H. (2000) Life at the oxic–anoxic interface: Microbial activities and adaptations. *Fems Microbiol Rev* **24**(5): 691–710.

Burke R. A. (1993) Possible influence of H concentration on microbial methane stable H isotopic composition. *Chemosphere* **26**: 55–67.

Burke R. A., Martens C. S. and Sackett W. (1988a) Seasonal variations of D/H and 13C/12C ratios of microbial methane in surface sediments. *Nature* **332**: 829–831.

Burke R. A., Barber T. and Sackett W. (1988b) Methane flux and stable H and C isotope composition of the sedimentary methane from the Florida Everglades. *Global Biogeochem Cycles* **2**: 329–340.

Chanton J. P. and Martens C. S. (1988) Seasonal variations in the isotopic composition and rate of methane bubble flux from a tidal freshwater estuary. *Global Biogeochem Cycles* **2**: 289–298.

Chanton J. P. and Dacey J. W. H. (1991) Effects of vegetation on methane flux, reservoirs and carbon isotopic composition. In *Trace Gas Emissions from Plants* (H. Mooney, E. Holland and T. Sharkey, eds) pp. 65–92. Academic Press.

Chanton J. and Smith L. (1993) Seasonal variations in the isotopic composition of methane associated with aquatic emergent macrophytes of the central Amazon basin. In *The Biogeochemistry of Global Change: Radiative Trace Gases* (R. S. Oremland, ed.) pp. 619–633. Chapman and Hall, NY.

Chanton J. P., Whiting G., Happell J. and Gerard G. (1993) Contrasting rates and diurnal patterns of methane emission from different types of vegetation. *Aquatic Botany* **46**: 111–128.

Chanton J. P., Bauer J., Glaser P., Siegel D., Ramonowitz E., Tyler S., Kelley C. and Lazrus A. (1995) Radiocarbon evidence for the substrates supporting methane formation within northern Minnesota peatlands. *Geochim Cosmochim Acta* **59**: 3663–3668.

Chanton J. P., Rutkowski C. M. and Mosher B. M. (1999) Quantifying methane oxidation from landfills using stable isotope analysis of downwind plumes. *Environ Sci Technol* **33**: 3755–3760.

Chanton J. P. and Liptay K. (2000) Seasonal variation in methane oxidation in a landfill cover soil as determined by an in situ stable isotope technique. *Global Biogeochem Cycles* **14**: 51–60.

Chanton J. P., Rutkowski C. M., Schwartz C. C., Ward D. E. and Boring L. (2000) Factors influencing the stable carbon isotope signature of CH_4 from combustion and biomass burning. *J Geophys Res Atomospheres* **105**: D2, 1867–1877.

Chasar L. S., Chanton J. P., Glaser P. H., Siegel D. I. and Rivers J. S. (2000a) Radiocarbon and stable carbon isotopic evidence for transport and transformation of DOC, DIC and CH_4 in a northern Minnesota peatland. *Global Biogeochem Cycles* **14**: 1095–1103.

Chasar L. S., Chanton J. P., Glaser P. H. and Siegel D. I. (2000b) Methane concentration and stable isotope distribution as evidence of rhizospheric processes: Comparison of a fen and bog in the Glacial Lake Agassiz peatland complex. *Ann Botany* **86**: 655–663.

Coleman D. D., Risatte J. B. and Schoell M. (1981) Fractionation of carbon and hydrogen isotopes by methane-oxidizing bacteria. *Geochim Cosmochim Acta* **45**: 1033–1037.

Conrad R., Klose M. and Claus P. (2002) Pathway of CH_4 formation in anoxic rice field soil and rice roots determined by ^{13}C-stable isotope fractionation. *Chemosphere* **47**: 797–806.

Crill P. M. and Martens C. S. (1986) Methane production from bicarbonate and acetate in an anoxic marine sediment. *Geochim Cosmochim Acta* **50**: 2089–2097.

Dacey J. W. H. (1981) How aquatic plants ventilate. *Oceanus* **24**: 43–51.

Daniels L., Fulton G., Spencer R. W. and Orme-Johnson W. H. (1980) Origin of H in methane produced by *Methanobacterium thermoautotrohicum*. *J Bacteriol* **141**: 694–698.

de Graaf W., Wellsbury P., Parkes R. J. and Cappenberg T. E. (1996) Comparison of acetate turnover in methanogenic and sulphate reducing sediments by radiolabeling and stable isotope labeling and by use of specific inhibitors: Evidence for isotopic exchange. *Appl Environ Microbiol* **62**: 772–777.

Duddleston K. N., Kinney M. A., Kiene R. P. and Hines M. E. (2002) Seasonal anaerobic biogeochemistry in a northern ombrotrophic bog: Acetate as a dominant metabolic end product. *Global Biogeochem Cycles* **16**(4): 1063, doi:10.29/2001GB001402.

Gelwicks T. T., Risatti J. B. and Hayes J. M. (1994) Carbon isotope effects associated with aceticlastic methanogenesis. *Appl Environ Microbiol* **60**: 467–472.

Happell J., Chanton J. P. and Showers W. (1994) The influence of methane oxidation on the stable isotopic composition of methane emitted from Florida Swamp forests. *Geochim Cosmochim Acta* **58**: 4377–4388.

Hines M. E., Duddleston K. N. and Kiene R. P. (2001) Carbon flow to acetate and C-1 compounds in northern wetlands. *Geophys Res Lett* **8**(22): 4251–4254

Hornibrook E. R. C., Longstaffe W. S. and Fyfe W. S. (1997) Spatial distribution of microbial methane production pathways in temperate zone wetland soils: Stable carbon and hydrogen isotope evidence. *Geochim Cosmochim Acta* **61**(4): 745–753.

Hornibrook E. R. C., Longstaffe F. J. and Fyfe W. S. (2000a) Evolution of stable carbon isotope compositions for methane and carbon dioxide in freshwater wetlands and other anaerobic environments. *Geochim Cosmochim Acta* **64**: 1013–1027.

Hornibrook E. R. C., Longstaffe F. J. and Fyfe W. S. (2000b) Factors influencing stable isotope ratios in CH_4 and CO_2 within subenvironments of freshwater wetlands: Implications for δ-signatures of emissions. *Isotopes Environ Health Stud* **36**: 151–176.

Kelley C. A., Dise N. and Martens C. S. (1992) Temporal variations in the stable carbon isotopic composition of methane emitted from Minnesota peatlands. *Global Biogeochem Cycles* **6**(3): 263–269.

Kelley C. A., Martens C. S. and Ussler W. (1995) Methane dynamics across a tidally flooded riverbank margin. *Limnol Oceanogr* **40**: 1112–1129.

King G. M. (1992) Ecological aspects of methane oxidation, a key determinant of global methane dynamics. *Adv Microbiol Ecol* **12**: 431–468; (K. C. Marshall, ed.) Plenum, New York.

Kruger M., Eller G., Conrad R. and Frenzel P. (2002) Seasonal variation in pathways of CH_4 production and in CH_4 oxidation in rice fields determined by stable isotopes and specific inhibitors. *Global Change Biol* **8**: 265–280.

Lansdown J. M., Quay P. D. and King S. L. (1992) CH_4 production via CO_2 reduction in a temperate bog: A source of ^{13}C depleted CH_4. *Geochim Cosmochim Acta* **56**: 3493–3503.

Liptay K., Chanton J., Czepiel P. and Mosher B. (1998) Use of stable isotopes to determine methane oxidation in landfill cover soils. *JGR-Atmospheres* **103**D: 8243–8250.

Lubina C., Bergamaschi P., Konigstedt R. and Fisher J. (1996a) Isotope studies of methane oxidation in landfill cover soils. Paper presented at IGBP Joint North American–European Workshop on Measurements and Modeling of Methane Fluxes from Landfills. Argonne, Ill.

Lubina C., Bergamaschi P., Konigstedt R. and Fisher J. (1996b) Isotope studies of methane oxidation in landfill cover soils (abstract). *Eos Trans* AGU 77, West Pac Geophys Meet (Suppl 3).

Martens C. S. and Klump J. V. (1980) Biogeochemical Cycling in an organic rich coastal marine basin. 1. methane sediment–water exchange processes. *Geochim Cosmochim Acta* **44**(3): 471–490.

Martens C. S., Chanton J. P. and Paull C. K. (1991) Fossil biogenic methane at the Florida Escarpment. *Geology* **19**: 851–854.

Martens C. S., Kelley C. A., Chanton J. P. and Showers W. J. (1992) Carbon and hydrogen isotopic composition of methane from wetlands and lakes of the Yukon–Kuskokwim Delta of the Alaskan tundra. *J Geophys Res* **97**: 16689–16701.

Miller J. B. (2004) The carbon isotopic composition of atmospheric methane and its constraint on the global methane budget. This volume.

Paull C. K., Martens C. S., Chanton J. P., Neumann A. C., Coston J., Jull A. T. and Toolin L. J. (1989) Fossil methane carbon in living chemosynthetic organisms and young carbonate cements at the Florida Escarpment. *Nature* **342**: 166–168.

Pine M. J. and Barker H. A. (1956) Studies on methane fermentation XII. The pathway of H in acetate fermentation. *J Bacteriol* **7**: 644–648.

Popp T. J. (1998) The methane stable isotope distribution and an evaluation of rhizospheric methane oxidation at a fen in North Central Alberta. Florida State University MS thesis, 166 p.

Popp T. J., Chanton J. P., Whiting G. J. and Grant N. (1999) The methane stable isotope distribution at a *Carex* dominated fen in North Central Alberta. *Global Biogeochem Cycles* **13**: 1063–1077.

Popp T. J., Chanton J. P., Whiting G. J. and Grant N. (2000) Evaluation of methane oxidation in the rhizosphere of a *Carex* dominated fen in North Central Alberta. *Biogeochemistry* **51**: 259–281.

Reeburgh W. S. (1996) 'Soft spots' in the global methane budget. In *Microbial Growth on C1 compounds* (M. E. Lidstrom and F. R. Tabia, eds) pp. 334–342. Kluwer Acad., Norwell, Mass.

Sansone F. J. (1986) Depth distribution of short-chain organic acid turnover in Cape Lookout Bight Sediments. *Geochim Cosmochim Acta* **50**: 99–105.

Sansone F. J. and Martens C. S. (1982) Volatile fatty acid cycling in organic-rich marine sediments. *Geochim Cosmochim Acta* **46**: 1575–1589.

Schoell M. (1988) Multiple Origins of methane in the earth. *Chem Geol* **71**: 1–10.

Shannon R. D. and White J. R. (1994) 3-year study of controls on methane emissions from 2 Michigan Peatlands. *Biogeochemistry* **27**: 36–60.

Shannon R. D., White J. R., Lawson J. E. and Gilmore B. S. (1996) Methane efflux from emergent vegetation in peatlands. *J Ecol* **84**: 239–246.

Siegel D. I., Chanton J. P., Glaser P. H., Chasar L. C. and Rosenberry D. O. (2001) Estimating methane production rates in bogs and landfills by deuterium enrichment of pore-water. *Global Biogeochem Cycles* **15**: 967–977.

Smith L. K., Lewis W. M., Chanton J. P., Cronin G. and Hamilton S. K. (2000) Methane emissions from the Orinoco River Floodplain, Venezuela. *Biogeochemistry* **51**: 113–140.

Stevens C. M. and Engelkemeir A. (1988) Stable carbon isotopic composition of methane from some natural and anthropogenic sources. *J Geophys Res* **93**: 725–733.

Sugimoto A. and Wada E. (1993) Carbon isotopic composition of bacterial methane in a soil incubation experiment: Contribution of acetate and CO_2/H_2. *Geochim Comochim Acta* **57**: 4015–4027.

Sugimoto A. and Wada E. (1995) Hydrogen isotopic composition of bacterial methane: CO_2/H_2 reduction and acetate fermentation. *Geochim Cosmochim Acta* **59**: 1329–1337.

Tyler S. C. (1991) The global methane budget. In *Microbial Production and Consumption of Greenhouse Gases: Methane, Nitrogen Oxides, and Halomethanes* (J. E. Rogers and W. B. Whitman, eds) pp. 7–38. American Society of Microbiology, Washington, DC.

Wahlen M., Tanaka N., Henry R., Deck B., Zeglen J., Vogel J. S., Southon J., Shemesh A., Fairbanks R. and Broecker W. (1989) Carbon-14 in methane sources and in atmospheric methane: The contribution from fossil carbon. *Science* **245**: 286–290.

Waldron S., Watson-Craik I. A., Hall A. J. and Falllick A. E. (1998a) The carbon and hydrogen stable isotope composition of bacteriogenic methane: A laboratory study using a landfill inoculum. *Geomicrobiology* **15**: 157–169.

Waldron S., Fallick A. and Hall A. (1998b) Comment on 'Spatial distribution of microbial methane production pathways in temperate zone wetland soils: Stable carbon and hydrogen evidence.' *Geochim Cosmochim Acta* **62**: 369–372.

Waldron S., Lansdown J. M., Scott E. M., Fallick A. E. and Hall A. J. (1999) The global influence of H isotopic composition of water on that of bacteriogenic methane from shallow freshwater environments. *Geochim Cosmochim Acta* **63**: 2237–2245.

Whiticar M. J. (1993) Stable isotopes and global budgets. In *Atmospheric Methane: Sources, Sinks, and Role in Global Climate Change* (M. A. K. Kahlil, ed.) NATO ASI Series, Vol. 113, pp. 138–167. Springer-Verlag, Berlin.

Whiticar M. J. (1999) Carbon and hydrogen isotope systematics of bacterial formation and oxidation of methane. *Chem Geol* **161**: 291–314.

Whiticar M. J. and Faber E. (1986) Methane oxidation in sediment and water column environments—isotope evidence. *Org Geochem* **10**: 759–768.

Whiticar M. J., Faber E. and Schoell M. (1986) Biogenic methane formation in marine and freshwater environments: CO_2 reduction vs acetate fermentation—isotope evidence. *Geochim Cosmochim Acta* **50**: 693–709.

Whiting G. J. and Chanton J. P. (1993) Primary production control of methane emissions from wetlands. *Nature* **364**: 794–795.

Woltemate I., Whiticar M. J. and Schoell M. (1984) Carbon and hydrogen isotopic composition of bacterial methane in a shallow freshwater lake. *Limnol Oceanogr* **29**: 985–992.

Part II

Ecosystem Scale Processes

7

Theoretical Examination of Keeling-plot Relationships for Carbon Dioxide in a Temperate Broadleaved Forest with a Biophysical Model, CANISOTOPE

Dennis D. Baldocchi, David R. Bowling

Introduction

Respiration by the soil microbes, roots, and leaves and photosynthesis by leaves alter the CO_2 mixing ratio (C) of the atmosphere and its isotopic composition (Yakir and Sternberg, 2000; Bowling *et al.*, 2001). Consequently, stable isotope $^{13}CO_2$ enables biogeoscientists to trace the flow of carbon among plants, the soil, and the atmosphere (Flanagan and Ehleringer, 1998; Yakir and Sternberg, 2000).

'Keeling plots' provide an experimental means of inferring the isotopic composition of respiring sources. The method assumes that the transfer of stable isotopes between a respiring source, such as an ecosystem, and the atmosphere can be represented as a two-end-member, mixing model. On the basis of this assumption, the intercept of the linear regression between the independent ($1/C$) and dependent, $\delta^{13}C$, variables equals isotopic composition of the respiring source (Keeling, 1958; Yakir and Sternberg, 2000; Pataki *et al.*, 2003).

In practice, the assumption that plant canopies are 'well-stirred mixing tanks' is not always true. Sources and sinks of CO_2 and atmospheric turbulence vary vertically in a plant canopy and assimilatory and respiratory carbon fluxes differ greatly on sunlit and shaded leaves (Baldocchi and Harley, 1995; Katul and Albertson, 1999). Furthermore, the turbulent transfer of trace gases, such as carbon dioxide, between plants and the atmosphere is very intermittent and non-local. In particular, wind shear at the plant atmosphere interface generates large-scale and intermittent coherent structures that produce counter-gradient mass transfer (Finnigan, 2000). At night, when the 'Keeling plot' approach is most valid, the stable thermal stratification of the atmosphere can cause periods when turbulent mixing inside a canopy is decoupled from that above it.

Over the past 20 years we have been conducting a series of field studies on the carbon, water, and energy exchange of a temperate broadleaved forest (Verma *et al.*, 1986; Bowling *et al.*, 1999; Baldocchi and Wilson, 2001). Through that work a biophysical model, CANOAK, was developed for computing fluxes of carbon and water between the forest and the atmosphere. The model's use of Lagrangian, random-walk diffusion theory to compute turbulent diffusion of trace gases within and above the canopy enables it to consider counter-gradient transport in the vicinity of canopy sources and sinks. Recently, we adapted this model to investigate environmental and plant factors that control the flux density and discrimination of the stable isotope $^{13}CO_2$ (Baldocchi and Bowling, 2003), producing a new model called CANISOTOPE. In this chapter we use the micrometeorological-based CANISOTOPE model to analyze the isotopic contents of a forest canopy produced by 'Keeling plots'. Our objective is to study how hypothetical changes in turbulent mixing, photosynthesis and respiration, canopy structure, and physiological capacity alter Keeling plot intercepts.

Model Framework

Model: Theory, Implementation, Inputs, and Parameters

The biophysical isotope model, CANISOTOPE, computes scalar profiles and flux densities of water vapor, CO_2, $^{13}CO_2$, and sensible heat within and above a broadleaf forest. The model, a member of the CANOAK/CANVEG family, consists of coupled micrometeorological and ecophysiological modules. The micrometeorological modules compute leaf and soil energy exchange, turbulent diffusion, and radiative transfer through the canopy. Environmental variables, computed with the micrometeorological module, drive the physiological modules that compute leaf photosynthesis, stomatal conductance, transpiration, and leaf, bole, and soil/root respiration. The model resolves vertical heterogeneity by dividing the canopy into multiple layers. Horizontal complexity is considered by evaluating mass and energy exchange of sunlit and shaded leaves separately. The model has been described and tested for summer-length studies by Baldocchi and Harley (1995) and over the course of several years by Baldocchi and Wilson (2001), and for $^{13}CO_2$ by Baldocchi and Bowling (2003). A brief overview of key components is provided later.

To compute isotopic flux densities and concentration profiles, we apply the conservation budget for a passive scalar to compute scalar fluxes (F) and their local ambient concentrations (C) (e.g., Raupach, 2001). By assuming that the canopy is horizontally homogeneous and environmental conditions are steady, the scalar conservation equation is reduced to an equality between the change, with height, of the vertical turbulent flux density (F) and the diffusive source/sink strength, $S(C, z)$. For $^{13}CO_2$, the conservation budget is expressed as a function of the ratio $^{13}C/(^{12}C + ^{13}C)$ in the air, R_a:

$$\frac{d^{13}C}{dt} = \frac{d(R_a \cdot C)}{dt} = -\left\{\frac{\partial[R_a \cdot F(C, z)]}{\partial z} + S(^{13}C, z)\right\} \tag{7.1}$$

The source term is a function of ^{13}C assimilated and ^{13}C respired by the vegetation and soil. Photosynthesis of $^{13}CO_2(^{13}A)$ is defined as a function of leaf photosynthesis (A) and the ratio of $^{13}C/(^{12}C + ^{13}C)$ in the photosynthetic products (R_{plant}):

$$^{13}A = A \cdot R_{plant} = \frac{A \cdot R_{air}}{1 + \Delta} \tag{7.2}$$

In Eq. 7.2, the isotopic discrimination factor due to photosynthesis is defined as:

$$\Delta = \left(\frac{R_{air}}{R_{plant}} - 1\right) \cdot 1000 \tag{7.3}$$

We calculated the photosynthetic discrimination against ^{13}C with a linear model that is a function of the CO_2 mixing ratio in the substomatal cavity of the leaf (C_i), relative to that in the atmosphere, C_a (Farquhar *et al.*, 1982):

$$\Delta = a + (b - a)\frac{C_i}{C_a} \tag{7.4}$$

In Eq. 7.4, a is the fractionation during diffusion (4.4‰); b is the net fractionation due to carboxylation (27.5‰). This algorithm is a simpler version of the more expansive model produced later by Farquhar *et al.* (1989), which includes additional terms for fractionation across the laminar boundary layer, an equilibrium fractionation as CO_2 enters solution, and a fractionation term representing $^{13}CO_2$ diffusion in water. The internal CO_2 concentration, C_i, is computed using set Fickian diffusion equations for the transfer of CO_2 between the atmosphere (C_a) and the leaf interior (C_i) that represent a balance between the supply of and demand for CO_2. One equation considers the transfer of CO_2 between the free atmosphere and leaf surface, which passes through the leaf boundary layer. The other equation considers the diffusion through the leaf stomata into the substomatal cavity:

$$A = g_s(C_s - C_i) = g_b(C_a - C_s) \tag{7.5}$$

In Eq. 7.5, g_b represents the boundary layer conductance and g_s the stomatal conductance. The solution to these sets of equations was performed with a coupled analytical model (Baldocchi, 1994).

Soil respiration of ^{13}C, $^{13}F_{soil}$, is represented as a multiplicative function of soil respiration, F_{soil}, and the isotope ratio of total ecosystem respired CO_2, R_{soil},

$$^{13}F_{soil} = R_{soil} \cdot F_{soil} \tag{7.6}$$

For simplicity, we assumed that the isotopic signal of the CO_2 respired by the soil was equal to the signal of the previous day's photosynthesis (Hogberg *et al.*, 2001). Specific information on model parameters, site location, weather and climate, and canopy structure are reported in papers by Bowling *et al.* (1999), Baldocchi and Wilson (2001), and Baldocchi and Bowling (2003).

Results and Discussion

Model Validation

Before we discuss the sensitivity of 'Keeling plot' intercepts to variations in factors such as time, soil respiration, turbulent mixing, leaf area, and photosynthetic capacity, we examine how well the model reproduces Keeling plot intercepts under field conditions. Figure 7.1 shows that the CANISOTOPE model is able to compute a Keeling plot intercept, for the summer growing season, that is indistinguishable from the one measured in the field; the measured intercept was -26.13 ± 0.56‰; and the computed intercept was -26.26 ± 0.01‰. We also report that the model was able to reproduce measured carbon isotope discrimination value in leaves and that computed values of isotopic flux densities agreed within 30% of measured values (Baldocchi and Bowling, 2003). With an acceptable level of agreement between measured and computed Keeling plot intercepts, we next use the model to diagnose the behavior of Keeling plot intercepts with respect to variations in controlling variables.

Model Sensitivity Studies

Two lines of evidence exist regarding the temporal variation of Keeling plot intercepts for $^{13}CO_2$. One body of data shows that the intercept does not change from night to day (Buchmann and Ehleringer, 1998;

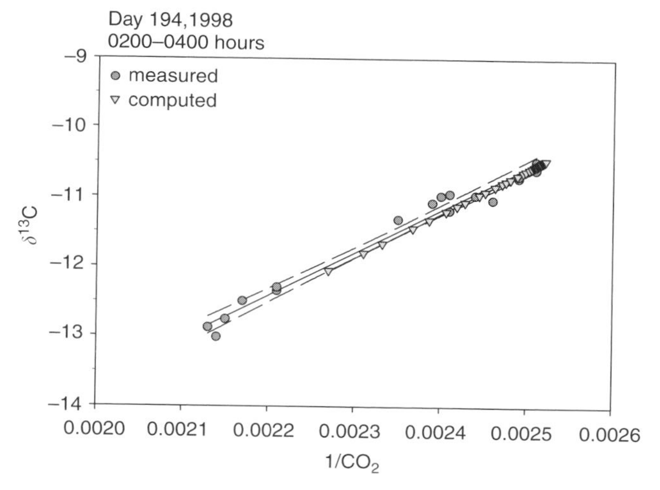

Figure 7.1 A test of the ability of the CANISOTOPE model to compute Keeling plot intercepts within a broadleaved deciduous forest. The measured intercept was -26.13 ± 0.56‰; the computed intercept was -26.26 ± 0.01‰ (after Baldocchi and Bowling, 2003). The geometric mean regression method was used to fit the data.

Mortazavi and Chanton, 2002), and another set of evidence suggests that it does (Bowling *et al.*, 1999; Pataki *et al.*, 2003).

Two theoretical Keeling plots are compared in Fig. 7.2. One is computed for daytime conditions and the other for nighttime conditions. The linear regressions through the two sets of data are offset from one another since

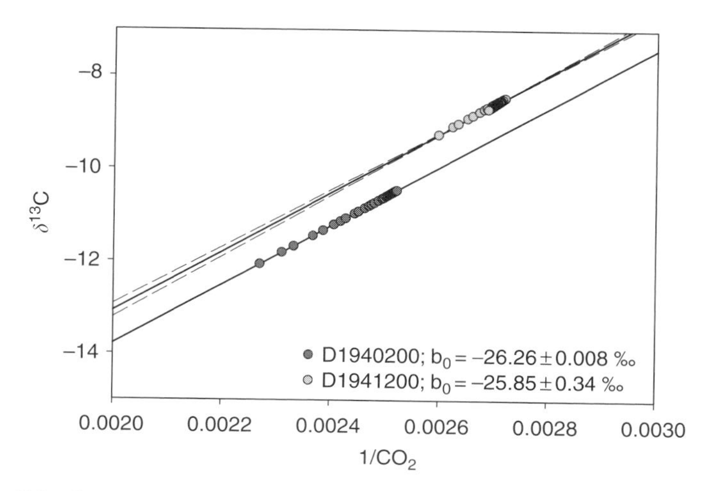

Figure 7.2 Computed Keeling plots for a daytime and nighttime period on day 194 during the 1998 growing season; the data were fit with geometric mean regressions.

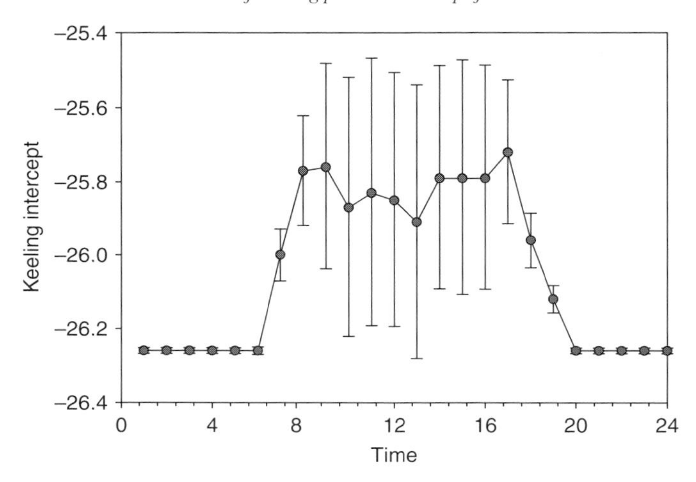

Figure 7.3 Diurnal pattern of Keeling plot intercepts computed with the CANISOTOPE model. These results are for a typical summer day, D194, 1998. The geometric mean regression technique was used to evaluate the regression intercepts and the error bars represent the statistical error associated with that estimate.

daytime CO_2 concentrations are lower than nighttime values. The nighttime regression produces an intercept of $-26.26 \pm 0.01‰$ and the regression intercept during the daytime is $-25.85 \pm 0.34‰$, values that are significantly different from one another.

Computing Keeling plot intercepts over the course of a typical summer day reveals that a diurnal pattern exists (Fig. 7.3). In general, Keeling plot intercepts are on the order of $-26.3‰$ during the night; during the day the intercepts fluctuate between $-25.4‰$ and $-25.9‰$. An important feature to notice in Fig. 7.3 is that the standard deviation of the calculated intercept is negligible at night ($<0.01‰$) but relatively large during the day ($\sim 0.3‰$).

Figure 7.4 provides an explanation why there is a diurnal pattern in the Keeling plot intercept and why the standard error of the regression intercept is large during the day. A daytime Keeling plot is composed of two intersecting lines, rather than one continuous line. One line is associated with air above the canopy, the other line corresponding with air inside the canopy. The Keeling plot intercept derived for inside the canopy equals $-26.2 \pm 0.01‰$, a value close to the intercept value derived from nighttime measurements that represents the respiring ecosystem. The Keeling plot intercept derived from air measured above the canopy yields an intercept of $-30.36 \pm 0.01‰$. The isotope discrimination value associated with air above the upper canopy is lower than the value that is associated with the nighttime Keeling plot because the photosynthetic sink in the upper portion of the canopy is probably respired air recycling from the lower levels

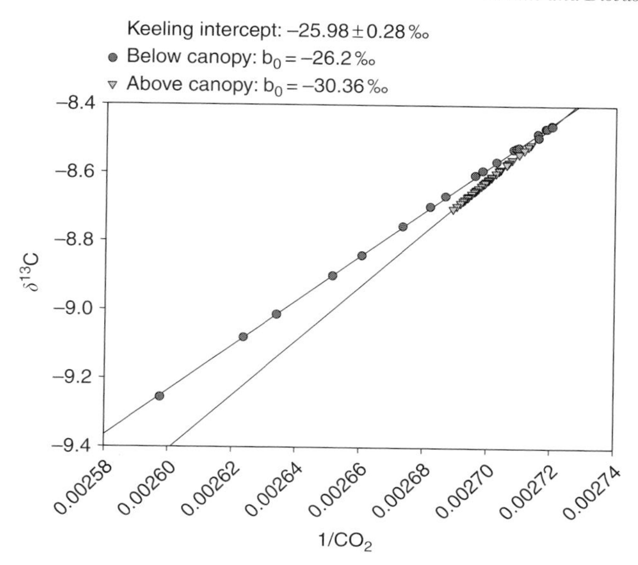

Figure 7.4 Keeling plots computed on separate profiles of air, from above and within the canopy. These theoretical estimates are derived from the CANISOTOPE model. D194, 1200 hours.

of the canopy depleted in $^{13}CO_2$; the isotopic content of the recycled air has a more negative isotopic ratio ($\sim -25.0‰$) than that of air originating from above the canopy ($\sim -8.0‰$) (e.g., Sternberg, 1989). Ironically, the linear regression applied to both sets of data produces a less-negative Keeling plot intercept.

The split in the two regression lines is either due to the differential carbon sources and sinks, as discussed earlier, or is an artefact of counter-gradient transfer that is occurring at a zone within the canopy, as is demonstrated by the calculations presented in Fig. 7.5. For example, between the canopy–atmosphere interface (~24 m) and 21 m, CO_2 concentrations decrease as one moves into the canopy, reflecting down-gradient diffusion and the uptake of carbon dioxide by the canopy. Below 21 m the CO_2 concentration gradient changes sign and CO_2 concentrations increase with further depth into the canopy (Fig. 7.5A) even though the canopy is a strong sink for CO_2 down to 16 m (Fig. 7.5B). At the lowest levels, the source strength of CO_2 released by the soil outpaces the vegetated sink (not shown); so, in this region, down-gradient transfer is occurring.

We emphasize that the subtle differences in the dependency of $\delta^{13}C$ on $1/C$ for air inside and above the canopy may be overlooked when examining field data. First, sampling errors can be large (10–40%) due to the interrelation between the timescales of turbulence and the frequency of sampling

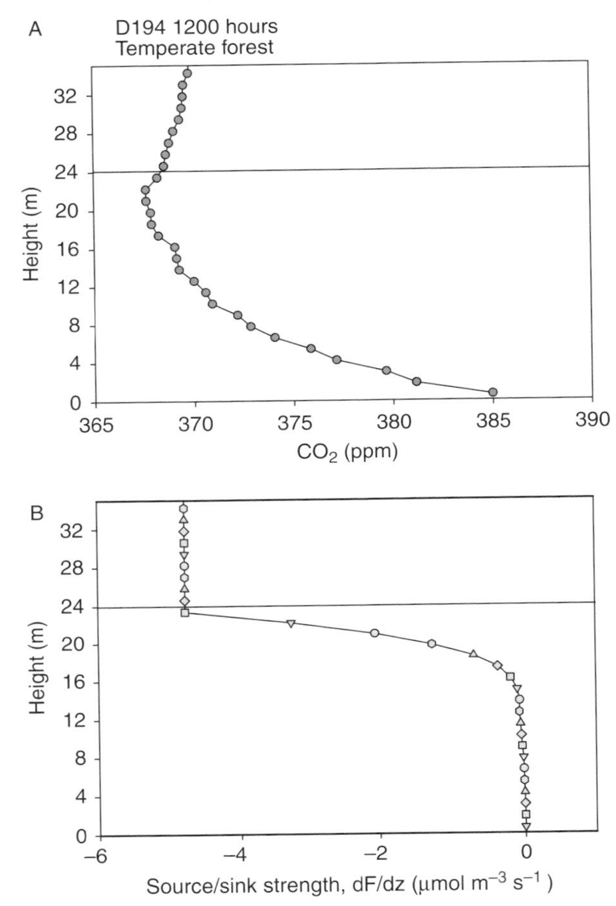

Figure 7.5 Computed vertical profiles of CO_2 concentrations (A) and source sink strength (B) for Day 194 1200. Counter-gradient transfer is identified in the region between 16 and 21 m.

air at a given level (Baldocchi and Bowling, 2003). Second, the finite sampling period required at a given level in the atmosphere and the cost and labor requirements of analyzing such air with a mass spectrometer restricts the number of levels that can be sampled for a representative period of time. Model calculations, on the other hand, have the luxury of producing information for many discrete levels throughout the canopy.

Next we focus on soil, plant, and atmospheric factors that can theoretically modulate the Keeling plot intercept. First, let us examine the impact of changing rates of soil respiration on Keeling plot intercepts. By turning off soil respiration, theoretically, we observe a narrower range of CO_2

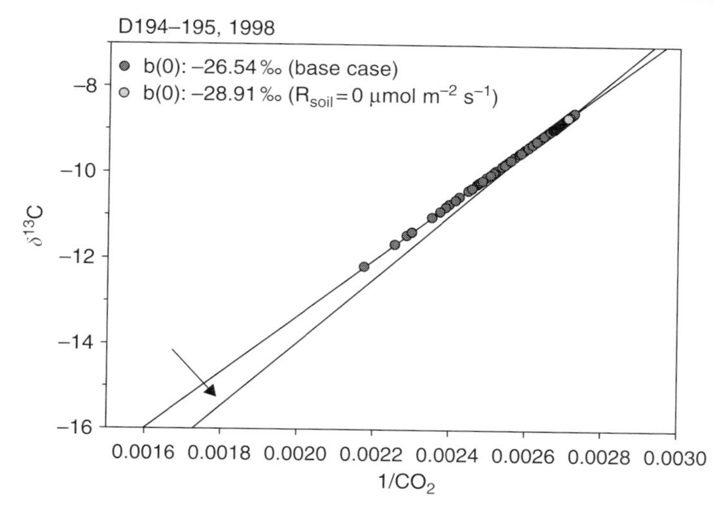

Figure 7.6 Theoretical computations of Keeling plot intercepts for cases with and without soil respiration. The arrow direction and magnitude indicates the tendency of change.

at night and a Keeling plot intercept that is 2.37‰ lower than the reference case, a difference of 9% (Fig. 7.6). This lower (more negative) value reflects leaf respiration of the previous day's photosynthesis, which has a lower isotopic discrimination value than that of carbon decomposing in the soil; the isotopic signal of soil respiration can reflect isotopic enrichment by Basidiomycete fungi (Henn and Chapela, 2001) and the isotopic signal of last year's leaves and carbon in older soil pools, in addition to root respiration that is reflecting recent photosynthesis (Hogberg *et al.*, 2001; Bowling *et al.*, 2002).

Doubling the base rate of soil respiration produces greater concentrations of CO_2 near the soil (Fig. 7.7). But this modification has a negligible effect on the Keeling plot intercept. In this circumstance, doubling the base rate of soil respiration causes the Keeling plot intercept to decrease by only 0.15‰. So we conclude that Keeling plot intercepts are independent of the rate of respiration.

How variations in leaf area index affect isotopic discrimination is a question of much interest to isotope biogeochemists (Buchmann *et al.*, 1997). In principle, variations in leaf-area index will have direct and indirect effects on the Keeling plot by changing the sink and source strengths for leaf photosynthesis and respiration, by altering the fraction of sunlit and shaded leaves, by altering turbulent mixing in the canopy, and by changing the pool size of decomposing litter and the amount of respiring roots. Data in Fig. 7.8 represent results from a numerical experiment where we doubled leaf-area index of the forest. The influence this change had on the Keeling plot

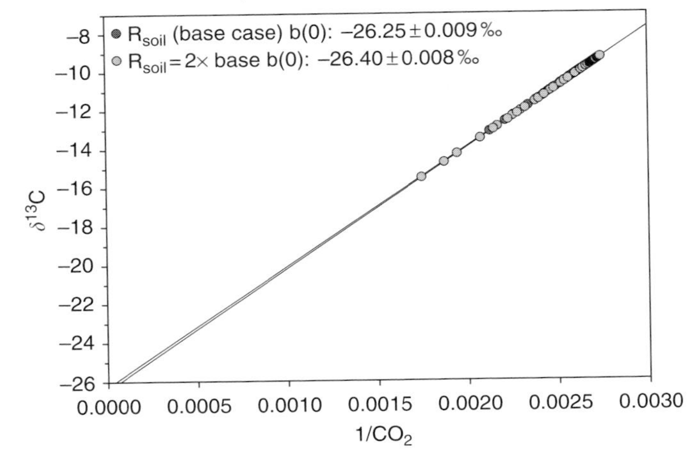

Figure 7.7 Theoretical Keeling plots for conditions of 2 times the base respiration case, which equalled 0.61 μmol m^{-2} s^{-1}. D194, 0300 to 0400 hours.

intercept was minor (a decrease of 0.12‰), a result that is consistent with the experimental findings of Buchmann *et al.* (1997). In general, calculated values of δ^{13}C and CO_2 moved up and down the regression line without greatly altering the slope or intercept.

Since we are using a Lagrangian-based turbulence scheme, these computations do not consider how changes in leaf area will alter turbulent mixing; for example, turbulent mixing deep within a canopy increases as leaf area

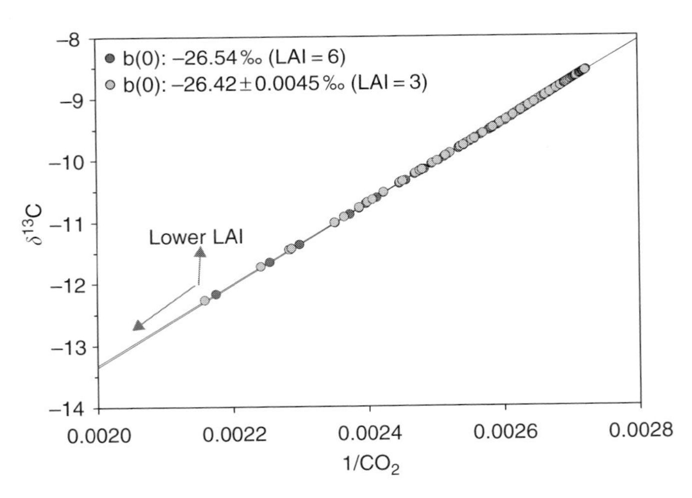

Figure 7.8 Impact of doubling leaf-area index on the Keeling plot for a nighttime period during day 194, 1998. The arrow directions and magnitudes identify the tendency of change.

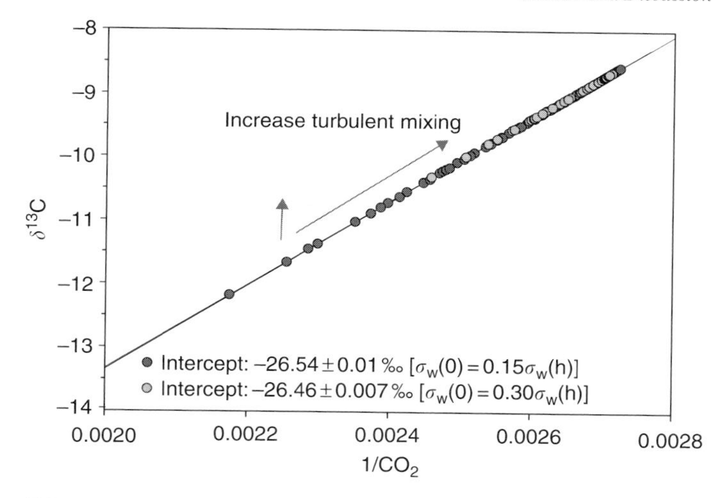

Figure 7.9 The impact of altering turbulent mixing in the canopy on the Keeling plot. The arrow directions and magnitudes identify the tendency of change.

decreases (Dwyer *et al.*, 1997; Massman and Weil, 1999). In contrast, a higher order closure Eulerian model can accommodate this feature because it calculates explicitly the impact of canopy drag on the kinetic energy and momentum exchange budgets of the canopy (see Meyers and Paw U, 1987; Katul and Albertson, 1999). Despite this limitation, we can mimic the interaction between turbulence and leaf-area index by prescribing a change in turbulent mixing within the canopy. Doubling turbulent mixing in the understorey—as quantified by the standard deviation of vertical velocity fluctuations near the soil—increases the Keeling plot intercept by only 0.10‰ (Fig. 7.9). The biggest change that occurs is a shifting of CO_2 concentrations up and down the linear regression line; as turbulent mixing increases, the range of CO_2 and $^{13}CO_2$ concentrations encountered diminishes.

Despite the results shown in Figs 7.8 and 7.9, we do not imply that leaf-area index has no impact on carbon isotope discrimination of a forest canopy, Δ_{canopy}. Theoretical calculations, presented in Fig. 7.10, show that the canopy isotopic discrimination value is about 10% less negative with a 50% decrease in leaf-area index. Additional model calculations (not shown) also indicate that the average value of C_i/C_a for sunlit leaves is lower than that for shaded leaves (~0.75 vs ~0.89). Reducing canopy leaf-area index increases the proportion of sunlit leaves and thereby forces the canopy isotopic discrimination value to be lower during daylight hours. It must also be recognized that changes in photosynthesis and isotopic discrimination, imposed by changes in leaf area, will affect the signature of carbon that is respired later, a feedback that is not satisfactorily simulated in our

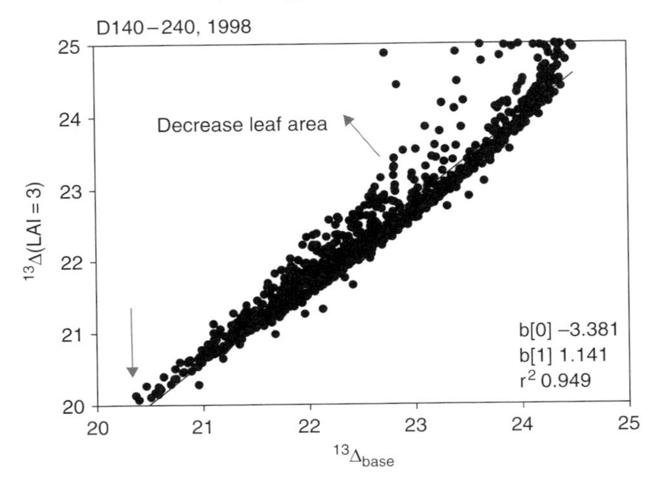

Figure 7.10 The impact of leaf-area index on the canopy isotopic discrimination for a temperate forest over the course of the summer growing season. The arrows identify the tendency of the change.

computations; at present we respire carbon from the litter and roots with the signal of the previous day's photosynthesis.

Next we examine how physiological capacity (Fig. 7.11) and stomatal conductance (Fig. 7.12) alters in the Keeling plot. The Keeling plot intercept is weakly associated with V_{cmax} and the stomatal conductance proportionality constant because changes in photosynthetic and stomatal capacity do not change C_i/C_a appreciably. In practice, sunlit leaves tend to open and close their stomata in concert with changes in photosynthesis to keep C_i/C_a rather conservative, near 0.7 (Wong *et al.*, 1979; Baldocchi, 1994). In addition, because these Keeling plots are computed at night, they do not reflect the impact that long-term changes in physiological capacity may have on the isotopic signature of litter and roots; in these calculations only dark respiration scales directly with photosynthetic capacity.

We computed stomatal conductance by coupling it to photosynthesis (A), relative humidity (rh) and CO_2, $g_s = k(A \cdot rh/C_a)$. When plants experience soil water deficits, it is common to see a decline in the proportionality constant, k (Sala and Tenhunen, 1996). Our model calculations, shown in Fig. 7.12, indicate that an alteration in the stomatal conductance proportionality constant, k, may not alter the Keeling plot intercept by much, despite the fractionation of carbon that occurs when it diffuses through the stomata. On the other hand, a change in k does affect and lower canopy isotope discrimination (Fig. 7.13). On the basis of this finding, we could expect long-term drought to affect the isotopic signal of decomposed leaves and root respiration that would be measured later in the year.

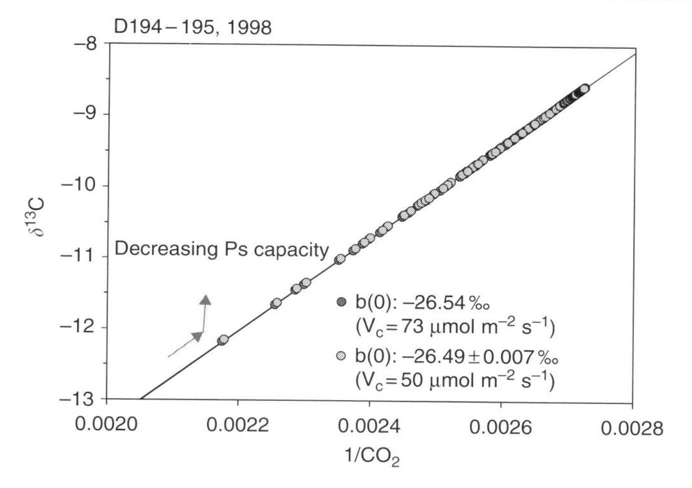

Figure 7.11 Impact of changing photosynthetic capacity on Keeling plots. In these calculations dark respiration scales with maximum carboxylation velocity, V_{cmax}. The arrow direction and magnitude indicates the tendency of change.

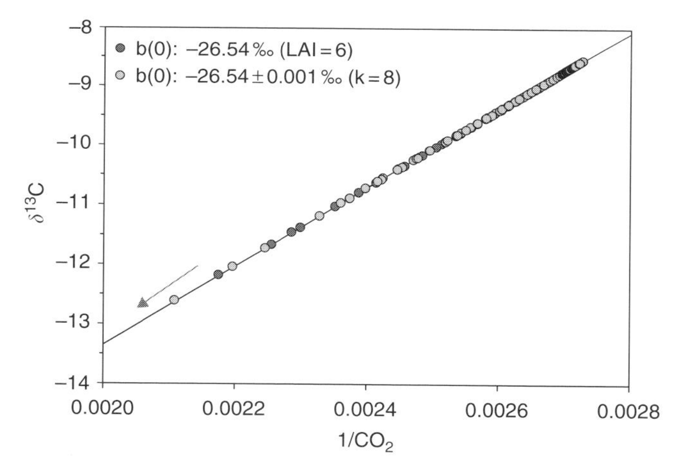

Figure 7.12 Impact of the stomatal conductance proportionality constant (e.g., the Ball-Berry stomatal conductance parameter) on the ^{13}C-CO_2 Keeling plot. The arrow indicates the tendency for change in direction and magnitude of the Keeling plot.

Conclusions

The biophysical models, such as CANOAK, have the potential to provide insight on the behavior of Keeling plots within ecosystems because

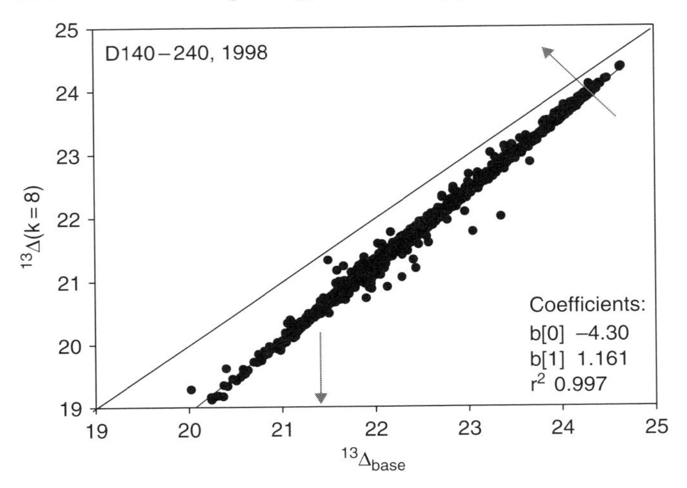

Figure 7.13 The impact of varying stomatal conductance on carbon isotope discrimination of a forest canopy. The arrows indicates the tendency of change in direction and magnitude.

biophysical models quantify interactions between canopy structure, leaf energy budgets, turbulent diffusion and physiological capacity of forests. For the cases studied here, we find that changes in soil respiration have the greatest impact on the Keeling plot intercept. Variations in photosynthetic capacity, leaf-area index, and stomatal conductance do not affect the Keeling plot intercept on the basis of budgeting near-term carbon fluxes. But variations in these variables do affect the isotopic discrimination of the canopy. Hence, we cannot discount their impact in the long term. In other words, if our model had been able correctly to capture how modified inputs of carbon to the soil pool altered the isotopic composition of its respiration as this material decayed and respired over the following year (e.g., Ehleringer *et al.*, 2000), we may have reported different results. Another modification to CANOAK which may improve its fidelity and universality in the near term and long term would involve using a higher-order closure model to assess the impact of leaf-area index on turbulent mixing and the computed dispersion matrix, *a la* Massman and Weil (1999).

Acknowledgments

D. R. Bowling was supported by a grant from NSF (ATM-9905717) to J. R. Ehleringer (University of Utah) and R. K. Monson (University of Colorado). D. D. Baldocchi was supported by the California Agricultural

Experiment Station, by the US Department of Energy's Terrestrial Carbon Project, and NASA's EOS Validation and Carbon Cycle Science Programs (FLUXNET). Any opinions, findings, and conclusions or recommendations expressed in this publication are those of the authors and do not necessarily reflect the views of the supporting government agencies.

References

Baldocchi D. D. (1994) An analytical solution for coupled leaf photosynthesis and stomatal conductance models. *Tree Physiol* **14**: 1069–1079.

Baldocchi D. D. and Harley P. C. (1995) Scaling carbon dioxide and water vapor exchange from leaf to canopy in a deciduous forest: model testing and application. *Plant Cell Environ* **18**: 1157–1173.

Baldocchi D. D. and Wilson K. B. (2001) Modeling CO_2 and water vapor exchange of a temperate broadleaved forest across hourly to decadal time scales. *Ecol Modeling* **142**: 155–184.

Baldocchi D. D. and Bowling D. R. (2003) Modeling the discrimination of $^{13}CO_2$ above and within a temperate broad-leaved forest canopy on hourly to seasonal time scales. *Plant Cell Environ* **26**: 231–244.

Bowling D. R., Baldocchi D. D. and Monson R. K. (1999) Dynamics of isotope exchange of carbon dioxide in a Tennessee deciduous forest. *Global Biogeochem Cycles* **13**: 903–921.

Bowling D. R., Monson R. and Tans P. P. (2001) Partitioning net ecosystem carbon exchange with isotopic fluxes of CO_2. *Global Change Biol* **7**: 127–145.

Bowling D. R., McDowell N. G., Bond B. J., Law B. E. and Ehleringer J. R. (2002) ^{13}C content of ecosystem respiration is linked to precipitation and vapor pressure deficit. *Oecologia* **131**: 113–124.

Buchmann N. and Ehleringer J. R. (1998) CO_2 concentration profiles and carbon and oxygen isotopes in C_3 and C_4 crop canopies. *Agric Forest Meteorol* **89**: 45–58.

Buchmann N., Kao W. Y. and Ehleringer J. R. (1997) Influence of stand structure on carbon-13 of vegetation, soils and canopy air within deciduous and evergreen forests in Utah, United States. *Oecologia* **110**: 109–119.

Dwyer M. J., Patton E. G. and Shaw R. H. (1997) Turbulent kinetic energy budgets from a largy-eddy simulation of airflow above and within a forest canopy. *Boundary Layer Meteorol* **84**: 23–43.

Ehleringer J. R., Buchmann N. and Flanagan L. B. (2000) Carbon isotope ratios in belowground carbon cycle processes. *Ecol Appl* **10**: 412–422.

Farquhar G. D., O'Leary M. H. and Berry J. A. (1982) On the relationship between carbon isotope discrimination and the intercellular carbon dioxide concentration in leaves. *Aust J Plant Physiol* **9**: 121–137.

Farquhar G. D., Ehleringer J. R. and Hubick K. T. (1989) Carbon isotope discrimination and photosynthesis. *Annu Rev Plant Physiol Molecular Biol* **40**: 503–537.

Finnigan J. J. (2000) Turbulence in plant canopies. *Annu Rev Fluid Mech* **32**: 519–571.

Flanagan L. B. and Ehleringer J. R. (1998) Ecosystem–atmosphere CO_2 exchange: interpreting signals of change using stable isotope ratios. *Trends Ecol Evol* **13**: 10–14.

Henn M. R. and Chapela I. H. (2001) Ecophysiology of C-13 and N-15 isotopic fractionation in forest fungi and the roots of the saprotrophic-mycorrhizal divide. *Oecologia* **128**: 480–487.

Hogberg P., *et al.* (2001) Large-scale forest girdling shows that current photosynthesis drives soil respiration. *Nature* **411**: 789–792.

Katul G. G. and Albertson J. D. (1999) Modeling CO_2 sources, sinks, and fluxes within a forest canopy. *J Geophys Res* **104**: 6081–6091.

Keeling C. D. (1958) The concentration and isotopic abundances of atmospheric carbon dioxide in rural areas. *Geochim Cosmochim Acta* **13**: 322–334.

Massman W. J. and Weil J. C. (1999) An analytical one-dimensional second-order closure model of turbulence statistics and the Lagrangian timescale within and above plant canopies of arbitrary structure. *Boundary Layer Meteorol* **91**: 81–107.

Meyers T. P. and Paw U. K. T. (1987) Modeling the plant canopy micrometeorology with higher order closure principles. *Agric Forest Meteorol* **41**: 143–163.

Mortazavi B. and Chanton J. P. (2002) Carbon isotopic discrimination and control of nighttime canopy $\delta^{18}O$-CO_2 in a pine forest in the southeastern United States. *Global Biogeochem Cycles* **16**: 1008.

Pataki D. E., Ehleringer J. R., Flanagan L. B., Yakir D., Bowling D. R., Still C. J., Buchmann N., Kaplan J. O. and Berry J. A. (2003) The application and interpretation of Keeling plots in terrestrial carbon cycle research. *Global Biogeochem Cycles* **17**: doi: 10.1029/2001GB001850.

Raupach M. R. (2001) Inferring biogeochemical sources and sinks from atmospheric concentrations: general consideration and application in vegetation canopies. In *Global Biogeochemical Cycles in the Climate System* (E. D. Schulze *et al.*, eds) pp. 41–60. Academic Press, San Diego.

Sala A. and Tenhunen J. D. (1996) Simulations of canopy net photosynthesis and transpiration in *Quercus ilex* L. under the influence of seasonal drought. *Agric Forest Meteorol* **78**: 203–222.

Sternberg L. S. L. O. (1989) A model to estimate carbon dioxide recycling in forests using $^{13}C/^{12}C$ ratios and concentrations of ambient carbon dioxide. *Agric Forest Meteorol* **48**: 163–173.

Verma S. B., Baldocchi D. D., Anderson D. E., Matt D. R. and Clement R. E. (1986) Eddy fluxes of CO_2, water vapor and sensible heat over a deciduous forest. *Boundary Layer Meteorol* **36**: 71–91.

Wong S. C., Cowan I. R. and Farquhar G. D. (1979) Stomatal conductance correlates with photosynthetic capacity. *Nature* **282**: 424–426.

Yakir D. and Sternberg L. S. L. O. (2000) The use of stable isotopes to study ecosystem gas exchange. *Oecologia* **123**: 297–311.

8

Partitioning Ecosystem Respiration Using Stable Carbon Isotope Analyses of CO_2

Kevin Tu, Todd Dawson

Introduction

Measurements of ecosystem respiration (R_{eco}) provide an estimate of carbon losses, a critical component of the ecosystem carbon balance. These measurements most often integrate respiratory activity from a wide range of sources including aboveground leaves and stems and belowground roots and microorganisms. Each of these sources can potentially respond to environmental conditions in unique ways and at different timescales. As the greater part of carbon dioxide (CO_2) fixed during photosynthesis is eventually returned to the atmosphere through respiration, partitioning what contributes to R_{eco} is central to understanding the environmental and biological controls on each of these respiratory sources and their roles in the dynamics of ecosystem carbon cycling. Partitioning ecosystem respiration is also needed to convert measurements such as net ecosystem CO_2 exchange (NEE) into net primary productivity (NPP) for testing remote-sensing products and ecosystem models.

Recent studies have shown that natural abundance stable carbon isotopes can be used to partition measurements of net ecosystem CO_2 exchange into gross fluxes of photosynthesis and respiration (Yakir and Wang, 1996; Bowling *et al.*, 2001). This chapter explores the related application of the natural abundance stable carbon isotope ratio of respired CO_2 to partitioning components of ecosystem respiration. For the use of labeled and radioactive carbon isotopes, the reader is directed to reviews by Schimel (1993), Coleman and Fry (1991), Hanson *et al.* (2000), and Trumbore (2000). Although the exploitation of the natural variation in stable isotope abundance in respired CO_2 holds the most promise for widespread application among existing measurement networks (e.g., BASIN, FLUXNET, SIBAE) and for comparative studies focused on ecosystem C balance, the reliance on natural variation may also have limitations from the standpoint that such variation is typically small. Recent evidence suggests, however, that variation can be large and significant such that measurable differences among respiration signatures can allow partitioning of ecosystem respiration and therefore inform ecosystem C-balance studies.

Keeling plots are now commonly used to characterize the isotopic signature of ecosystem respiration (Flanagan and Ehleringer, 1998; Pataki *et al.*, 2003). Studies to date indicate significant spatial and temporal variation among ecosystem respiration signatures (Pataki *et al.*, 2003). The cause of such variation has largely been attributed to the environment (Bowling *et al.*, 2002), yet different ecophysiological responses of each ecosystem component can also lead to variation in the isotopic signature of the ecosystem as a whole. That is, Keeling plot intercepts can mask significant variation among the contributing sources, such as between overstory and understory

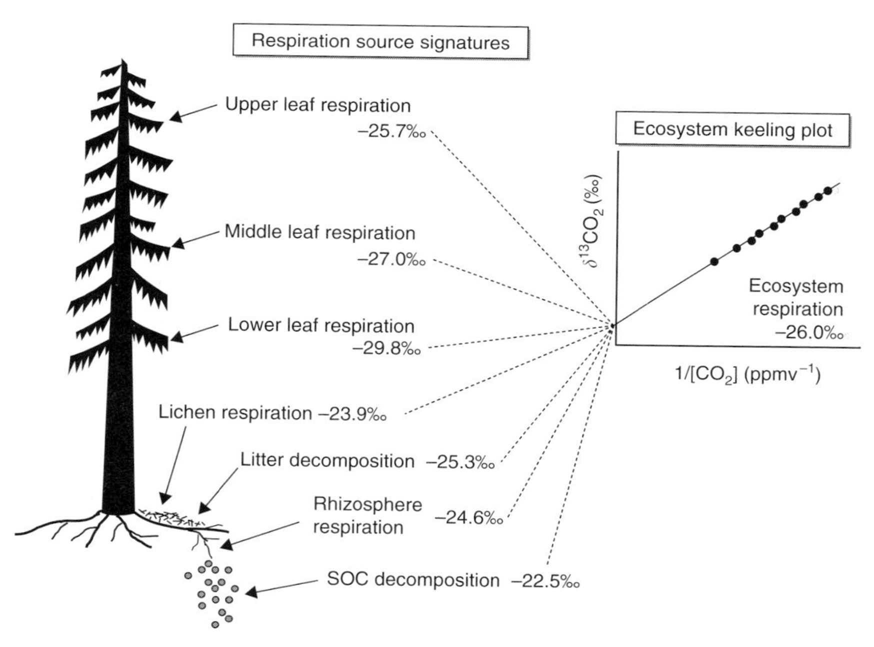

Figure 8.1 The Keeling plot intercept is a composite signal of all respiratory sources in an ecosystem, from canopy leaves to soil microorganisms. Samples were collected in a redwood forest (*Sequoia sempirvirens*) near Occidental, California (unpublished results) using the syringe method described in the section Determining Source Signatures.

species or between plants and microbes. Figure 8.1 for example shows data that demonstrate the range of variation in the $\delta^{13}C$ of respired CO_2 from a redwood forest ecosystem in California. The measured range for seven different sources was 7.3‰. Questions arise from such differences as to the relative contribution of each source to the Keeling plot intercept and how the source contributions change over space or time and under different environmental conditions. Resolving these issues will not only enable the partitioning of ecosystem respiration but will improve our understanding of the causes underlying variation in ecosystem isotope signatures.

After first outlining the theoretical approach to partitioning, we review empirical evidence for isotopic variation among respiration signatures, explore some sampling considerations when determining source signatures, and discuss unresolved issues that need to be addressed with future studies.

Partitioning Approach

Ecosystem respiration has typically been partitioned using chamber flux measurements on each of the components (roots, soils, stems, leaves, etc.) and solving by difference for those components (e.g., microbes) that are difficult to measure or isolate with a chamber (Law *et al.*, 2001; Xu *et al.*, 2001). Isotope approaches on the other hand can provide an alternative measure that additionally holds the promise of in-situ partitioning without disturbance effects common to most chamber methods. Realization of this promise has been limited to ecosystems that have experienced a change in the dominant vegetation with different photosynthetic pathways such as from C_3 to C_4 grasses or vice versa (Robinson and Scrimgeour, 1995; Rochette and Flanagan, 1997; Rochette *et al.*, 1999) or that have been exposed to labeled CO_2 that is either highly enriched or depleted in ^{13}C (Hungate *et al.*, 1997; Lin *et al.*, 1999; Andrews *et al.*, 1999). Beyond these applications, natural abundance stable isotopes can in theory be used to partition respiration sources when the isotopic difference between ecosystem components can be resolved. We argue and then show that many (most) of these components are not often, if ever, measured and therefore past studies provide an incomplete picture of the C-dynamics within the system under study. More detailed partitioning is therefore likely to yield new insights and allow for a more complete understanding of carbon inputs, sequestration, and losses.

Ecosystem respiration (R_{eco}) can be expressed as the sum of aboveground (R_{above}) and belowground (R_{below}) respiration:

$$R_{eco} = R_{above} + R_{below} \tag{8.1}$$

Belowground respiration (R_{below}) can be further partitioned between root (R_{root}) and microbial (R_{mic}) sources:

$$R_{below} = R_{root} + R_{mic} \tag{8.2}$$

Expanding the above two expressions following the conservation of mass for carbon isotopes (Bowling *et al.*, 2001) gives:

$$\delta_{eco} R_{eco} = \delta_{above} R_{above} + \delta_{below} R_{below} \tag{8.3}$$

$$\delta_{below} R_{below} = \delta_{root} R_{root} + \delta_{mic} R_{mic} \tag{8.4}$$

rearranging gives the aboveground fraction of ecosystem respiration ($F_{above} = R_{above}/R_{eco}$) and the root fraction of belowground respiration ($f_{root} = R_{root}/R_{below}$), respectively:

$$F_{above} = \frac{(\delta_{eco} - \delta_{below})}{(\delta_{above} - \delta_{below})} \tag{8.5}$$

$$f_{root} = \frac{(\delta_{below} - \delta_{mic})}{(\delta_{root} - \delta_{mic})} \tag{8.6}$$

From these two expressions, the fractions of ecosystem respiration contributed by belowground (F_{below}), root (F_{root}), and microbial (F_{mic}) sources can be determined as:

$$F_{below} = 1 - F_{above} \tag{8.7}$$

$$F_{root} = f_{root} F_{below} \tag{8.8}$$

$$F_{mic} = 1 - F_{above} - F_{root} \tag{8.9}$$

Clearly, application of the above approach requires measurable differences among the respiratory signatures in question. In the following section we review the theoretical and empirical basis for such natural variation, show that it can be in some instances very large, and because of this can contain useful information for C-dynamics research.

Variation in Respiration Signatures Among Ecosystem Components

The isotopic composition of ecosystem respiration is generally believed to reflect that of the dominant vegetation (Flanagan and Ehleringer, 1998). However, soil respiration (root plus microbial) is often enriched in ^{13}C relative to ecosystem respiration whereas aboveground respiration (leaf plus stem) is often depleted (Fig. 8.2). While the isotopic composition of leaf organic matter is sometimes a poor indicator of the isotopic composition of

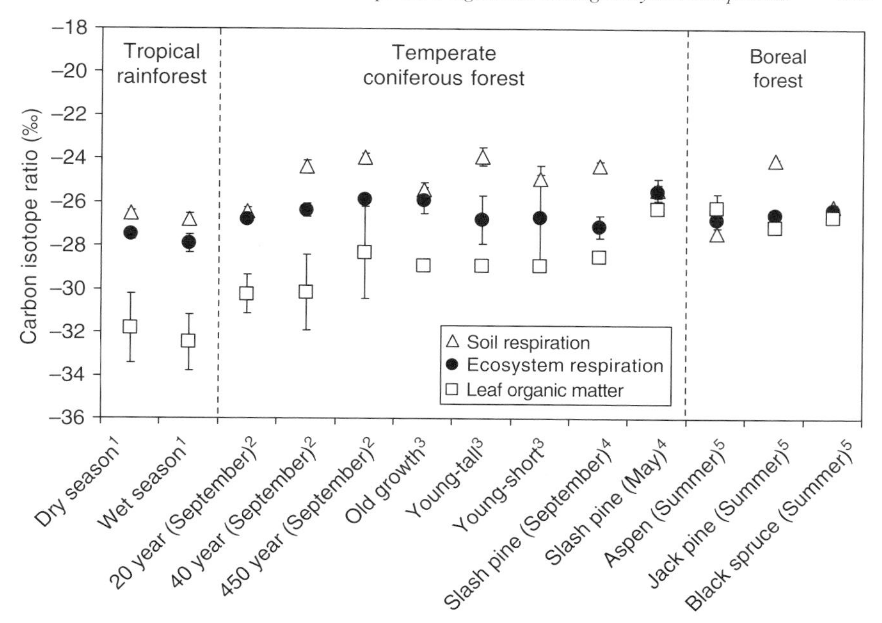

Figure 8.2 Relationship between carbon isotope signatures of leaf organic matter, and ecosystem and soil respiration. ([1]Buchmann *et al.*, 1997; [2]Fessenden and Ehleringer, 2002; [3]Buchmann *et al.*, 1998; [4]Mortavazi and Chanton, 2002; [5]Flanagan *et al.*, 1996). It should be noted that although the isotopic composition of leaf organic matter is sometimes a poor indicator of the isotopic composition of recent photosynthates (Pate and Arthur, 1998) and leaf respiration (Duranceau *et al.*, 1999; Ghashghaie *et al.*, 2001), the data indicate the potential for differences between nighttime soil, ecosystem and aboveground plant respiration.

recent photosynthates (e.g., Pate and Arthur, 1998) or plant respiration (Duranceau *et al.*, 1999; Ghashghaie *et al.*, 2001), the data depicted in Fig. 8.2 indicate the potential for differences between aboveground and belowground respiration.

Recent studies indicate a general pattern of ^{13}C enrichment in microbial (largely fungal) biomass relative to plant organic matter (Fig. 8.3). Although early studies suggested there is significant fractionation during microbial respiration (Blair *et al.*, 1985; Mary *et al.*, 1992) recent studies indicate small (Santruckova *et al.*, 2000; Högberg *et al.*, this volume) or negligible fractionation (Henn and Chapela, 2000; Ekblad *et al.*, 2001). Nevertheless, the isotopic differences in microbial biomass indicated in Fig. 8.3 are clearly suggestive of the potential for isotopic differences in respired CO_2, both between microbial functional groups (particularly saprotrophic vs mycorrhizal fungi) and between plants and microbes (autotrophic vs heterotrophic).

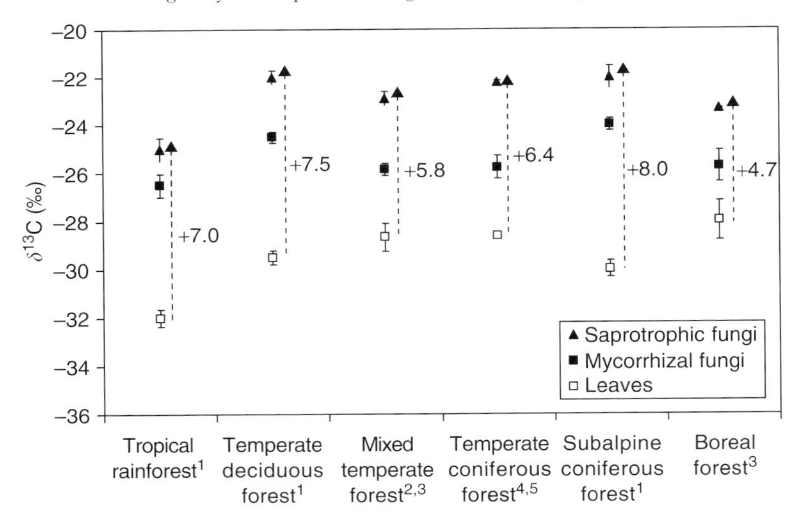

Figure 8.3 Relationship between carbon isotope signatures of leaf and mycorrhizal and saprotrophic fungal biomass across biome types. Error bars represent the standard error. ([1]Kozhu *et al.*, 1999; [2]Hobbie *et al.*, 1999; [3]Högberg *et al.*, 1999; [4]Henn and Chapela, 2001; [5]Hobbie *et al.*, 2001.)

Causes of Variation in Respiration Signatures

Differences in respiration signatures among ecosystem components (Figs 8.2 and 8.3) can result from (1) temporal lags in the movement of carbon sisotope signals through various ecosystem pools, (2) metabolic fractionation, (3) anaplerotic CO_2 fixation, (4) compound-specific effects, and (5) kinetic fractionation. Each of these factors is discussed below.

Temporal Lags Variation in photosynthetic discrimination combined with temporal lags in the movement of photosynthates through plant and soil pools caused by different turnover rates of each pool can result in disequilibrium among substrates and the resulting respiration. For example, soil respiration signatures can lag behind that of the dominant vegetation in temperate grasslands with seasonal transitions from C_3 to C_4 plants (Still *et al.*, 2003). Similar lags can occur as a result of environmental rather than plant-related effects. For example, the decrease in the carbon isotope ratio of atmospheric CO_2 of ~1.4‰ over the last 250 years, the so-called Suess effect (Ehleringer *et al.*, 2000), could cause an equivalent change in photosynthates over this same time period. The temporal lag of this carbon moving through the soil is believed to contribute to the ^{13}C enrichment commonly observed with depth (Ehleringer *et al.*, 2000). Additionally, a change in humidity can cause a change in stomatal conductance and photosynthetic discrimination which, in turn, can cause a change in the carbon isotope ratio of photosynthates and plant respiration before the photosynthates reach the soil microbial community and appear in heterotrophic respiration (Ekblad and Högberg, 2001). Lags on the scale of decades or even centuries might also be observed under special circumstances such as when organic matter, like peat, thaws after years of being frozen in permafrost. The CO_2 released from such a C source may in turn have a unique $\delta^{13}C$.

Metabolic Fractionation Kinetic isotope effects (see Kinetic Fractionation below) as well as isotope effects associated with C moving preferentially in particular directions at metabolic branching points during synthesis (Gleixner *et al.*, 1993; Schmidt and Gleixner, 1998) can cause variation in the carbon isotope signature of CO_2 released during the biosynthesis of secondary compounds (e.g., lipids, lignin, cellulose). For example, lignin and lipids can be depleted in ^{13}C by as much as 3–6‰ relative to bulk organic matter whereas carbohydrates such as glucose and sucrose and related polymers such as starch and cellulose can be enriched by 1–3‰ (Fig. 8.4; O'Leary, 1981; Schmidt and Gleixner, 1998). By mass balance, any depletion (or enrichment) of ^{13}C in a synthesized compound, must necessarily complement an enrichment (or depletion) of ^{13}C in some other compound. This may be a solid or gas such as respired CO_2

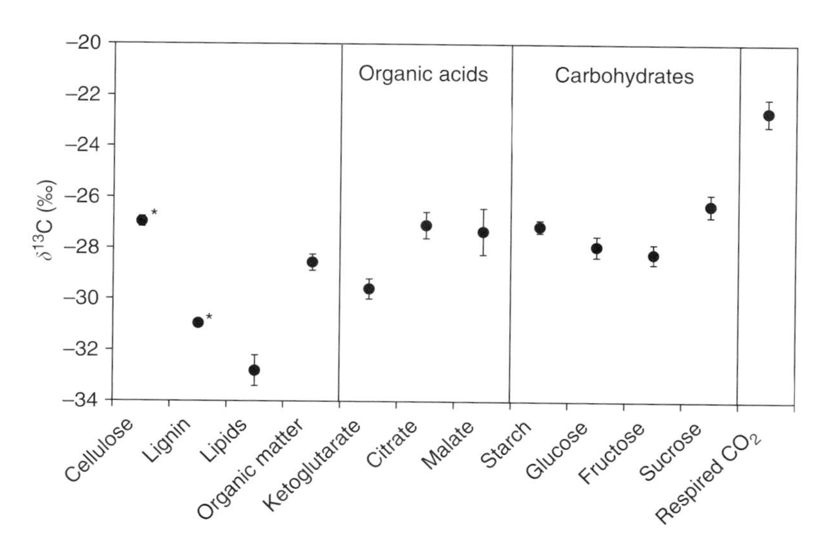

Figure 8.4 Variation in carbon isotope ratios of various leaf compounds. From Ghashghaie *et al.* (2001). Points with an asterix (*) are from Nadelhoffer and Fry (1988) and were adjusted so that their organic matter value equaled that of Ghashghaie *et al.* (2001).

(Park and Epstein, 1961; DeNiro and Epstein, 1977; Rossmann *et al.*, 1991). Accordingly, Park and Epstein (1961) suggest that CO_2 evolved during lipid synthesis can be enriched by as much as 8‰ relative to sugars. Those sorts of effects have been largely ignored in studies conducted at the ecosystem scale.

Anaplerotic CO_2 Fixation Heterotrophic CO_2 fixation (Wood *et al.*, 1941) can occur when CO_2 is fixed by PEP carboxylase in roots and microbes to replace the carbon that was removed from the tricarboxylic acid cycle (TCA) during the biosynthesis of secondary compounds such as amino acids (Wingler *et al.*, 1996, Dunn, 1998). As soil CO_2 has an isotopic signature between that of the atmosphere (e.g., −8‰) and soil respiration (e.g., −27‰), fixation of soil CO_2 will result in the ^{13}C enrichment of microbial biomass relative to plants (e.g., −27‰). Fixation contributing to only a small fraction of the microbial biomass (e.g., 5%), may cause a significant shift (1–1.5‰) in microbial carbon isotope composition (Ehleringer *et al.*, 2000). Subsequent decomposition of the enriched microbial biomass will then lead to enriched $^{13}CO_2$ released during respiration.

Compound-Specific Effects Selective use of specific compounds as substrates for respiration that differ in their isotope composition as a result of fractionation during their formation (see Metabolic Fractionation above) can lead to isotopic differences in respired CO_2. That is, although it may be true that microbes 'are what they eat,' not all microbes eat the same things. For example, isotopic differences among microbial functional groups (Fig. 8.5) are consistent with the isotopic variation among their respective substrates; wood decay fungi are more enriched than litter decay fungi, which are more enriched than mycorrhizal fungi (Kohzu *et al.*, 1999; Hobbie *et al.*, 2001). Cellulose appears to be the preferred substrate for saprotrophic fungi, regardless of whether or not they are capable of decomposing lignin (Gleixner *et al.*, 1993; Hobbie *et al.*, 2001). In fact, ^{14}C tracer studies indicate lignin-derived C is not assimilated into microbial biomass (Hoffrichter *et al.*, 1999) but is mostly solubilized and, to a lesser extent, mineralized to CO_2 (Hackett *et al.*, 1977). Thus, the typical ^{13}C-enrichment of wood cellulose relative to leaf cellulose of 3‰ (Leavitt and Long, 1982) can explain the observed differences between wood decay and litter decay fungi (Hobbie *et al.*, 2001). Mycorrhizal fungi feed on soluble root sugars (Finlay and Söderström, 1992) and, correspondingly, their isotope signature is closer to that of intact leaves and roots than either litter or wood decay fungi. The mechanisms underlying the offsets between plants and mycorrhizae (Fig. 8.3) are not known but may stem from the fact that most ectomycorrhizal fungi engage in saprotrophic activity to some extent (Gadgil and Gadgil, 1987) and thereby incorporate ^{13}C-enriched cellulose signatures,

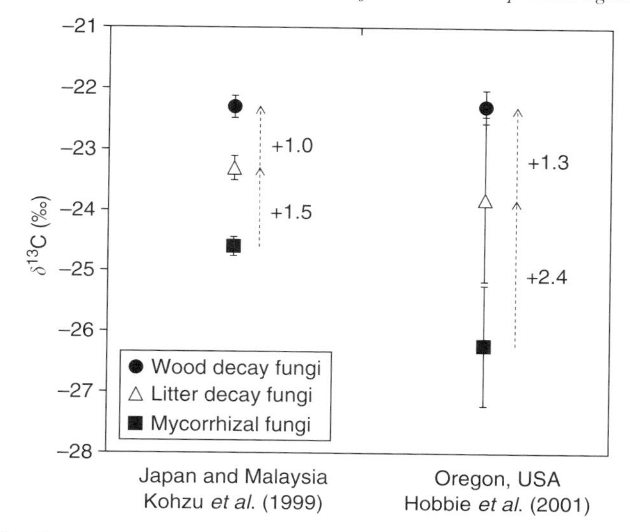

Figure 8.5 Site-specific differences among the carbon isotope ratios of wood decay, litter decay, and mycorrhizal fungi. Error bars represent the standard error.

in addition to potential fractionation during the transfer of carbon from the root to the fungi (Högberg *et al.*, 1999; Henn and Chapela, 2000).

Kinetic Fractionation Kinetic fractionation refers to mass-dependent isotope effects and principally involves fractionation during respiration or fractionation during the incorporation of sugars by microbes. Lin and Ehleringer (1997) found minimal fractionation during autotrophic respiration in isolated plant protoplasts. As discussed above (see Metabolic Fractionation), further research is needed to determine if fractionation during auto- or heterotrophic respiration occurs and to what extent it may occur under different substrate and environmental conditions. Fractionation can occur during sugar uptake by fungi because hydrolysis of sucrose can yield ^{13}C-enriched glucose and ^{13}C-depleted fructose fractions if hydrolysis is not allowed to reach completion (Gonzalez *et al.*, 1999) (see also Henn and Chapela, 2000). Fungi do not appear to take up sucrose directly (Chen and Hampp, 1993; Buscot *et al.*, 2000) and they have a higher affinity for glucose than fructose (Chen and Hampp, 1993) such that incomplete hydrolysis and fractionation may occur during uptake. This may explain some of the isotopic shift between roots and mycorrhizae (Fig. 8.3). Bacteria on the other hand, do not appear to rely on glucose as the primary carbon source (Abraham *et al.*, 1998). They exhibit no clear trend in the isotopic difference between substrate and biomass, as fractionation effects are highly substrate-specific (Abraham *et al.*, 1998). Sugar amendments to intact soil organic horizons, where bacterial activity generally predominates, indicate no significant microbial fractionation during respiration (Ekblad and Högberg, 2000; Ekblad *et al.*, 2002).

Implications of Variation in Respiration Signatures on Soil C-Dynamics

It should be noted that ^{13}C enrichment of microbial respiration relative to plant substrates (Figs 8.3 and 8.5) that in turn leads to ^{13}C enrichment of soil respiration (root and microbial) relative to plant substrates (Figs 8.2 and 8.4), implies the selective loss of enriched carbon that should eventually cause ^{13}C depletion of soil organic carbon (SOC). This contrasts with the typical pattern of ^{13}C enrichment by 1–3‰ observed with depth in soil profiles (Nadelhoffer and Fry, 1988; Ehleringer *et al.*, 2000). However, as noted by Ehleringer *et al.* (2000), ^{13}C enrichment with depth may be explained by a combination of (1) the historical trend of ^{13}C depletion in atmospheric CO_2 and incorporation of this signal in photosynthates and plant litter inputs over time (Suess effect) and (2) incorporation of ^{13}C enriched carbons through anaplerotic CO_2 fixation (see above) during microbial biosynthesis (Ehleringer *et al.*, 2000) with selective preservation of microbial products as a progressively more significant component of soil organic matter (SOM) over time. The latter will cause a progressive enrichment of resynthesized microbial C, and is thus consistent with an enrichment of both soil microbial biomass (Fig. 8.3) and soil respiration (Fig. 8.2) relative to plant organic matter. The extent of microbial recycling and overall influence of microbial products on SOC with depth is evidenced by the overall decrease in C content (Nadelhoffer and Fry, 1988; Bird and Pousai, 1997), decrease in C/N ratio (Nadelhoffer and Fry, 1988) towards that of microbes themselves (fungi 5–15 : 1, bacteria 3–6), decrease in particle size (Bird and Pousai, 1997), and compound-specific shifts in SOC reflecting products of microbial origin (Huang *et al.*, 1996; Marseille *et al.*, 1999; Kracht and Gleixner, 2000).

Enrichment of ^{13}C with depth in the soil profile also indicates that preferential preservation of ^{13}C-depleted lignin does not occur during decomposition (Nadelhoffer and Fry, 1988). Although enrichment with depth is unlikely to result from preferential preservation of ^{13}C-enriched compounds such as sugars, amino acids, and cellulose, for example via Maillard reactions (Maillard, 1917) or abiotic recondensation (see discussion by Burdon, 2001), the ^{13}C-enriched signatures of these relatively labile compounds can be transferred to humus through consumption by microorganisms and subsequent biosynthesis of resistant secondary compounds such as aliphatic biopolymers (Lichtfouse *et al.*, 1995, 1998) and proteinaceous materials (Gliexner *et al.*, 1999) including derivatives of malanins and other polyketides (Burdon, 2001). Since the greater part of plant litter is composed of either polysaccharides (e.g., cellulose and hemicellulose) or lignin (Aber and Melillo, 1991), the balance between lignin-derived and polysaccharide-derived products retained in soils may ultimately determine its carbon isotope ratio. The extent of isotopic enrichment or depletion may therefore depend on the initial isotopic difference between cellulose and lignin, and on the factors that influence their rates of decomposition including temperature, pH, moisture, oxygen tension, soil texture, microbial community composition, nutrient availability, and litter quality (e.g., ratio of lignin to cellulose). For example, anaerobic conditions (e.g., peat soils) that limit lignin degradation can lead to ^{13}C depletion whereas aerobic conditions conducive to lignin degradation can lead to ^{13}C enrichment. In theory, isotope effects caused by discrimination against ^{13}C during respiration (Blair *et al.*, 1985; Mary *et al.*, 1992; Santruckova *et al.*, 2000) can lead to depleted respiration and enriched microbial biomass and SOC, as shown by Agren *et al.* (1996). However, it is possible that other processes including compound-specific discrimination (e.g., preferential use of cellulose over lignin), anaplerotic CO_2 fixation and the Suess effect can lead to enriched microbial biomass and SOC without fractionation during respiration.

Figure 8.6 depicts the results of a model we have developed for describing the variation that can result in the $\delta^{13}C$ of decomposing litter solely from compound-specific effects alone. The model follows the fate of key C-pools through the microbial metabolic processes and into SOC while maintaining $\delta^{13}C$ mass balance and tracking the subsequent fate of the dissolved organic carbon (DOC) flux. The aforementioned discussion and the figure caption provide the details underlying time-dependent C-transformations and their specific isotope effects. The results of the model reveal that such effects can be quite large and therefore must be accounted for if robust partitioning is to be accomplished.

Mass balance dictates that enrichment with depth can only occur if the ^{12}C losses exceed those of ^{13}C (or conversely, if ^{13}C gains exceed those of ^{12}C, for example as a result of anaplerotic CO_2 fixation). For this reason, observations of soils enriched in ^{13}C often lead to the expectation that microbial respiration is depleted in ^{13}C (Nadelhoffer and Fry, 1988; Agren *et al.*, 1996). The latter is consistent with the notion that 'light' ^{12}C is preferentially metabolized whereas 'heavy' ^{13}C is polymerized (Schmidt and Gleixner, 1998; Santruckova *et al.*, 2000). Further, laboratory ^{14}C tracer studies suggest lignin mineralization efficiencies can reach 75% (Hoffrichter *et al.*, 1999; Tuomela *et al.*, 2000). However, rates in natural soils are generally low except at high temperatures conducive to thermophilic microbes (>35°C; Hackett *et al.*, 1977) such that ^{13}C-depleted lignin losses via respiration

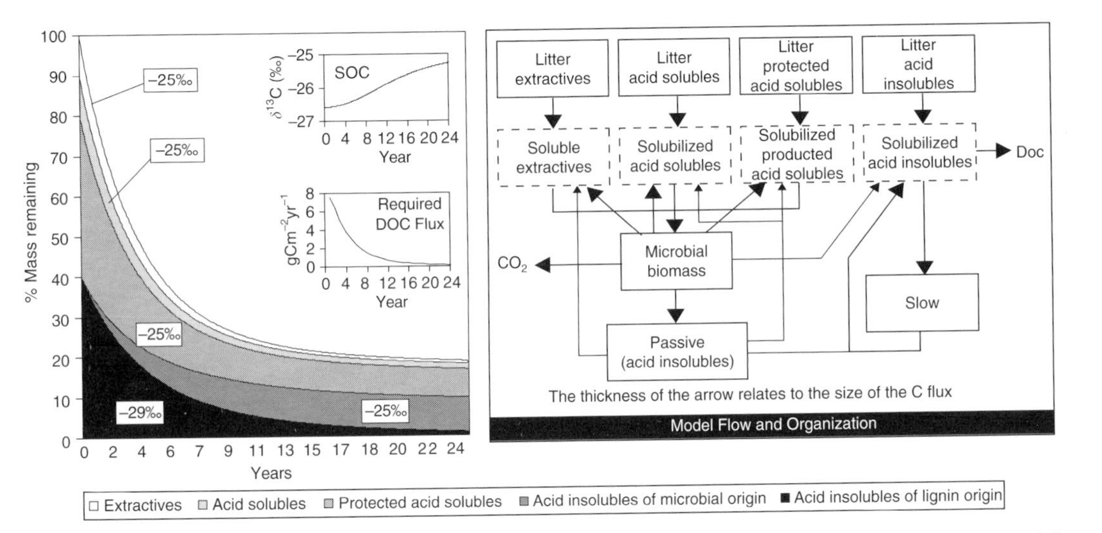

Figure 8.6 Proposed model for describing carbon isotope variation during litter decomposition and humus formation due solely to compound-specific decomposition effects ('microbes are what they eat, but they eat different things') and the hypothesis that DOC derives primarily from the by-products of ^{13}C-depleted lignin ('microbes degrade but don't eat lignin'). In actuality, a combination of compound-specific decomposition, anaplerotic CO_2 fixation, and the Suess effect may explain much of the enrichment in δ^{13}C values with depth in aerobic soils. For simplicity, all microbially active substrates (extractives and cellulose) are given the same δ^{13}C value (−25‰), which contrasts with the δ^{13}C value of the microbially undesirable lignin (−29‰). Initially (year = 0), litter has a carbon isotope ratio of −26.5‰ (top inset graph) due to the abundance of the ^{13}C-depleted acid insolubles (e.g., lignin) relative to the ^{13}C-enriched compounds such as extractives (e.g., sugars, amino acids), acid solubles (cellulose), and protected acid solubles (lignin-bound cellulose). As decomposition progresses, microbes preferentially incorporate the ^{13}C-enriched cellulose and extractives into their biomass. A fraction of this assimilated C is used to synthesize secondary compounds, which are ultimately converted to resistant acid insoluble substances after they are exuded or released upon death and degradation of the cells (the rest is recycled back to the extractive, acid soluble, and protected acid-soluble pools). These lignin-like acid-insoluble substances of microbial origin carry the isotopic signature of the enriched microbial biomass (e.g., −25‰), which reflects their enriched substrate (extractives and cellulose). As lignin is progressively degraded, solubilized, and lost as DOC, these resistant microbial compounds increasingly dominate the mass balance of remaining organic matter and therefore its overall isotopic composition. The DOC flux required to sustain the lignin losses (bottom inset graph) and, in this case, the eventual 1.5‰ enrichment of the remaining organic matter (top inset graph) is within the range cited in the literature (1–84 g C m^{-2} yr^{-1}; Neff and Asner, 2001).

should be small (<20%). Moreover, field measurements indicate that soil surface respiration is typically enriched in ^{13}C relative to plant litter inputs (Fig. 8.2), contrary to the depleted-respiration hypothesis.

If both SOM and respired CO_2 are enriched in ^{13}C in aerobic soils, a mechanism such as ^{13}C-depleted DOC leaching is required to close the carbon isotope mass balance. Measurements of carbon isotope ratios in DOC (Schiff *et al.*, 1990; Trumbore *et al.*, 1992; Ludwig *et al.*, 2000) suggest that DOC is indeed more depleted in ^{13}C than both surface litter inputs and SOC with depth in soils where SOC is also enriched (Fig. 8.7). Further, radiocarbon dating (Trumbore *et al.*, 1992; Schiff *et al.*, 1997) suggests that the age of the DOC increases with depth consistent with its autochthonous production from in-situ microbial activity rather than by transfer from overlying layers. Thus, with increasing depth in the soil profile, each soil layer is progressively more isolated from fresh litter inputs at the surface (aside from C inputs from deep roots and some leaching). At more mature stages of decomposition deeper in the soil profile, the significance of ^{13}C-enriched polysaccharide-derived microbial products that are effectively recycled by the microbial community increases relative to ^{13}C-depleted lignin-derived products that are continuously degraded and lost as DOC (Fig. 8.6). Although ^{13}C-CO_2 losses through respiration greatly exceed ^{12}C-DOC losses, the longer effective retention time of microbial resynthesized ^{13}C allows the cumulative effect of small DOC fluxes (see bottom inset graph in Fig. 8.6) to eventually cause significant shifts in soil isotope ratios (see top inset graph in Fig. 8.6). The cumulative nature of the DOC effect

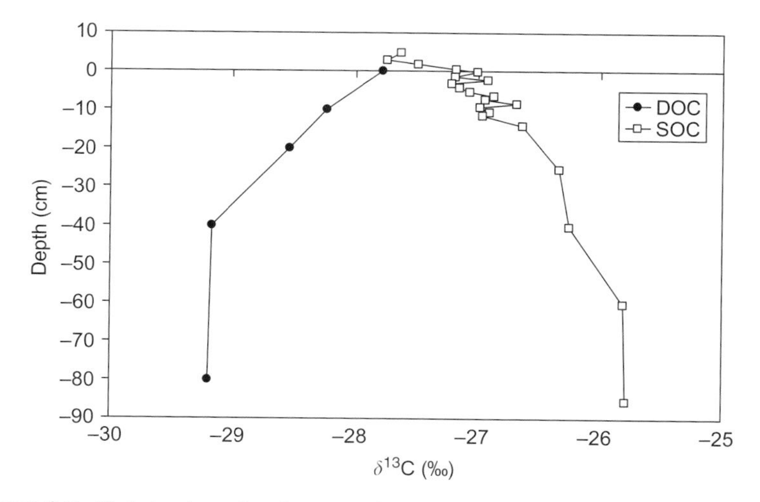

Figure 8.7 Variation in carbon isotope ratios of dissolved organic carbon (DOC) and soil organic carbon (SOC) with depth in the soil profile. From Ludwig *et al.* (2000).

is evidenced by the greater ^{13}C-enrichment observed with depth in the soil profile where enough time has elapsed to degrade lignin and its derivatives.

Resolving the complexities of root–microbe–soil stable isotope interactions and their application to partitioning ecosystem respiration will undoubtedly benefit from studies aimed at closing the ecosystem carbon isotope budget. Such studies should include compound-specific ^{13}C measurements of the major carbon fractions (e.g., soluble carbohydrates, amino acids, alkanes, cellulose, lipids, and lignin) and ^{14}C analyses to better understand the fate and transformations of stable carbon isotopes as they move from leaves to roots to litter and among various soil microbial, DOC, and SOC pools.

Determining Source Signatures

What to Measure

Partitioning CO_2 fluxes using carbon isotope ratios requires direct measurement of carbon isotope ratios in CO_2. That is, temporal, metabolic and kinetic fractionation effects can cause unpredictable isotopic differences between bulk organic matter and respired CO_2 such that the adequacy of using *only* organic matter signatures and assuming they can act as surrogates for the respired CO_2 signatures will depend on site conditions and in many cases will not work (Fig. 8.1). For example, Cheng (1996) found minimal difference between root respiration and $\delta^{13}C$ values of bulk root C under controlled growth conditions. However, in a naturally dynamic growth environment, it is not uncommon to find isotopic disequilibrium between bulk tissues and respired CO_2. For example, during drought, significant differences can arise between $\delta^{13}C$ values of leaf organic matter and leaf respired CO_2 (Duranceau *et al.*, 1999; Ghashghaie *et al.*, 2001), presumably as a result of transient reductions in discrimination caused by reduced stomatal conductance. Similar disequilibrium can occur with respect to whole plant respiration as the $\delta^{13}C$ values of metabolically active C in phloem can differ from the $\delta^{13}C$ values of leaf dry matter (Fig. 8.8; Pate and Arthur, 1998). This may in part stem from the fact that organic matter reflects C fixed at an earlier date (Smedley *et al.*, 1991; Terwilliger, 2001) whereas respired CO_2 reflects recent photosynthates (Pate, 2001). For example, leaf-to-phloem isotope differences reported by Pate and Arthur (1998) apparently arose because leaf dry matter was synthesized during relatively unstressed wet season conditions whereas leaf respiration during peak summer conditions (Fig. 8.8) reflected recent water stress.

Owing to the dynamic nature of carbon allocation among plant sources and sinks, the CO_2 respired by the leaves may not reflect the CO_2 respired by the roots. Aside from the frequently cited isotopic differences between

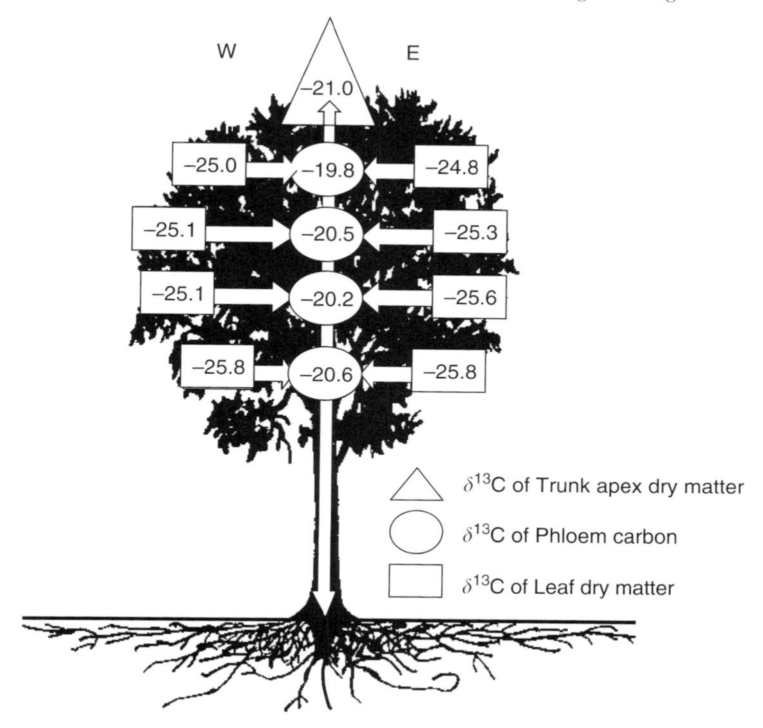

Figure 8.8 Differences between carbon isotope composition of leaf dry matter and phloem carbon during peak summer. Leaf dry matter was presumably synthesized during less-stressful wet season conditions. From Pate and Arthur (1998).

leaf and root organic matter (Gleixner *et al.*, 1993; Schweizer *et al.*, 1999; Arndt and Wanek, 2002), Terwilliger and Huang (1996) found a 1–3‰ enrichment of heterotrophic plant parts adjacent to photoautotrophic leaves. Similar results from other studies (Gleixner *et al.*, 1998; Terwilliger, 2001) suggest that isotope effects related to C translocation (e.g., de-novo sucrose synthesis) and anaplerotic CO_2 fixation can contribute to isotopic differences among organs and their metabolically active C pools. Anaplerotic CO_2 fixation by roots or mycorrhizae may be particularly important during nitrogen assimilation (Wingler *et al.*, 1996; Dunn, 1998).

It is not known if isotopic differences exist between fungal hyphae and sporocarps yet fruiting bodies are widely used to assess hyphal isotopic signatures (Hobbie *et al.*, 1999; Högberg *et al.*, 1999; Kohzu *et al.*, 1999). Thus, although there does not appear to be fractionation during fungal respiration (Henn and Chapela, 2000), further studies are needed to determine if sporocarp isotope signatures can be used to provide an accurate assessment of fungal $\delta^{13}C$-CO_2 values. It should be noted that Chapela *et al.* (2001) reported no isotopic difference between mycorrhizal root tips and the host plant, whereas fruit bodies of the same fungal species exhibited a 3‰ difference relative to the host plant. In this specific case, unusually high rates of fungal productivity suggested that this isotopic gradient between fungal tissues adjacent to roots and their more distant fruit bodies resulted from the incorporation of ^{13}C-enriched SOC as a result of significant saprotrophic activity. This notion was supported by radiocarbon dating that indicated the presence of older carbon in the fruit bodies, consistent with different C sources between the fungal tissues adjacent to the roots and their associated fruit bodies. Saprotrophic activity of ectomycorrhizal fungi (Gadgil and Gadgil, 1987; Durall *et al.*, 1994) in addition to long-distance transport of C from interconnected mycelia (Simard *et al.*, 1997) may be the cause of isotopic differences between ectomycorrhizal fungi and their apparent plant host (Fig. 8.3).

Clearly, the potential for isotopic disequilibrium between plant and fungal parts necessitates careful consideration as to what to measure in order to achieve a specific partitioning goal (e.g., root vs microbial, aboveground vs belowground).

When to Measure Source Signatures

Determination of *what* to measure also requires a consideration of *when* to measure, as carbon isotope signatures of respired CO_2 can exhibit significant variation on daily (Duranceau *et al.*, 1999; Ghashghaie *et al.*, 2001; Tcherkez *et al.*, 2003) to seasonal timescales (Pate and Arthur, 1998). Seasonal variations in the isotopic composition of plant respiration are primarily driven by changes in photosynthetic discrimination, which are in turn driven by moisture limitations to stomatal conductance (Pate and Arthur, 1988) in addition to other factors such as air temperature, genotype, leaf age, etc. (see Dawson *et al.*, 2002). On diel timescales, leaf respiration can be enriched in ^{13}C during the day relative to its value near the end of the night (Duranceau *et al.*, 1999; Tcherkez *et al.*, 2003). This pattern is most pronounced in variable environments where daytime photosynthetic discrimination is limited due to water or temperature stress, and the resulting isotopic signature of the labile C pool (e.g., sucrose) is continuously changing relative to the more stable pool of C reserves which are respired near the end of the night after the labile pool is exhausted. The turnover time of this labile pool appears to be on the order of 4 hours (Fig. 8.9), implying that the best time to sample plant respiration for the purpose of partitioning nighttime respiration is at least 4 hours *after* sunset. Sooner than this, significant isotopic differences between plant parts may exist (Fig. 8.10) as a result of differential allocation and thus different amounts of labile C

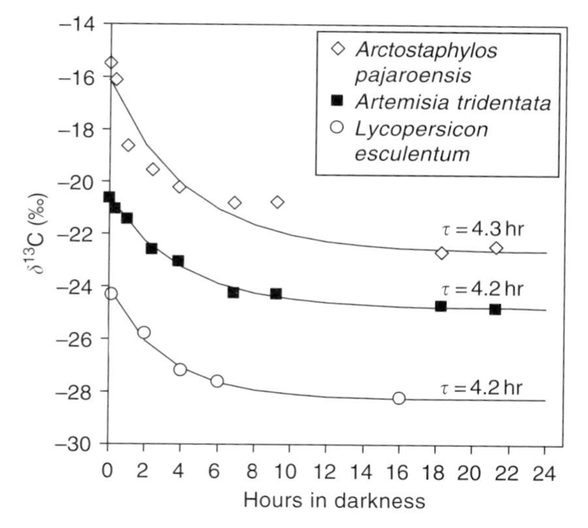

Figure 8.9 Change in carbon isotope composition of leaf respiration with respect to duration in darkness. Initially ($t = 0$ hr), the $\delta^{13}C$ value of respired CO_2 presumably reflects light-dependent and photosynthate-driven reactions involved in growth and biosynthesis of secondary compounds such as lipids (Park and Epstein, 1961). By mass balance, synthesis of such ^{13}C-depleted compounds will result in the ^{13}C-enriched CO_2 signals shown at $t = 0$. As the labile photosynthate pool becomes exhausted, the biosynthetic reactions will decrease, allowing the respired CO_2 signal to reflect that of the more stable pool of C reserves. The turnover times ($\tau = 1/k$) were determined by fitting first order decay functions ($Y = Ae^{-kt}$) to the $\delta^{13}C$ values that were first adjusted so that values at hour 24 equaled 0‰ (r^2 ranged from 0.92 to 0.99). The values for *Lycopersicon esculentum* are from Park and Epstein (1961) whereas those for *Arctostaphylos pajaroensis* and *Artemisia tridentata* are unpublished results. All CO_2 samples were collected in chambers filled with CO_2-free air (see also Tcherkez *et al.*, 2003).

and associated anabolic reactions in the various plant organs (e.g., synthesis of ^{13}C-depleted lipids; Park and Epstein, 1961).

How to Measure Source Signatures

Arguably the most difficult aspect of determining source signatures is not *what* or *when* to sample but *how* to sample without introducing artefacts while isolating and collecting CO_2 from the various respiration sources. Foliage respiration CO_2 is relatively easy to collect using, for example, the chamber method described by Duranceau *et al.* (1999) in which CO_2 is collected in a CO_2-free chamber connected to an infra-red gas analyzer (IRGA) in a closed loop to monitor CO_2 concentration. Whereas this approach has the possible disadvantage of introducing artefacts related to respiration into

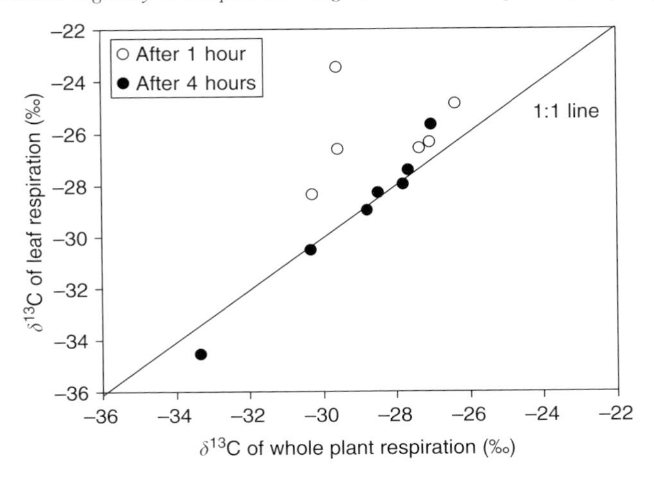

Figure 8.10 Relationship between carbon isotope composition of leaf and whole plant respiration after 1 and 4 h in darkness (unpublished results). The two sets of samples were collected on different dates (after 1 hr—January 11, 2002; after 4 hr—December 21, 2001).

CO_2-free air, it has the advantage of requiring only one isotope analysis per sample.

In contrast, Keeling plot approaches, such as those described by Fessenden and Ehleringer (2002) for measuring soil surface respiration (see also Högberg and Ekblad, 1996; Ekblad and Högberg, 2000) generally require a minimum of five samples to determine one source signature (over a substantial CO_2 concentration range of >300 ppm). This type of system is generally configured in a closed loop between a sample and the IRGA and can be modified for use with detached or attached leaves, or on boles, branches, and roots (with the appropriate chamber). Because Keeling plot approaches require a range of CO_2 concentrations, the sample is thus typically subjected to elevated CO_2 concentrations in contrast to the aforementioned CO_2-free method. Particular attention must be given to maintaining the volume and background isotopic signature. The latter is crucial so as not to introduce a third end-member into the system and confounding the interpretation. In general, care must be exercised with chamber approaches to avoid leaks and pressure-related effects on rates of CO_2 efflux (Lund *et al.*, 1999) as soil air that is ^{13}C-enriched relative to the source (by up to $\sim$4.4‰) can be drawn into the chamber airspace if the chamber pressure is less than ambient. In addition to chambers placed on the soil surface, soil gas probes (Rochette *et al.*, 1999) and wells (Andrews *et al.*, 1999) provide practical methods for characterizing the isotopic composition of soil-respired CO_2 (see also Amundson *et al.*, 1998).

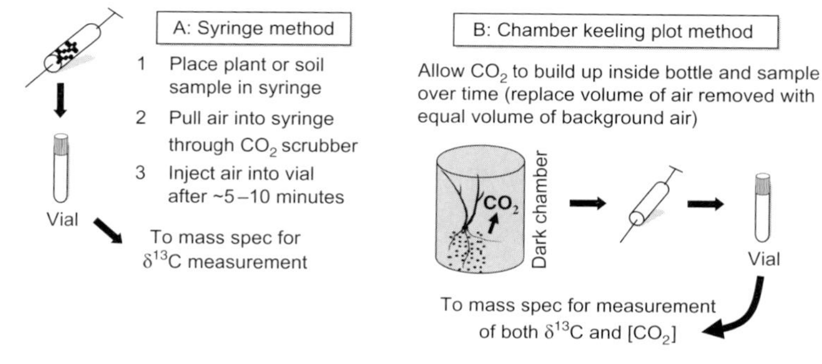

Figure 8.11 Syringe and chamber Keeling plot methods for determining respiration signatures. A: Syringe method—samples are placed within a large syringe (e.g., 60 mL), CO_2-free air is then drawn into the syringe; and finally, the headspace is injected into a sample vial after the appropriate concentration is reached (e.g., 5 to 10 min); B: Chamber method—samples (soil, roots, leaves, etc.) are placed within a closed chamber (e.g., 1000 mL) and air is withdrawn at regular intervals (e.g., every 2 min), transferred to a sample vial (e.g., 10 mL) or flask, and subsequently analyzed for stable isotope ratios and CO_2 concentration. Alternatively, CO_2 can be initially scrubbed and an air sample can be withdrawn after the respired CO_2 has reached a suitable level for analysis (e.g., 400 ppm after ~10 min, depending on respiration rate).

As a further modification of the Keeling plot approach, samples (e.g., leaves, roots, stems, and soil) can be placed within a closed, static chamber that can either be scrubbed and used in a CO_2-free fashion or air samples can be withdrawn at regular intervals as CO_2 concentrations build up as in a typical Keeling plot approach (Fig. 8.11B: Chamber Method). Alternatively, samples can be placed within airtight syringes, scrubbed of CO_2, the respired CO_2 then being injected into septum-capped vials or transferred to flasks for later isotopic analysis (Fig. 8.11A: Syringe Method). The main drawback to these methods is the disturbance effect as samples must be detached (e.g., leaves, roots, stems) or removed from the soil. On the other hand, the principal advantages are (1) the flexibility in chamber sizes that ensures minimal sampling time as a result of rapid build-up of CO_2 (or rapid scrubbing of initial CO_2), (2) the materials are inexpensive (e.g., plastic bottles and syringes), and (3) leaks are easy to minimize due to the simplicity of the system. We now regularly employ a method similar to this (e.g., used for collecting respired CO_2 as in Fig. 8.1) and have obtained repeatable results that can be statistically replicated within a relatively short period of time (Tu *et al.*, 2001).

Microbial respiration is clearly the most difficult signature to determine because of the difficulty in isolating microbes from the soil without artefacts related to disturbance effects (e.g., changing the oxygen concentration, removal of rhizosphere C inputs, disruption of hyphal networks). One promising approach for determining microbial respiration signatures involves careful sieving and removal of root material from the soil (e.g., under

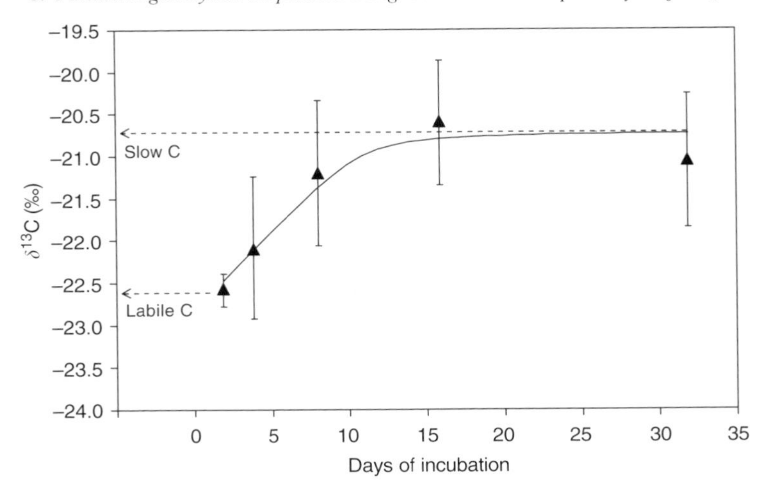

Figure 8.12 Estimation of carbon isotope ratios of microbial respiration by incubating root-free soil. From Andrews *et al.* (1999).

a dissecting microscope), followed by incubation of the root-free soil at field temperature and moisture conditions for several days or longer (Fig. 8.12; Andrews *et al.*, 1999, 2000). The headspace can then be collected (e.g., after flushing with CO_2-free air) and analyzed. Clearly, this approach precludes determination of in-situ respiration rates and isotope fluxes but offers arguably the most robust approach to isolating microbial respiration under field-like conditions. Moreover, it affords the opportunity to assess the respiration signatures of both the labile 'active' SOC pool from the initial $\delta^{13}C$ values and the resistant and stable 'slow' SOC pool from the final steady-state $\delta^{13}C$ values (Fig. 8.12).

Although no single method will work equally well for all sites and will undoubtedly require modification as the specific site conditions dictate (i.e., due to plant height, leaf size, rooting depth, understory growth, etc.), some degree of standardization—perhaps through inter-comparison methods is needed, particularly in the context of comparative studies as part of measurement networks such as FLUXNET, BASIN, or SIBAE. Chamber methodologies (open, closed, static, pressure-regulated, small, large, etc.) have come under particular scrutiny given the potential for errors and apparent inconsistencies between the relatively recent observations of enriched soil respiration (Fig. 8.2) and the older well-established patterns of enrichment with depth in the soil profile, as their co-occurrence is often difficult to reconcile from a mass balance perspective (although see Causes of Variation in Respiration Signatures, earlier).

Using Variation in Respiration Signatures to Partition Ecosystem Respiration

To illustrate the use of stable carbon isotopes in CO_2 for partitioning respiration fluxes between plant and microbial sources, we refer to the data in Fig. 8.1 with Eqs (5)–(9) outlined earlier in Partitioning Approach. The signature of ecosystem respiration (−26.0‰, SE = 0.01‰) was determined with the Keeling plot approach using within-canopy air collected at night (see Bowling *et al.*, 2002 for determining Keeling plot intercepts). Source signatures of the contributing component fluxes were determined using the syringe method at night as well (see Fig. 8.11). As a first approximation, the carbon isotope signature of plant respiration was assumed to be equal to that of leaf respiration, the latter of which was, for simplicity, taken as the average of the upper, middle, and lower values (−27.5‰, SE = 0.28‰). The microbial respiration signature was taken as the average of litter and SOC decomposition (−23.9‰, SE = 1.4‰). The soil respiration signature was measured using the chamber Keeling plot method (Fig. 8.11) similar to that described earlier (see How to Measure Source Signatures), with the exception that a chamber (~18 L volume) was placed on the soil surface (−25.5‰, SE = 0.1‰). Applying Eq. 8.5 gives aboveground plant respiration (leaf + stem) as a fraction of total ecosystem respiration:

$$F_{\text{above}} = \frac{(\delta_{\text{eco}} - \delta_{\text{below}})}{(\delta_{\text{above}} - \delta_{\text{below}})} = \frac{(-26.0 + 25.5)}{(-27.5 + 25.5)} = 0.25 \pm 0.10$$

Further, Eq. 8.6 gives root respiration as a fraction of total belowground respiration:

$$f_{\text{root}} = \frac{(\delta_{\text{below}} - \delta_{\text{mic}})}{(\delta_{\text{root}} - \delta_{\text{mic}})} = \frac{(-25.5 + 23.9)}{(-27.5 + 23.9)} = 0.44 \pm 0.20$$

From these two values, the fractions of ecosystem respiration contributed by belowground (F_{below}), root (F_{root}), and microbial (F_{mic}) sources can be determined as:

$$F_{\text{below}} = 1 - F_{\text{above}} = 1 - 0.25 = 0.75 \pm 0.10$$

$$F_{\text{root}} = f_{\text{root}}\, F_{\text{below}} = 0.44 \cdot 0.25 = 0.33 \pm 0.20$$

$$F_{\text{mic}} = 1 - F_{\text{above}} - F_{\text{root}} = 1 - 0.25 - 0.33 = 0.42 \pm 0.22$$

Figure 8.13 summarizes the partitioning of ecosystem respiration between above- and belowground components as well as between plant and microbial sources for a mature coastal California redwood forest. For determining the uncertainty associated with these partitioning values, the

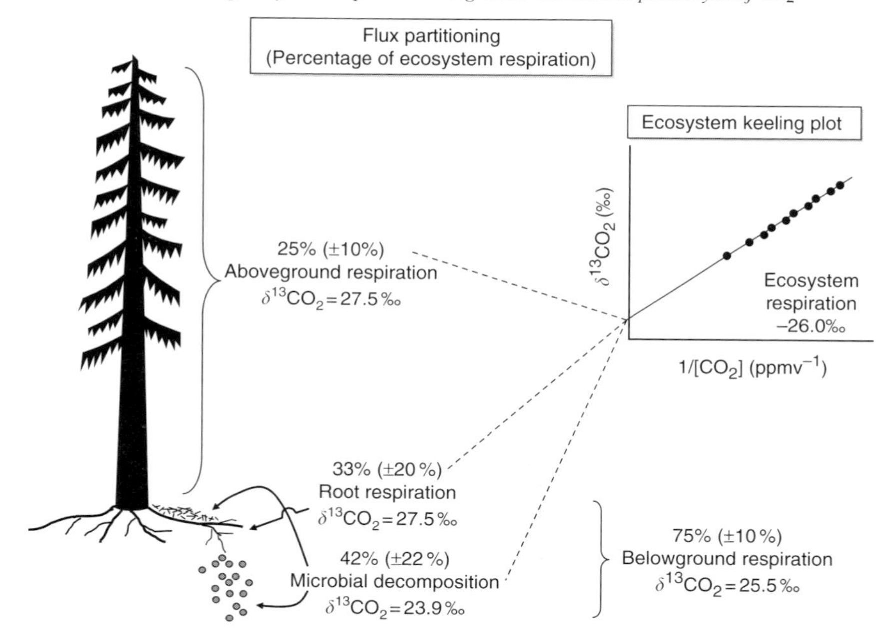

Figure 8.13 Partitioning ecosystem respiration using stable carbon isotope signatures. Samples were collected in a redwood forest (*Sequoia sempirvirens*) near Occidental, California (unpublished results) using the syringe method described in the section Determining Source Signatures for determining the isotopic signature of aboveground (leaf), root, and microbial respiration. The Keeling plot method was used for belowground and ecosystem respiration by sampling air within a chamber on the soil surface and from within the canopy airspace at night, respectively. Note that roots and microbes contribute ~44% and ~56% to belowground respiration respectively (root = 33/75, microbial = 42/75). The values in parentheses are the standard errors of the partitioning estimates as determined following Phillips and Gregg (2001).

reader is directed to Phillips and Gregg (2001). In practice, such uncertainty should be minimized by increasing the number and frequency of samples.

Conclusions

Partitioning R_{eco} is central to understanding the controls on different respiratory sources and their roles in the dynamics of ecosystem carbon cycling. Recent evidence suggests that isotopic differences among ecosystem sources can at times be large enough to be resolved. These differences can be exploited to partition ecosystem respiration and improve our understanding of the causes underlying variation in Keeling plot intercepts and biosphere–atmosphere carbon isotope exchange.

Isotopic differences between plant and microbial biomass are clearly suggestive of the potential for isotopic differences in respired CO_2, both between microbial functional groups (saprotrophic vs mycorrhizal) and between plants and microbes (autotrophic vs heterotrophic). Such differences in respiration signatures among ecosystem components can result from (1) temporal lags in the movement of carbon isotope signals through various ecosystem pools, (2) metabolic fractionation, (3) anaplerotic CO_2 fixation, (4) compound-specific effects, and (5) kinetic fractionation.

Differences in respiration signatures must be reconciled with respect to known patterns of soil C and C isotope dynamics. Although isotope effects caused by discrimination against ^{13}C during respiration can lead to depleted-CO_2, enriched microbial biomass, and ^{13}C-enrichment of SOC, isotope effects related to compound-specific decomposition, in addition to anaplerotic CO_2 fixation and the Suess effect, can result in similar ^{13}C enrichments in the absence of fractionation during respiration. Further, if both SOM and respired CO_2 are enriched in ^{13}C in aerobic soils, a mechanism such as ^{13}C-depleted DOC leaching is required to close the carbon isotope mass balance.

Temporal, metabolic, and kinetic fractionation effects can cause unpredictable isotopic differences between bulk organic matter and respired CO_2, precluding the use of bulk organic matter as a surrogate for direct measurement of carbon isotope ratios of CO_2. Determination of *what* to measure also requires a consideration of *when* to measure, as carbon isotope signatures of respired CO_2 can exhibit significant variation on daily to seasonal timescales. Arguably the most difficult aspect of determining source signatures is not *what* or *when* to sample but *how* to sample without introducing artefacts while isolating and collecting CO_2 from the various respiration sources. Existing methods include CO_2-free and Keeling plot chamber approaches for collecting leaf, stem, root, and soil respiration products, soil gas probes for characterizing soil respiration, and root-free soil incubations for assessing microbial heterotrophic respiration. Microbial respiration is clearly the most difficult signature to determine because of the difficulty in isolating microbes from the soil without artefacts related to both physical and biological disturbances.

Future studies should focus on the dynamics of stable isotope interactions among roots, microbes, and soil with the goal of closing the ecosystem carbon isotope budget. These studies should encompass compound-specific ^{13}C measurements of the major carbon fractions (e.g., soluble carbohydrates, amino acids, alkanes, cellulose, lipids, and lignin) along with ^{14}C analyses to better understand the fate and transformations of stable carbon isotopes as they move from leaves to roots to litter and among various soil microbial, DOC, and SOC pools. Once this has been achieved, future investigations may also be in the position to manipulate particular components of a system in order to enhance our understanding of the underlying biophysical controls on ecosystem C exchange (inputs, storage, and loss). In the area of global change research, we feel such a pursuit has tremendous promise in helping to resolve open issues in global C balance research.

In summary, partitioning ecosystem respiration using natural abundance stable carbon isotopes would benefit from future studies directed at:

(1) kinetic and metabolic fractionation effects during respiration in both microbes and plants;
(2) the significance of compound-specific isotope effects during decomposition (e.g., fate of lignin and role of microbial resynthesis of cellulose products);
(3) the isotopic composition of DOC and its effect on SOC ^{13}C-enrichment;
(4) the relationship between the carbon isotope ratios in mycorrhizal root tips, mycelia, and fruiting bodies and the influence of saprotrophic activity;
(5) closing the ecosystem carbon isotope budget;
(6) developing standardization and inter-comparison of methods, particularly in the context of comparative studies as part of measurement networks such as BASIN, SIBAE, and FLUXNET;
(7) capitalizing on new and emerging technologies such as tunable diode lasers and cavity ring-down lasers for instantaneous and continuous partitioning of eddy flux measurements.

Acknowledgments

Funding for research related to the topics presented here came from the A. W. Mellon Foundation and the University of California. We thank Larry Flanagan, Jim Ehleringer, and Diane Pataki for the invitation to attend the Banff meetings and to contribute our views to this volume.

References

Aber J. D. and Melillo J. M. (1991) *Terrestrial Ecosystems.* Saunders College Publishing, Philadelphia, USA.

Abraham W. R., Hesse C. and Pelz O. (1998) Ratios of carbon isotopes in microbial lipids as an indicator of substrate usage. *Appl Environ Microbiol* **64**: 4202–4209.

Agren G. J., Bosatta E. and Balesdent J. (1996) Isotope discrimination during decomposition of organic carbon: A theoretical analysis. *Soil Sci Soc Am J* **60**: 1121–1126.

Amundson R., Stern L., Baisden T. and Wang Y. (1998) The isotopic composition of soil and soil-respired CO_2. *Geoderma* **82**: 83–114.

Andrews J. A., Harrison K. G., Roser M. and Schlesinger W. H. (1999) Separation of root respiration from total soil respiration using carbon-13 labeling during free-air carbon dioxide enrichment (FACE). *Soil Sci Soc Am J* **63**: 1429–1435.

Andrews J. A., Matamala R., Westover K. M. and Schlesinger W. H. (2000) Temperature effects on the diversity of soil heterotrophs and the $\delta^{13}C$ of soil-respired CO_2. *Soil Biol Biochem* **32**: 699–706.

Arndt S. K. and Wanek W. (2002) Use of decreasing foliar carbon isotope discrimination during water limitation as a carbon tracer to study whole plant carbon allocation. *Plant Cell Environ* **25**: 609–616.

Bird M. I. and Pousai P. (1997) Variations of $\delta^{13}C$ in the surface soil organic carbon pool. *Global Biogeochem Cycles* **11**: 313–322.

Blair N., Leu A., Munoz E., Olsen J., Kwong E. and Desmarais D. (1985) Carbon isotopic fractionation in heterotrophic microbial-metabolism. *Appl Environ Microbiol* **50**: 996–1001.

Bowling D. R., Tans P. R. and Monson R. K. (2001) Partitioning net ecosystem carbon exchange with isotopic fluxes of CO_2. *Global Change Biol* **7**: 127–145.

Bowling D. R., McDowell N. G., Bond B. J., Law B. E. and Ehleringer J. R. (2002) ^{13}C content of ecosystem respiration is linked to precipitation and vapor pressure deficit. *Oecologia* **131**: 113–124.

Buchmann N., Guehl J.-M., Barigah T. S. and Ehleringer J. R. (1997) Interseasonal comparison of CO_2 concentrations, isotopic composition, and carbon dynamics in an Amazonian rainforest (French Guiana). *Oecologia* **110**: 120–131.

Buchmann N., Hinckely T. M. and Ehleringer J. R. (1998) Carbon isotope dynamics in *Abies amabils* stands in the Cascades. *Can J For Res* **28**: 808–819.

Burdon J. (2001) Are the traditional concepts of the structures of humic substances realistic? *Soil Sci* **166**: 752–769.

Buscot F., Munch J. C., Charcosset J. Y., Gardes M., Nehls U. and Hampp R. (2000) Recent advances in exploring physiology and biodiversity of ectomycorrhizas highlight the functioning of these symbioses in ecosystems. *FEMS Microbiol Rev* **24**: 601–614.

Chapela I. H., Osher L. J., Horton T. R. and Henn M. R. (2001) Ectomycorrhizal fungi introduced with exotic pine plantations induce soil carbon depletion. *Soil Biol Biochem* **33**: 1733–1740.

Chen X.-Y. and Hampp R. (1993) Sugar uptake by protoplasts of the ectomycorrhizal fungus *Amanita muscaria* (L. ex Fr.) Hooker. *New Phytol* **125**: 601–608.

Cheng W. (1996) Measurement of rhizosphere respiration and organic matter decomposition using natural ^{13}C. *Plant Soil* **183**: 263–268.

Coleman D. C. and Fry B. (1991) *Carbon Isotope Techniques.* Academic Press, San Diego, USA.

Dawson T. E., Mambelli S., Plamboeck A. H., Templer P. H. and Tu K. P. (2002) Stable isotopes in plant ecology. *Annu Rev Ecol Sys* **33**: 507–559.

DeNiro M. J. and Epstein S. (1977) Mechanism of carbon isotope fractionation associated with lipid synthesis. *Science* **197**: 261–263.

Dunn M. F. (1998) Tricarboxylic acid cycle and anaplerotic enzymes in rhizobia. *FEMS Microbiol Rev* **22**: 105–123.

Duranceau M., Ghashghaie J., Badeck F., Deleens E. and Cornic G. (1999) $\delta^{13}C$ of CO_2 respired in the dark in relation to $\delta^{13}C$ of leaf carbohydrates in *Phaseolus vulgaris* L. under progressive drought. *Plant Cell Environ* **22**: 515–523.

Durall D. M., Todd A. W. and Trappe J. M. (1994) Decomposition of C14-labeled substrates by ectomycorrhizal fungi in association with douglas-fir. *New Phytol* **127**: 725–729.

Ehleringer J. R., Buchmann N. and Flanagan L. B. (2000) Carbon and oxygen isotope ratios in belowground carbon cycle processes. *Ecol Appl* **10**: 412–422.

Ekblad A. and Högberg P. (2000) Analysis of $\delta^{13}C$ of CO_2 distinguishes between microbial respiration of added C_4 sucrose and other soil respiration in a C_3 ecosystem. *Plant Soil* **219**: 197–209.

Ekblad A. and Högberg P. (2001) Natural abundance of ^{13}C in CO_2 respired from forest soils reveals speed of link between tree photosynthesis and root respiration. *Oecologia* **127**(3): 305–308.

Ekblad A., Nyberg G. and Högberg P. (2002) ^{13}C-discrimination during microbial respiration of added C_3, C_4, and ^{13}C-labelled sugars to a C_3 forest soil. *Oecologia* **131**: 245–249.

Fessenden J. E. and Ehleringer J. R. (2002) Age-related variations in ^{13}C of ecosystem respiration across a coniferous forest chronosequence in the Pacific Northwest. *Tree Physiol* **22**: 159–167.

Finlay R. D. and Söderström B. (1992) Mycorrhiza and carbon flow to soil. In *Mycorrhizal Functioning* (M. F. Allen, ed.) Chapman & Hall, London.

Flanagan L. B., Brooks J. R., Varney G. T., Berry S. C. and Ehleringer J. R. (1996) Carbon isotope discrimination during photosynthesis and the isotope ratio of respired CO_2 in boreal forest ecosystems. *Global Biogeochem Cycles* **10**: 629–640.

Flanagan L. B. and Ehleringer J. R. (1998) Ecosystem-atmosphere CO_2 exchange: Interpreting signals of change using stable isotope ratios. *Trends Ecol Evol* **13**: 10–14.

Gadgil R. L. and Gadgil P. D. (1987) Mycorrhiza and litter decomposition. *Nature* **233**: 133.

Ghashghaie J., Duranceau M., Badeck F. W., Cornic G., Adeline M. T. and Deleens E. (2001) $\delta^{13}C$ of CO_2 respired in the dark in relation to $\delta^{13}C$ of leaf metabolites: Comparison between *Nicotiana sylvestris* and *Helianthus annuus* under drought. *Plant Cell Environ* **24**: 505–515.

Gleixner G., Danier H.-J., Werner R. A. and Schmidt H.-L. (1993) Correlations between the ^{13}C content of primary and secondary plant products in different cell compartments and that in decomposing basidiomycetes. *Plant Physiol* **102**: 1287–1290.

Gleixner G., Scrimgeour C., Schmidt H. L. and Viola R. (1998) Stable isotope distribution in the major metabolites of source and sink organs of *Solanum tuberosum* L.: A powerful tool in the study of metabolic partitioning in intact plants. *Planta* **207**: 241–245.

Gleixner G., Bol R. and Balesdent J. (1999) Molecular insight into soil carbon turnover. *Rapid Commun Mass Spectrometry* **13**: 1278–1283.

Gonzalez J., Remaud G., Jamin E., Naulet N. and Martin G. (1999) Specific natural isotope profile studied by isotope ratio mass spectrometry (SNIP-IRMS): $^{13}C/^{12}C$ ratios of fructose, glucose, and sucrose for improved detection of sugar addition to pineapple juices and concentrates. *J Agric Food Chem* **47**: 2316–2321.

Hackett W. F., Connors W. J., Kirk T. K. and Zeikus J. G. (1977) Microbial decomposition of synthetic ^{14}C-labeled lignins in nature: Lignin biodegradation in a variety of natural materials. *Appl Environ Microbiol* **33**: 43–51.

Hanson P. J., Edwards N. T., Garten C. T. and Andrews J. A. (2000) Separating root and soil microbial contributions to soil respiration: A review of methods and observations. *Biogeochemistry* **48**: 115–146.

Henn M. R. and Chapela I. H. (2000) Differential carbon isotope discrimination by fungi during decomposition of C_3- and C_4-derived sucrose. *Appl Environ Microbiol* **66**: 4180–4186.

Henn M. H. and Chapela I. H. (2001) Ecophysiology of ^{13}C and ^{15}N, isotopic fractionation in forest fungi and the roots of the saprotrophic-mycorrhizal divide. *Oecologia* **128**: 480–487.

Hobbie E. A., Macko S. A. and Shugart H. H. (1999) Insights into nitrogen and carbon dynamics of ectomycorrhizal and saprotrophic fungi from isotopic evidence. *Oecologia* **118**: 353–360.

Hobbie E. A., Weber N. S. and Trappe J. M. (2001) Mycorrhizal vs saprotrophic status of fungi: The isotopic evidence. *New Phytol* **150**: 601–610.

Hoffrichter M., Vares T., Kalsi M., Galkin S., Scheibner K., Fritsche W. and Hatakka A. (1999) Production of manganese peroxidase and organic acids and mineralization of ^{14}C-labelled lignin (^{14}C-DHP) during solid-state fermentation of wheat straw with the white rot fungus *Nematoloma frowardii*. *Appl Environ Microbiol* **65**: 1864–1870.

Högberg P. and Ekblad A. (1996) Substrate induced respiration measured *in situ* in a C_3-plant ecosystem using additions of C_4-sucrose. *Soil Biol Biochem* **28**: 1131–1138.

Högberg P., Plamboeck A. H., Taylor A. F. S. and Fransson P. M. A. (1999) Natural ^{13}C abundance reveals trophic status of fungi and host-origin of carbon in mycorrhizal fungi in mixed forests. *Proc Natl Acad Sci USA* **96**: 8534–8539.

Huang Y. S., Bol R., Harkness D. D., Ineson P. and Eglinton G. (1996) Post-glacial variations in distributions, ^{3}C and ^{14}C contents of aliphatic hydrocarbons and bulk organic matter in three types of British acid upland soils. *Org Geochem* **24**: 273–287.

Hungate B. A., Jackson R. B., Chapin F. S., III, Mooney H. A. and Field C. B. (1997) The fate of carbon in grasslands under carbon dioxide enrichment. *Nature* **388**: 576–579.

Kohzu A., Yoshioka T., Ando T., Takahashi M., Koba K. and Wada E. (1999) Natural ^{13}C and ^{15}N abundance of field-collected fungi and their ecological implications. *New Phytol* **144**: 323–330.

Kracht O. and Gleixner G. (2000) Isotope analysis of pyrolysis products from Sphagnum peat and dissolved organic matter from bog water. *Org Geochem* **31**: 645–654.

Law B. E., Kelliher F. M., Baldocchi D. D., Anthoni P. M., Irvine J., Moore J. and Van Tuyl S. (2001) Spatial and temporal variation in respiration in a young ponderosa pine forest during a summer drought. *Ag For Met* **110**: 27–43.

Leavitt S. W. and Long A. (1982) Evidence for $^{13}C/^{12}C$ fractionation between tree leaves and wood. *Nature* **298**: 742–744.

Lichtfouse E., Berthier G., Houot S., Barriuso E., Bergheaud V. and Vallaeys T. (1995) Stable carbon isotope evidence for the microbial origin of C_{14}-C_{18} n-alkanoic acids in soils. *Org Geochem* **23**: 849–852.

Lichtfouse E., Leblond C., Da Silva M. and Behar F. (1998) Occurrence of biomarkers and straight-chain biopolymers in humin: Implication for the origin of soil organic matter. *Naturwissenschaften* **85**: 497–501.

Lin G. and Ehleringer J. R. (1997) Carbon isotopic fractionation does not occur during dark respiration in C_3 and C_4 plants. *Plant Physiol* **114**: 391–394.

Lin G., Ehleringer J. E., Rygiewicz P. T., Johnson M. G. and Tingey D. T. (1999) Elevated CO_2 and temperature impacts on different components of soil CO_2 efflux in Douglas-fir terracosms. *Global Change Biol* **5**: 157–168.

Ludwig B., Heil B., Flessa H. and Beese F. (2000) Dissolved organic carbon in seepage water–production and transformation during soil passage. *Acta Hydrochim Hydrobiol* **28**: 77–82.

Lund C. P., Riley W. J., Pierce L. L. and Field C. B. (1999) The effects of chamber pressurization on soil-surface CO_2 flux and the implications for NEE measurements under elevated CO_2. *Global Change Biol* **5**: 269–282.

Maillard L. C. (1917) Identite des matieres humiques de synthese avec les matieres humiques naturelles. *Ann Chim Paris* **7**: 113–152.

Marseille F., Disnar J. R., Guillet B. and Noack Y. (1999) n-Alkanes and free fatty acids in humus and A1 horizons of soils under beech, spruce and grass in the Massif-Central (Mont-Lozere), France. *Eur J Soil Sci* **50**: 433–441.

Mary B., Mariotti A. and Morel J. L. (1992) Use of ^{13}C variations at natural abundance for studying the biodegradation of root mucilage, roots and glucose in soil. *Soil Biol Biochem* **24**: 1065–1072.

Mortazavi B. and Chanton J. P. (2002) Carbon isotopic discrimination and control of nighttime canopy $\delta^{18}O$-CO_2 in a pine forest in the southeastern United States. *Global Biogeochem Cycles* **16**: 1390–1407.

Nadelhoffer K. J. and Fry B. (1988) Controls on natural ^{15}N and ^{13}C abundances in forest soil organic matter. *Soil Sci Soc Am J* **52**: 1633–1640.

Neff J. C. and Asner G. P. (2001) Dissolved organic carbon in terrestrial ecosystems: Synthesis and a model. *Ecosystems* **4**: 29–48.

O'Leary M. H. (1981) Carbon isotope fractionation in plants. *Phytochemistry* **20**: 553–567.

Park R. and Epstein S. (1961) Metabolic fractionation of ^{13}C and ^{12}C in plants. *Plant Physiol* **36**: 133–138.

Pataki D. E., Ehleringer J. R., Flanagan L. B., Yakir D., Bowling D. R., Still C. J., Buchmann N., Kaplan J. O. and Berry J. A. (2003) The application and interpretation of Keeling plots in terrestrial carbon cycle research. *Global Biogeochem Cycles* **17**(1): 1022.

Pate J. (2001) Carbon isotope discrimination and plant water-use efficiency: Case scenarios for C_3 Plants, In *Stable Isotope Techniques in the Study of Biological Processes and Functioning of Ecosystems* (M. Unkovich, J. Pate, A. McNeill and J. D. Gibbs, eds) pp. 19–36. Kluwer Academic.

Pate J. and Arthur D. (1998) $\delta^{13}C$ analysis of phloem sap carbon: Novel means of evaluating seasonal water stress and interpreting carbon isotope signatures of foliage and trunk wood of Eucalyptus globules. *Oecologia* **117**: 301–311.

Phillips D. L. and Gregg J. W. (2001) Uncertainty in source partitioning using stable isotopes. *Oecologia* **127**: 171–179.

Robinson D. and Scrimgeour C. M. (1995) The contribution of plant C to soil CO_2 measured using ^{13}C. *Soil Biol Biochem* **27**: 1653–1656.

Rochette P. and Flanagan L. B. (1997) Quantifying rhizosphere respiration in a corn crop under field conditions. *Soil Sci Soc Am J* **61**: 466–474.

Rochette P., Flanagan L. B. and Gregorich E. G. (1999) Separating soil respiration into plant and soil components using analyses of the natural abundance of ^{13}C. *Soil Sci Soc Am J* **63**: 1207–1213.

Rossmann A., Butzenlechner M. and Schmidt H. (1991) Evidence for a nonstatistical carbon isotope distribution in natural glucose. *Plant Physiol* **96**: 609–614.

Santruckova H., Bird M. I. and Lloyd J. (2000) Microbial processes and carbon-isotope fractionation in tropical and temperate grassland soils. *Funct Ecol* **14**: 108–114.

Schiff S. L., Aravena R., Trumbore S. E. and Dillon P. J. (1990) Dissolved organic carbon cycling in forested watersheds: A carbon isotope approach. *Water Resour Res* **26**: 2949–2957.

Schiff S. L., Aravena R., Trumbore S. E., Hinton M. J., Elgood R. and Dillon P. J. (1997) Export of DOC from forested catchments on the Precambrian Shield of Central Ontario: Clues from ^{13}C and ^{14}C. *Biogeochemistry* **36**: 43–65.

Schimel D. S. (1993) *Theory and Application of Tracers.* Academic Press, San Diego.

Schmidt H. L. and Gleixner G. (1998) Carbon isotope effects on key reactions in plant metabolism and ^{13}C-patterns in natural compounds. In *Stable Isotopes: Integration of Biological, Ecological and Geochemical Processes* (H. Griffiths ed.) pp. 13–26. BIOS, Oxford.

Schweizer M., Fear J. and Cadisch G. (1999) Isotopic (^{13}C) fractionation during plant residue decomposition and its implications for soil organic matter studies. *Rapid Commun Mass Spectrometry* **13**: 1284–1290.

Simard S. W., Durall D. M. and Jones M. D. (1997) Carbon allocation and carbon transfer between *Betula papyrifera* with *Pseudotsuga menziesii* seedlings using a ^{13}C pulse-labeling method. *Plant Soil* **191**: 41–55.

Smedley M. P., Dawson T. E. and Comstock J. P. (1991) Seasonal carbon isotope discrimination in a grassland community. *Oecologia* **85**: 314–320.

Still C. J., Berry J. A., Collatz G. J. and DeFries R. S. (2003) Global distribution of C_3 and C_4 vegetation: Carbon cycle implications. *Global Biogeochem Cycles* **17**: 1006.

Tcherkez G., Nogues S., Bleton J., Cornic G., Badeck F. and Ghashghaie J. (2003) Metabolic origin of carbon isotope composition of leaf dark-respired CO_2 in French Bean. *Plant Physiol* **131**: 237–244.

Terwilliger V. J. and Huang J. (1996) Heterotrophic whole plant tissues show more enrichment in ^{13}C than their carbon sources. *Phytochemistry* **43**: 1183–1188.

Terwilliger V. J. (2001) Intrinsic water-use efficiency and heterotrophic investment in tropical leaf growth of two Neotropical pioneer tree species as estimated from δ^{13}C values. *New Phytol* **152**: 267–281.

Trumbore S. E., Schiff S. L., Aravena R. and Elgood R. (1992) Sources and transformation of dissolved organic carbon in the Harp Lake forested catchment: The role of soils. *Radiocarbon* **34**: 626–635.

Trumbore S. (2000) Age of soil organic matter and soil respiration: Radiocarbon constraints on belowground C dynamics. *Ecol Appl* **10**: 399–411.

Tu K. P., Brooks P. D. and Dawson T. E. (2001) Using septum-capped vials with continuous-flow isotope ratio mass spectrometric analysis of atmospheric CO_2 for Keeling plot application. *Rapid Commun Mass Spectrometry* **15**: 952–956.

Tuomela M., Vikman M., Hatakka A. and Itavaara M. (2000) Biodegradation of lignin in a compost environment: A review. *Bioresource Technol* **72**: 169–183.

Wingler A., Wallenda T. and Hampp R. (1996) Mycorrhiza formation on Norway spruce (*Picea abies*) roots affects the pathway of anaplerotic CO_2 fixation. *Physiol Plantarum* **96**: 699–705.

Wood H. G., Werkman C. H., Hemingway A. and Nier A. O. (1941) Heavy carbon as a tracer in heterotrophic carbon dioxide assimilation. *J Biol Chem* **139**: 365–376.

Xu M., Debiase T. A., Qi Y., Goldstein A. and Liu Z. (2001) Ecosystem respiration in a young ponderosa pine plantation in the Sierra Nevada Mountains, California. *Tree Physiol* **21**: 309–318.

Yakir D. and Wang X.-F. (1996) Fluxes of CO_2 and water between terrestrial vegetation and the atmosphere estimated from isotope measurements. *Nature* **380**: 515–517.

9

Simulation of Ecosystem $C^{18}OO$ Isotope Fluxes in a Tallgrass Prairie: Biological and Physical Controls

Christopher Still, William J. Riley, Brent R. Helliker, Joseph A. Berry

Introduction

Spatial and temporal variations in the ^{18}O composition of atmospheric CO_2 (denoted $\delta^{18}O$-CO_2) are determined largely by terrestrial carbon exchanges, with small contributions from ocean gas exchange, biomass and fossil fuel burning, and stratospheric reactions (Francey and Tans, 1987; Farquhar *et al.*, 1993; Ciais *et al.*, 1997a,b; Peylin *et al.*, 1999). Francey and Tans (1987) first proposed that terrestrial carbon fluxes interacting with soil water and leaf water drive the large (~1.5‰) interhemispheric gradient in $\delta^{18}O$-CO_2. Friedli *et al.* (1987) simultaneously suggested that terrestrial fluxes were responsible for variations in $\delta^{18}O$-CO_2 they observed above Switzerland. Later work by Hesterburg and Siegenthaler (1991) elucidated the processes by which soil CO_2 fluxes influence $\delta^{18}O$-CO_2. Tans (1998) developed steady-state analytical solutions for the $\delta^{18}O$ value of soil-gas CO_2 for several CO_2 source profiles. Farquhar *et al.* (1993) and Farquhar and Lloyd (1993) confirmed the earlier inference by Francey and Tans (1987) and developed equations to describe the interaction of leaf fluxes with $\delta^{18}O$-CO_2. Ciais *et al.* (1997a,b) extended the global-scale study of Farquhar *et al.* (1993) by simulating surface fluxes with a land surface model and coupling them to a tracer-transport model to examine seasonal and latitudinal variations in $\delta^{18}O$-CO_2. Peylin *et al.* (1999) examined in further detail the influence of specific land regions, and showed the dominant influence of the Siberian taiga forests on northern hemisphere variations of $\delta^{18}O$-CO_2.

While the predominant influence of terrestrial exchanges on global-scale variations in $\delta^{18}O$-CO_2 has been established, many uncertainties remain, including our understanding of the mechanisms driving large interannual variations in $\delta^{18}O$-CO_2 and the observed downward excursion of $\delta^{18}O$-CO_2 during most of the 1990s (Gillon and Yakir, 2001; Stern *et al.*, 2001; Ishizawa *et al.*, 2002). It is unclear how well our understanding of the climatological processes affecting $\delta^{18}O$-CO_2 translates to interannual or decadal forcing. Resolving these uncertainties requires an improved basis for understanding the magnitude of, and mechanisms underlying, variations in terrestrial $C^{18}OO$ exchanges. To date, however, relatively few studies have examined controls on $\delta^{18}O$ in CO_2 at the ecosystem scale (see the review by Yakir and Sternberg, 2000). Additionally, most of these studies were conducted in forested ecosystems (e.g., Flanagan *et al.*, 1997; Harwood *et al.*, 1998, 1999) or in agro-ecosystems (e.g., Yakir and Wang, 1996; Buchmann and Ehleringer, 1998).

This chapter focuses on $C^{18}OO$ exchanges in a tallgrass prairie grassland located in north-central Oklahoma, USA. We use a mechanistic ecosystem model modified for simulating oxygen isotopes to explore the physiological and biophysical controls on $C^{18}OO$ exchanges in this ecosystem. Because these exchanges are strongly coupled to the ^{18}O composition of leaf and soil water, we also describe these quantities and their relation to $\delta^{18}O$-CO_2.

Background

Plant Oxygen Isotope Exchanges

Photosynthesis requires the transport of CO_2 along a concentration gradient from the atmosphere to the site of carboxylation in the chloroplast organelle. This diffusive flux is composed of gross bi-directional fluxes across the stomatal pore, which together sum to net photosynthesis (O'Leary, 1981). The net photosynthetic $C^{18}OO$ flux is dependent primarily on the relative magnitude of these bi-directional fluxes and on the $\delta^{18}O$ value of water in mesophyll cells adjacent to the stomatal pore (Farquhar and Lloyd, 1993; Farquhar *et al.*, 1993; Ciais *et al.*, 1997a; Gillon and Yakir, 2000; Yakir and Sternberg, 2000). The back- or retro-diffusion of CO_2 molecules out of leaves with a different $\delta^{18}O$ from when they entered the leaf occurs because only some of the CO_2 entering the leaf is fixed by photosynthesis (the net flux), while the remainder diffuses out after full or partial equilibration with leaf water. The difference in the $\delta^{18}O$ value occurs because CO_2 molecules dissolved in the mesophyll cells that line the stomatal cavity exchange their oxygen atoms with mesophyll water during a hydration reaction catalyzed by the enzyme carbonic anhydrase (CA) (Farquhar and Lloyd, 1993; Williams *et al.*, 1996; Gillon and Yakir, 2000).

$$CO_{2(g)} + H_2{}^{18}O_{(l)} \leftrightarrow H^+ + [HCO_2{}^{18}O]_{(aq)} \leftrightarrow C^{18}OO_{(g)} + H_2O_{(l)}$$

[Reaction 1]

The mesophyll water effectively sets the $\delta^{18}O$ value of the CO_2 molecules since the number of oxygen atoms in the water is several orders of magnitude greater than the number of oxygen atoms on dissolved CO_2 molecules. Equilibrium and kinetic isotope effects and interactions with atmospheric water vapor determine the isotopic composition of leaf water (Craig and Gordon, 1965; Dongmann *et al.*, 1974; Flanagan *et al.*, 1991; Roden and Ehleringer, 1999). As water molecules evaporate from the mesophyll cell wall surfaces adjacent to the stomatal cavity during transpiration, the lighter isotopologue ($H_2{}^{16}O$) evaporates more readily and diffuses through the stomatal pores more quickly, thus enriching the evaporating surface with $H_2{}^{18}O$ molecules. This enrichment of the water pool is then transferred to the hydrated CO_2 molecules (Farquhar and Lloyd, 1993; Farquhar *et al.*, 1993; Ciais *et al.*, 1997a).

These factors are captured in the following equation for the net leaf flux of $C^{18}OO$ (Farquhar and Lloyd, 1993; Farquhar *et al.*, 1993; Keeling, 1995; Ciais *et al.*, 1997a):

$$^{18}F_p = F_p \alpha_D R_a + F_p f \alpha_D (R_a - R_c) \tag{9.1}$$

where F_p is net leaf photosynthesis, f is the factor representing retrodiffusion of CO_2 molecules (i.e., a multiple of net leaf photosynthesis), and α_D represents the weighted mean diffusive fractionation that occurs during diffusion across the laminar leaf boundary layer, molecular diffusion through the stomatal pore, and equilibrium dissolution and liquid-phase diffusion from the bottom of the stomatal pore to the chloroplast ($\sim$0.9926 for typical conditions; Farquhar and Lloyd, 1993). The subscripts on R refer to the concentration and oxygen isotope ratio of CO_2 in equilibrium with chloroplast (leaf) water (R_c) or present in the canopy atmosphere (R_a). The first term corresponds to the net flux, and the second term captures the exchange, or 'disequilibrium,' component of the flux (Ciais *et al.*, 1997a).

Upon converting the terms in Eq. 9.1 to delta notation, a fractionation factor for the discrimination against atmospheric $C^{18}OO$ during net photosynthesis can be expressed (Farquhar and Lloyd, 1993; Farquhar *et al.*, 1993; Keeling, 1995; Ciais *et al.*, 1997a):

$$\Delta^{18} = -\varepsilon_D + \frac{C_c}{C_a - C_c}(\delta^{18}O_c - \delta^{18}O_a) \tag{9.2}$$

Soil Oxygen Isotope Exchanges

Analogous to leaf processes, the $\delta^{18}O$ value of the soil–atmosphere CO_2 flux is determined by equilibrium isotope exchanges with soil water and kinetic diffusive fractionation against $C^{18}OO$ relative to $C^{16}OO$. However, the processes are qualitatively different from those occurring in leaves. Carbon dioxide produced in the soil column by autotrophic (root) and heterotrophic (microbial and fungal) respiration presumably carries the oxygen isotope signatures of root water and the water of the bacterial medium, respectively. However, the original signature is rarely preserved, as the CO_2 sequentially re-equilibrates with soil water as it diffuses away from the site of production, taking on the new $\delta^{18}O$ water value after equilibration. Differential diffusion of $C^{18}OO$ and $C^{16}OO$ in air also contributes to the $\delta^{18}O$ of soil CO_2. The study of Hesterburg and Siegenthaler (1991) was the first to highlight this competition between diffusion and equilibration in determining the $\delta^{18}O$ of soil-respired CO_2. Our understanding of this competition was enhanced by the work of Tans (1998) and Miller *et al.* (1999). Examining extreme situations in the soil profile highlights the competing effects of diffusion and equilibration. If the timescale of equilibration in reaction 1 is very long relative to the timescale for diffusion, the $\delta^{18}O$ value of soil-respired CO_2 will approach the signature of the last water pool with which it equilibrated, effectively cancelling the diffusive fractionation. At the other extreme, with very fast equilibration (as is the case in plant leaves because of CA), the full diffusive fractionation will be expressed and the CO_2 will be in isotopic equilibrium with water from depth to the surface (Tans, 1998; Miller *et al.*, 1999). The depth-profile of CO_2 production is also important in affecting the diffusion–equilibration competition (Tans, 1998). Finally, gradients in the $\delta^{18}O$ value of soil water in the top few centimeters of soil will impact the $\delta^{18}O$ value of the soil-surface CO_2 flux (Riley *et al.*, Impact of the $\delta^{18}O$ value of near-surface soil water on the $\delta^{18}O$ value of the soil-surface CO_2 flux; submitted to *Geochim Cosmochim Acta*).

Another process that influences the $\delta^{18}O$ of soil-respired CO_2 fluxes is the 'invasion effect,' which occurs when CO_2 just above the soil–atmosphere interface diffuses into the soil column, equilibrates with soil water in the upper layers, and diffuses back out with a new isotope signature (Tans, 1998; Amundson *et al.*, 1998; Miller *et al.*, 1999; Stern *et al.*, 2001). (The invasion flux is also referred to as the 'abiotic' flux, as it is not associated with microbial or plant respiration within the soil column (Stern *et al.*, 1999, 2001)). Thus, the influence of soils on the $\delta^{18}O$ value of atmospheric CO_2 results from bi-directional fluxes that can be decomposed into net and 'disequilibrium' fluxes, analogous to ^{13}C in CO_2 (e.g., Ciais *et al.*, 1997a). The extent to which invasion occurs is a function of several factors, including soil temperature and water content. In certain environments, such as those with low respiration rates, the invasion flux can dominate the $\delta^{18}O$ signature of soil-respired CO_2 (Tans, 1998; Stern *et al.*, 2001).

Ecosystem Model Simulations

We modified an existing 'big leaf' ecosystem model, LSM1.0 (hereafter, LSM), to capture the aforementioned plant and soil isotope exchange processes. LSM is a widely used single-canopy model that simulates CO_2,

H_2O, and energy exchanges between terrestrial ecosystems and the atmosphere on sub-hourly timesteps (Bonan, 1996). The canopy is divided into sunlit and shaded fractions; the partitioning of measured shortwave radiation into direct and diffuse components is described in Riley *et al.* (2002). The model is capable of simulating twenty-eight different surface types comprising different fractional covers of thirteen different vegetation types. In addition to calculating exchanges of CO_2 and H_2O between the surface and the atmosphere, LSM predicts aboveground fluxes of radiation, sensible and latent heat, and momentum. LSM also has modules for predicting belowground fluxes of energy and H_2O. We prescribed seasonal dynamics such as LAI from measurements taken at the tallgrass site (e.g., Suyker and Verma, 2001; Suyker *et al.*, 2003).

We modified LSM to predict isotopic values for H_2O and CO_2 pools and fluxes. Specifically, we developed modules to compute the following quantities: the $H_2{}^{18}O$ content of canopy water vapor, leaf water, and vertically resolved soil water; CO_2 isotopic equilibration with ^{18}O in leaf and soil water; bi-directional photosynthetic CO_2 and $C^{18}OO$ fluxes; and soil CO_2 and $C^{18}OO$ fluxes. We call this modified model ISOLSM. Model development and sensitivity testing is described in Riley *et al.* (2002) and model testing is described in Riley *et al.* (2003). The isotope pools and exchange fluxes captured in ISOLSM are detailed in Fig. 9.1.

ISOLSM is driven with measured air temperature, wind speed, CO_2 concentration, water vapor pressure and its isotopic ratio, short- and long-wave radiation, amount of precipitation and its isotopic ratio, and CO_2 concentration at the soil surface. The non-isotopic driving quantities were collected using equipment mounted on an eddy flux tower (Suyker and Verma, 2001). The isotopic quantities were either measured in a mostly continuous fashion (e.g., the isotopic ratio of precipitation) or were estimated using common assumptions (e.g., the isotopic ratio of above-canopy water vapor was set to be 7‰ lighter than predicted stem water at the site). The isotope data and collection methods are described elsewhere (Helliker *et al.*, 2002; Riley *et al.*, 2003). In this section, we discuss model simulations of vertically resolved soil water $H_2{}^{18}O$ composition and leaf $H_2{}^{18}O$ composition for select days in the 2000 growing season.

Soil Water Isotopic Composition

As described in Riley *et al.* (2002), soil water $H_2{}^{18}O$ composition was simulated by ISOLSM at finely resolved depth increments (2.5 cm) in the upper 30 cm of soil and at lower resolution for the remainder of the soil column. The simulated values agree fairly well with measurements over the course of the growing season and throughout the top 20 cm of soil (Riley *et al.*, 2003). Figure 9.2 (top panel) shows simulated soil water $\delta^{18}O$

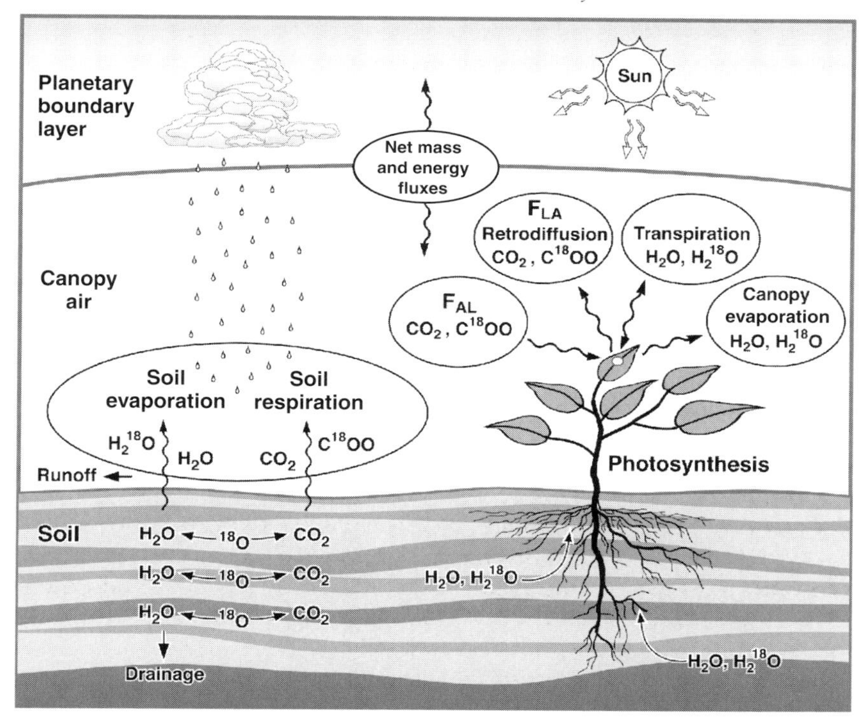

Figure 9.1 Simplified schematic of the isotopic pools and exchange fluxes simulated by ISOLSM. F_{AL} and F_{LA} represent gross CO_2 fluxes from the atmosphere to the leaf and from the leaf to the atmosphere, respectively.

values for four depth increments over a 3-day period (May 7–9) in 2000. The strong enrichment and extreme variability of $H_2{}^{18}O$ in the uppermost soil water layer (0–2.5 cm) predicted by the model is striking. However, we did not take soil water measurements throughout a day or over several days to compare with the predicted extreme diurnal variability in the upper soil layer (Riley *et al.*, 2003). The modeled variability is dampened in the layers below the top layer (Fig. 9.2, top panel). The model predicts much smaller variation in the ^{18}O composition of soil water layers below 20 cm (not shown).

Leaf Water Isotopic Composition

ISOLSM calculates the $\delta^{18}O$ value of leaf water using the modified Craig–Gordon equation developed by Flanagan *et al.* (1991). This equation requires the $\delta^{18}O$ value of stem or source water as well as $\delta^{18}O$ of water vapor at the leaf surface and in the canopy atmosphere. We estimated the stem water isotopic signature from modeled $\delta^{18}O$ in upper soil water

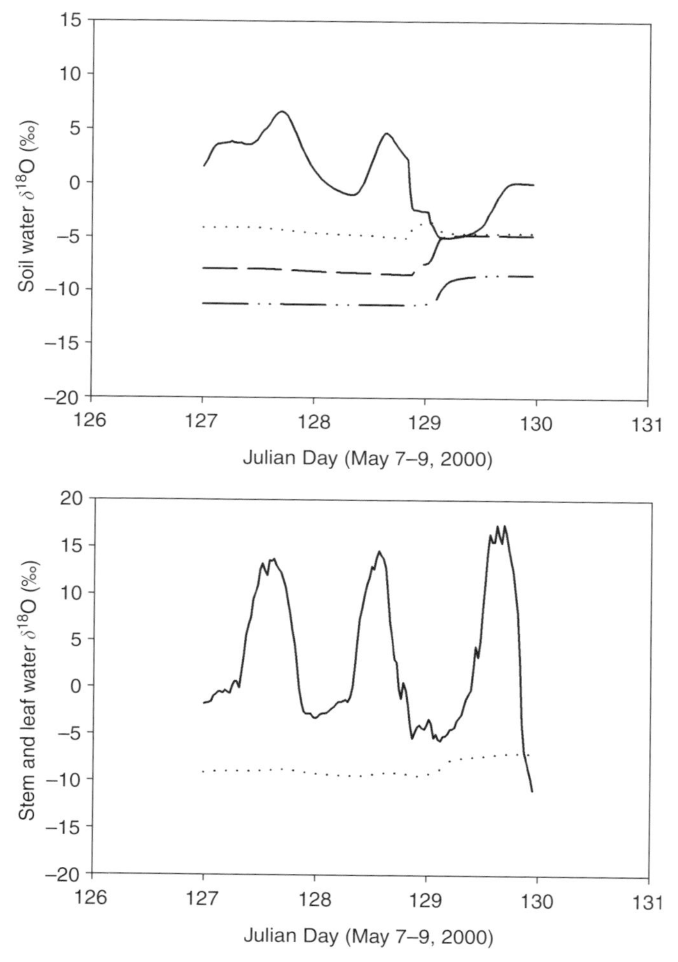

Figure 9.2 Simulated isotopic composition of soil, stem, and leaf water in the tallgrass prairie ecosystem. Values are in ‰, expressed relative to V-SMOW. The oxygen isotope composition of soil water during May 7–9, 2000 is displayed in the top panel. The depth increments are 0–2.5 cm (solid line), 2.5–5 cm (dotted line), 5–10 cm (dashed line), and 10–20 cm (dash–dot lines). The simulated oxygen isotope composition of tallgrass stem (source) and leaf water during May 7–9, 2000 is plotted in the bottom panel. The solid line represents leaf water, the dotted line stem water.

layers weighted by the vertical rooting mass distribution (Riley *et al.*, 2002). The $\delta^{18}O$ value of canopy water vapor was calculated based on a balance between fluxes from the leaves and soil surface, and between the canopy air and atmosphere. Predicted leaf water (solid line) and stem (source) water values are plotted for May 7–9 in Fig. 9.2 (bottom panel). As with soil water, predicted leaf water values are quite variable over the 3-month measurement period (not shown), but sufficient measurements over this entire period to confirm this modeled variability are not available.

Ecosystem Isofluxes

ISOLSM was used to calculate leaf and soil ^{18}O fluxes in CO_2 (isofluxes) over the May–July period in 2000. The isoflux ($\mu mol\, m^{-2}\, s^{-1}$ ‰) is calculated as the CO_2 flux magnitude multiplied by its $\delta^{18}O$ value. The net ecosystem isoflux is the sum of soil and leaf isofluxes, ignoring the stem respiratory isoflux, i.e.,

$$I = F_{SA}\delta_{SA} + F_{NL}\delta_{NL} \tag{9.3}$$

where F_{SA} and δ_{SA} represent the soil–atmosphere flux magnitude and its $\delta^{18}O$ value (which includes the influence of soil invasion), and F_{NL} and δ_{NL} represent the magnitude and $\delta^{18}O$ value of the net leaf CO_2 flux. The gross fluxes across the stomata (atmosphere-to-leaf, F_{AL}, and leaf-to-atmosphere, F_{LA}) are calculated following Ciais *et al.* (1997a). At night, the leaf flux consists of an outgoing flux only (F_{LA}), which is equal to the leaf maintenance respiration. The isotopic ratios of these fluxes are determined by the processes described previously in Background, this chapter, and in Riley *et al.* (2002). The soil flux is calculated from the sum of root autotrophic respiration and soil heterotrophic respiration and the isotopic composition of the soil–atmosphere CO_2 flux (Riley *et al.*, 2002). Figure 9.3 displays predicted leaf, soil, and net (leaf-plus-soil) isofluxes for the period May 7–9 from the ISOLSM simulation (C_3 model). The photosynthetic isoflux is the overwhelming component during the daytime; relative to this isoflux, the soil respiratory component is small and fairly invariant, as is the nighttime leaf isoflux. The daily leaf isoflux closely follows the profile of photosynthetic uptake and the leaf water oxygen isotope composition. Maximum values of photosynthesis and leaf water roughly coincide at midday, reinforcing the isoflux signal.

There are several reasons why isofluxes in this tallgrass ecosystem are dominated by the leaf contribution. First, precipitation and soil water $\delta^{18}O$ values are not strongly depleted relative to the atmospheric background $\delta^{18}O$-CO_2 value (as compared to, for example, interior boreal regions at high latitudes); see Fig. 9.2(top panel). Second, this grassland is characterized by fast growth and large photosynthetic fluxes, with daily net ecosystem exchange rates approaching 35 $\mu mol\, m^{-2}\, s^{-1}$ (Suyker and Verma, 2001). Finally, leaf water $\delta^{18}O$ values predicted by the model are quite enriched,

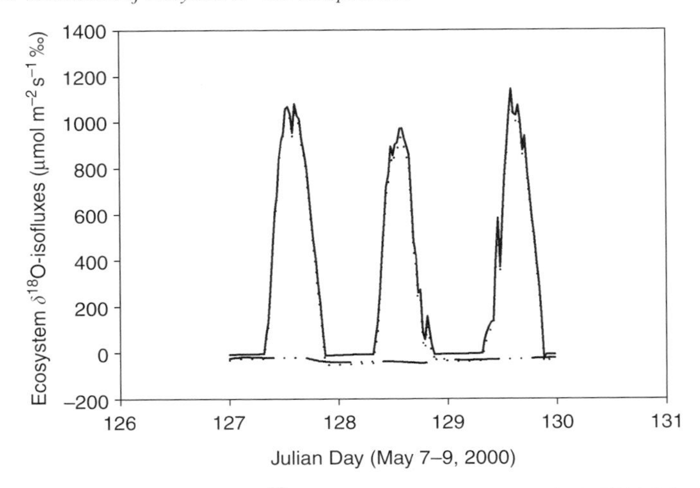

Figure 9.3 Simulated ecosystem $^{18}O\text{-}CO_2$ isofluxes during May 7–9, 2000. The dash–dot line represents the soil isoflux; the solid line is the leaf isoflux; and the dotted line gives the net (soil plus leaf) ecosystem isoflux, ignoring the stem respiration isoflux. This simulation was conducted with the C_3 photosynthesis model.

often approaching 20–25‰ at midday, reflecting the low relative humidity and relatively enriched soil water at the site.

Large positive isofluxes can result from both enriched water pools and large carbon exchange fluxes. The study by Flanagan (see Chapter 10) in a mixed grass prairie showed that the largest isofluxes took place during drought years. These large isofluxes took place because upper soil water layers and leaf water became enriched in ^{18}O, while carbon fluxes were relatively constant. This result agrees with earlier work at the leaf level showing the dramatic effect that changes in leaf water isotopic composition have on discrimination against $C^{18}OO$ (Flanagan *et al.*, 1994). In our simulations, the largest net ecosystem isofluxes occurred during periods of moderately high leaf water enrichment and large leaf exchange fluxes. Variations in the soil isoflux were small by comparison.

C_3 and C_4 Isofluxes

To explore physiological differences between the dominant plant types in this grassland (C_4 and C_3 plants), we ran ISOLSM over the entire measurement period with either the C_3 or C_4 photosynthesis model. (In reality, this prairie is a mixture of C_3 and C_4 plants whose abundance changes seasonally in response to climatic seasonality [Still *et al.*, 2003].) The effect of photosynthetic physiology on leaf and ecosystem isofluxes is dramatic. Modeled leaf isofluxes are separated into net and exchange component fluxes (see Eq. 9.1) in Fig. 9.4 for a 3-day period in May 2000. Overall leaf isofluxes are similar for the two simulations, but the exchange component is a much larger fraction of the total and larger in magnitude in the C_3 simulation (Fig 9.4, bottom panel). The net component is both a larger fraction of the leaf isoflux and larger in magnitude for the C_4 simulation (Fig. 9.4, top panel). Leaf water $\delta^{18}O$ is effectively identical in both the C_3 and C_4 simulations.

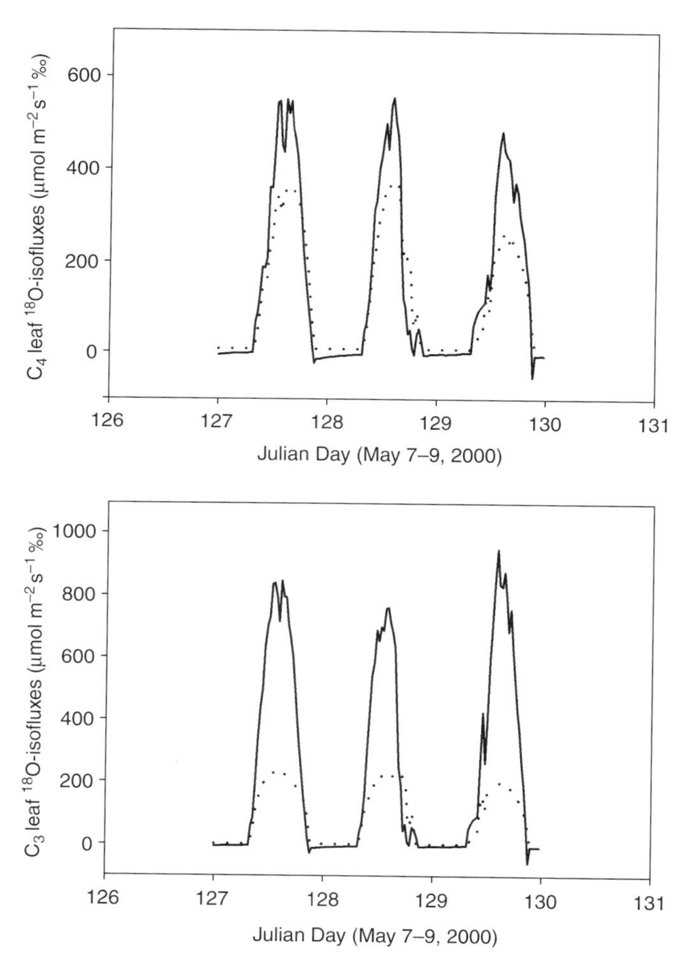

Figure 9.4 Modeled leaf isofluxes, expressed as positive quantities, partitioned into net and exchange terms $(F_{AL} - F_{LA})\delta_{AL}$ and $(\delta_{AL} - \delta_{LA})F_{LA}$, respectively, during May 7–9, 2000. Isofluxes from the C_4 simulation are in the top panel. The net isoflux is given by the dotted line, and the exchange isoflux by the solid line. Isofluxes from the C_3 simulation are in the bottom panel. The net isoflux is given by the dotted line, and the exchange isoflux by the solid line. Note the different y-axis scales in each panel.

The larger photosynthetic isoflux in the C_3 simulation is due to the larger leaf exchange flux in these plants (Farquhar *et al.*, 1993). This is illustrated by the factor, f, in Eq. 9.1. At similar flux rates, this factor can be fourfold higher in C_3 plants than in C_4 plants. This relative increase is due to higher stomatal conductance and internal CO_2 concentration in C_3 plants. For example, the flux-weighted average intercellular CO_2 concentration in the C_3 simulation was 261 ppm, whereas for the C_4 simulation it was 190 ppm. This difference in isotope exchange flux vs photosynthetic flux is highlighted in Fig. 9.5. In this figure, the leaf isoflux is plotted against the photosynthetic flux for all points in the C_3 and C_4 simulations (filled circles and open circles, respectively). For a given photosynthetic flux, the leaf isoflux is greater in the C_3 simulation, by as much as a factor of two or more, due to the much greater exchange flux in these plants.

The differing photosynthetic isofluxes strongly impact net (leaf photosynthetic plus soil respiratory) isofluxes. This is illustrated by Fig. 9.6, which displays predicted cumulative isofluxes from the C_4 and C_3 simulations in the top and bottom panels, respectively. The cumulative isoflux is calculated by summing the isoflux from each timestep over the 3-month simulation. The cumulative net isoflux is about 25% larger in the C_3 simulation (Fig. 9.6, bottom panel), even though the cumulative net C_3 carbon flux is smaller over this period due to the larger photosynthetic rates predicted by the C_4 photosynthesis model, particularly in the hotter months

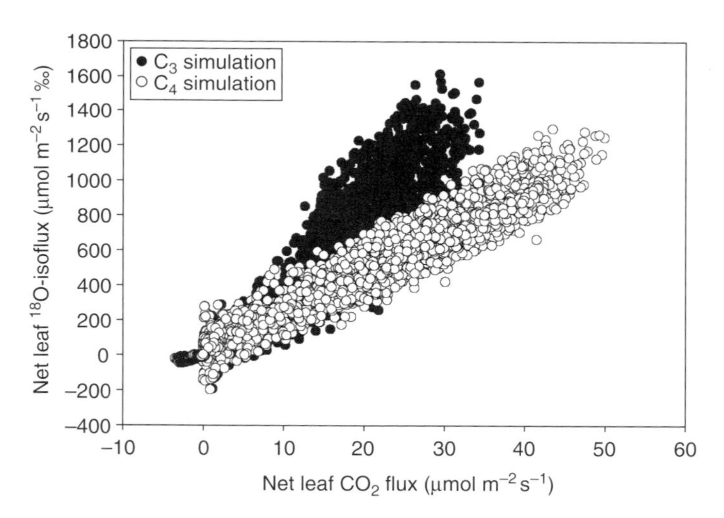

Figure 9.5 Modeled net leaf $^{18}O-CO_2$ isofluxes plotted against net leaf CO_2 fluxes for every timestep from May to July. Points from the C_3 simulation are given by solid circles, and points from the C_4 simulation by open circles.

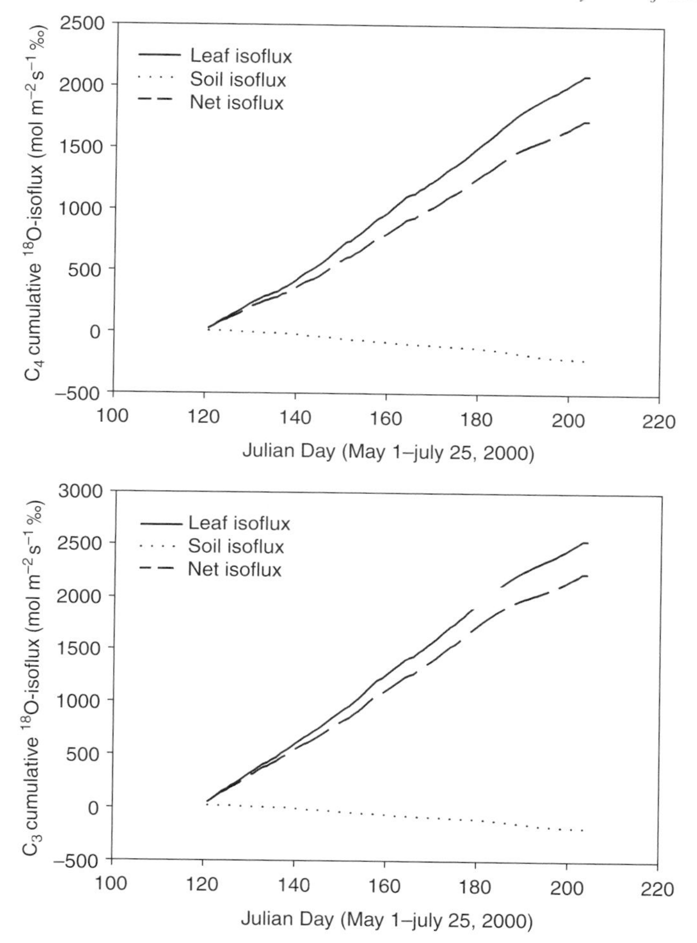

Figure 9.6 Modeled cumulative isofluxes from the beginning of May to late July 2000. The solid line represents the leaf isoflux; the dotted line is the soil isoflux; and the dashed line gives the net isoflux. Modeled cumulative isofluxes from the C_4 simulation are in the top panel; modeled cumulative isofluxes from the C_3 simulation are in the bottom panel. Note the different y-axis scales in each panel.

of June and July. The cumulative net isofluxes plotted in Fig. 9.6 show a continuous upward trend. This is deceiving, as the cumulative net flux should level off or decrease later in the growing season as respiration outpaces photosynthesis. Because we had isotopic forcing data only through late July of 2000, the influence of late-season respiration is not apparent. Also, we do not include stem respiration, which would reduce the net isoflux slightly.

However, despite these gaps, the net isoflux for the entire year would still be positive in this system.

Isofluxes from Grasses and Forbs

After the photosynthetic pathway, the most important functional type distinction in this ecosystem is probably that of grasses vs forbs. One of the primary isoflux differences between grasses and forbs is the rooting depth profile, with forbs expected to have deeper roots than grasses. Riley *et al.* (2002) explored the effect of rooting depths on the $\delta^{18}O$ signature of ecosystem CO_2 fluxes. The two primary effects are on leaf water $\delta^{18}O$ and the $\delta^{18}O$ value of the soil CO_2 flux. A deeper rooting distribution would tend to decrease leaf water $\delta^{18}O$ (by up to several per mil for realistic variations in rooting depth) and thus decrease the photosynthetic isoflux because deeper soil water layers are insulated from evaporative enrichment at the soil surface. A deeper rooting distribution would also decrease the $\delta^{18}O$ signature of the soil CO_2 flux because the respiration source profile would change: as a result, H_2O-CO_2 equilibration would occur at greater depths and be associated with more isotopically depleted water. The latter conclusion, however, depends very much on the soil moisture content. Riley *et al.* (2002) showed that a deeper rooting (and thus source) profile resulted in a more negative surface flux primarily in dry conditions, reflecting the competition between CO_2 diffusion and isotopic equilibration with soil water.

Other possible isoflux differences between grasses and forbs (not explored here) include stomatal conductance and photosynthetic capacity, seasonality of plant activity, and aboveground vs belowground carbon partitioning. It is also possible that differences in vegetation stature will affect the isofluxes, since vegetation height influences leaf temperature and the vapor pressure deficit that leaves experience. However, we did not attempt to simulate these other differences. It would be useful for future modeling studies to examine the complete array of structure and function differences between grasses and forbs and how these differences affect ecosystem isofluxes on seasonal and interannual bases.

Summary and Conclusions

Photosynthesis and respiration exert opposing influences on the oxygen isotope composition of atmospheric CO_2. These influences result from the labeling of CO_2 molecules with the oxygen isotope signatures of leaf and soil water ($\delta^{18}O$-H_2O) during photosynthesis and soil respiration. Typically, CO_2 in isotopic equilibrium with leaf and soil water is enriched and depleted in ^{18}O, respectively, relative to the background atmosphere $\delta^{18}O$-CO_2 value. To explore the biological and physical controls on isotope fluxes of ^{18}O in CO_2 (isofluxes) associated with photosynthesis and respiration and to assess their relative magnitudes, we modified an ecosystem model to simulate the important processes (Riley *et al.*, 2002). Simulated quantities (e.g., the $\delta^{18}O$ value of leaf water, soil water, and nighttime ecosystem-respired CO_2) compare well with measurements taken in a tallgrass prairie during discrete campaigns during the 2000 growing season (Riley *et al.*, 2003).

On diurnal and seasonal timescales, the leaf isoflux typically dominates the ecosystem isoflux, with soil isofluxes contributing a fairly small amount. These results are counter to global scale modeling studies, which infer a dominant contribution from boreal ecosystem soil respiratory isofluxes to match the atmospheric constraints imposed by mid- and high-latitude flask sampling stations (Ciais *et al.*, 1997a,b; Peylin *et al.*, 1999). The dominance of photosynthetic isofluxes in the tallgrass prairie ecosystem is likely to be true in other tropical and subtropical grassland ecosystems. In general, the net annual ecosystem isoflux in these ecosystems would be positive due to the relatively enriched ^{18}O content of precipitation, low relative humidity, and high net radiation typical of these regions. It is important to acknowledge the limited temporal extent of this study, which covers only about 3/4 of the growing season and does not include the net ecosystem respiration isofluxes occurring during the rest of the year. Also, it is possible that the 2000 growing season is unrepresentative of this ecosystem, and other climatically different years should be analyzed in future studies.

Studies of isofluxes in grassland ecosystems must recognize the important functional differences between C_3 and C_4 plants. The different physiologies impact the $\delta^{18}O$ value of atmospheric CO_2 in very different fashions. Plants with the C_3 photosynthetic pathway generate larger photosynthetic isofluxes than do C_4 plants. This is due to the much smaller leaf isotope exchange flux in C_4 plants at similar photosynthetic rates. One aspect of this difference is that respiratory fluxes should have a larger impact on $\delta^{18}O$-CO_2 in C_4-dominated grasslands. This offset in isofluxes provides an additional constraint in partitioning carbon fluxes between C_3 and C_4 plants, especially in combination with the $\delta^{13}C$ tracer.

The model simulations presented here demonstrate the utility of a land surface model that simulates the isotopic contents of H_2O and CO_2 ecosystem pools and fluxes. This utility is apparent because of the many non-linear processes that affect the ^{18}O of soil and leaf water as well as the dynamics of gross carbon exchanges between the land surface and the atmosphere. For example, the partitioning of net radiation into latent and sensible heat fluxes influences surface physical and biological processes that impact the $\delta^{18}O$ value of both H_2O and CO_2 exchanges. Future global simulations of $\delta^{18}O$-CO_2 should explicitly couple a land surface model like

ISOLSM with an atmospheric model capable of simulating ^{18}O throughout the hydrological cycle. With such a modeling system, other ecosystems could be investigated in detail to better understand biological and physical controls on isofluxes. For example, it would be valuable to contrast isofluxes from a grass–herbaceous understory with isofluxes from a tree overstory (such as occurs in savannas). It is also likely that large isoflux differences exist between evergreen and deciduous trees in boreal and temperate forests.

References

Amundson R., Stern L., Baisden T. and Wang Y. (1998) The isotopic composition of soil and soil-respired CO_2. *Geoderma* **82**: 83–114.

Bonan G. (1996) A land surface model (LSM version 1.0) for ecological, hydrological, and atmospheric studies: Technical description and user's guide, TN-417 + STR. National Center for Atmospheric Research, Boulder, Colorado.

Buchmann N. and Ehleringer J. R. (1998) CO_2 concentration profiles, and carbon and oxygen isotopes in C_3 and C_4 crop canopies. *Ag Forest Met* **89**: 45–58.

Ciais P., Denning A. S., Tans P. P., Berry J. A., Randall D. A., Collatz J. G., Sellers P. J., White J. W. C., Trolier M., Meijer H. A. J., Francey R. J., Monfray P. and Heimann M. (1997a) A three-dimensional synthesis study of $\delta^{18}O$ in atmospheric CO_2. 1. Surface fluxes. *J Geophys Res* **102**: 5857–5972.

Ciais P., Tans P. P., Denning A. S., Francey R. J., Trolier M., Meijer H. A. J., White J. W. C., Berry J. A., Randall D. A., Collatz J. G., Sellers P. J., Monfray P. and Heimann M. (1997b) A three-dimensional synthesis study of $\delta^{18}O$ in atmospheric CO_2. 2. Simulations with the TM2 transport model. *J Geophys Res* **102**: 5873–5883.

Craig H. and Gordon L. (1965) Deuterium and oxygen-18 variations in the ocean and marine atmosphere. In *Proceedings of the Conference on Stable Isotopes in Oceanographic Studies and Paleotemperatures* (E. Tongiorgi, ed.) pp. 9–130. Laboratory of Geology and Nuclear Science, Pisa.

Dongmann G., Nurnberg H. E., Forstel H. and Wagener K. (1974) On the enrichment of $H_2{}^{18}O$ in the leaves of transpiring plants. *Radiat. Environ. Biophys.* **11**: 41–52.

Farquhar G. D. and Lloyd J. (1993) Carbon and oxygen isotope effects in the exchange of carbon dioxide between terrestrial plants and the atmosphere. In *Stable Isotopes and Plant Carbon–Water Relations* (J. R. Ehleringer, A. E. Hall and G. D. Farquhar, eds) pp. 47–70. Academic Press, San Diego.

Farquhar G. D., Lloyd J., Taylor J. A., Flanagan L. B., Syvertsen J. P., Hubick K. T., Wong S. C. and Ehleringer J. R. (1993) Vegetation effects on the isotope composition of oxygen in atmospheric CO_2. *Nature* **363**: 439–443.

Flanagan L. B., Brooks J. R., Varney G. T. and Ehleringer J. R. (1997) Discrimination against $C^{18}O^{16}O$ during photosynthesis and the oxygen isotope ratio of respired CO_2 in boreal forest ecosystems. *Global Biogeochemical Cycles* **11**: 83–98.

Flanagan L. B., Comstock J. and Ehleringer J. R. (1991) Comparison of modeled and observed environmental influences on the stable oxygen and hydrogen isotope composition of leaf water in *Phaseolus vulgaris* L. *Plant Physiol* **96**: 588–596.

Flanagan L. B., Phillips S. L., Ehleringer J. R., Lloyd J. and Farquhar G. D. (1994) Effect of changes in leaf water oxygen isotopic composition on discrimination against $C^{18}O^{16}O$ during photosynthetic gas exchange. *Aust J Plant Physiol* **21**: 221–234.

Francey R. J. and Tans P. P. (1987) Latitudinal variation in oxygen-18 of atmospheric CO_2. *Nature* **327**: 495–497.

Friedli H., Siegenthaler U., Rauber D. and Oeschger H. (1987) Measurements of concentration, 13C/12C and 18O/16O ratios of tropospheric carbon dioxide over Switzerland. *Tellus* **39**B: 80–88.

Gillon J. S. and Yakir D. (2000) Naturally low carbonic anhydrase activity in C-4 and C-3 plants limits discrimination against $C^{18}OO$ during photosynthesis. *Plant Cell Environ* **23**: 903–915.

Gillon J. S. and Yakir D. (2001) Influence of carbonic anhydrase activity in terrestrial vegetation on the ^{18}O content of atmospheric CO_2. *Science* **291**: 2584–2587.

Harwood K. G., Gillon J. S., Griffiths H. and Broadmeadow M. S. J. (1998) Diurnal variation in $\Delta^{13}CO_2$, $\Delta C^{18}O^{16}O$ and evaporative site enrichment of $\delta H_2{}^{18}O$ in *Piper aduncum* under field conditions in Trinidad. *Plant Cell Environ* **21**: 269–283.

Harwood K. G., Gillon J. S., Roberts A. and Griffiths H. (1999) Determinants of isotopic coupling of CO_2 and water vapour within a *Quercus petraea* forest canopy. *Oecologia* **119**: 109–119.

Helliker B. R., Roden J. S., Cook C. and Ehleringer J. R. (2002) A rapid and precise method for sampling and determining the oxygen isotope ratio of atmospheric water vapor. *Rapid Commun Mass Spectrometry* **16**: 929–932.

Hesterburg R. and Siegenthaler U. (1991) Production and stable isotopic composition of CO_2 in soil near Bern, Switzerland. *Tellus* **43**B: 197–205.

Ishizawa M., Nakazawa T. and Higuchi K. (2002) A multi-box model study of the role of the biospheric metabolism in the recent decline of $\delta^{18}O$ in atmospheric CO_2. *Tellus* **54**B: 307–324.

Keeling R. (1995) The atmospheric oxygen cycle: The oxygen isotopes of atmospheric CO_2 and O_2 and the O_2/N_2 ratio. *Rev Geophys* **33**: 1253–1262.

Miller J. B., Yakir D., White J. W. C. and Tans P. P. (1999) Measurement of O-18/O-16 in the soil–atmosphere CO_2 flux. *Global Biogeochem Cycles* **13**: 761–774.

O'Leary M. H. (1981) Carbon isotope fractionation in plants. *Phytochemistry* **20**: 553–567.

Peylin P., Ciais P., Denning A. S., Tans P. P., Berry J. A. and White J. W. C. (1999) A 3-dimensional study of $\delta^{18}O$ in atmospheric CO_2: Contribution of different land ecosystems. *Tellus* **51**B: 642–667.

Riley W. J., Still C. J., Torn M. S. and Berry J. A. (2002) A mechanistic model of $H_2{}^{18}O$ and $C^{18}OO$ fluxes between ecosystems and the atmosphere: Model description and sensitivity analyses. *Global Biogeochem Cycles* **16**: doi: 10.1029/2002GB001878.

Riley W. J., Still C. J., Helliker B. R., Ribas-Carbo M. and Berry J. A. (2003) ^{18}O composition of CO_2 and H_2O ecosystem pools and fluxes: Simulations and comparisons to measurements. *Global Change Biol* **9**: 1567–1581.

Roden J. S. and Ehleringer J. R. (1999) Observations of hydrogen and oxygen isotopes in leaf water confirm the Craig–Gordon model under wide-ranging environmental conditions. *Plant Physiol* **120**: 1165–1173.

Stern L. A., Amundson R. and Baisden W. T. (1999) Processes controlling the oxygen isotope ratio of soil CO_2: Analytic and numerical modeling. *Geochim Cosmochim Acta* **63**: 799–814.

Stern L. A., Amundson R. and Baisden W. T. (2001) Influence of soils on oxygen isotope ratio of atmospheric CO_2. *Global Biogeochem Cycles* **15**: 753–759.

Still C. J., Berry J. A., Ribas-Carbo M. and Helliker B. R. (2003) The contribution of C_3 and C_4 plants to the carbon cycle of a tallgrass prairie: An isotopic approach. *Oecologia* **136**: 347–359.

Suyker A. E. and Verma S. B. (2001) Year-round observations of the net ecosystem exchange of carbon dioxide in a native tallgrass prairie. *Global Change Biol* **7**: 279–289.

Suyker A. E., Verma S. B. and Burba G. G. (2003) Interannual variability in net CO_2 exchange of a native tallgrass prairie. *Global Change Biol* **9**: 255–265.

Tans P. P. (1998) Oxygen isotopic equilibrium between carbon dioxide and water in soils. *Tellus* **50**B: 163–178.

Williams T. G., Flanagan L. B. and Coleman J. R. (1996) Photosynthetic gas exchange and discrimination against $^{13}CO_2$ and $C^{18}O^{16}O$ in tobacco plants modified by an antisense construct to have low chloroplastic carbonic anhydrase. *Plant Physiol* **112**: 319–326.

Yakir D. and Wang X.-F. (1996) Fluxes of CO_2 and water between terrestrial vegetation and the atmosphere estimated from isotope measurements. *Nature* **381**: 515–518.

Yakir D. and Sternberg L. D. L. (2000) The use of stable isotopes to study ecosystem gas exchange. *Oecologia* **123**: 297–311.

10

Ecosystem CO_2 Exchange and Variation in the $\delta^{18}O$ of Atmospheric CO_2

Lawrence B. Flanagan

Introduction

Changes in the composition of the atmosphere have important effects on the functioning of terrestrial ecosystems (Canadell *et al.*, 2000). Elevated carbon dioxide levels cause changes to ecosystem photosynthesis and water flux (Drake *et al.*, 1997), and in turn ecosystem activity influences the atmosphere by causing seasonal fluctuations and latitudinal gradients in the concentration of atmospheric CO_2 (Flanagan and Ehleringer, 1998; Fung, 2000). Atmospheric monitoring programs have developed as a tool to measure and document changes in the activity of terrestrial ecosystems on large spatial scales (Tans and White, 1998; Canadell *et al.*, 2000). A working hypothesis in earth system science is that variation in the uptake and release of carbon dioxide in terrestrial ecosystems is largely responsible for the substantial year-to-year variation in the annual rate of increase in atmospheric CO_2 concentration (Fung, 2000). Measurements of the stable carbon isotope composition of atmospheric CO_2 provide additional information to help separate out the relative contribution of oceanic and terrestrial processes to seasonal and interannual variation in atmospheric CO_2 concentration (Tans and White, 1998; Chapters 13 and 14). In addition, the oxygen isotope ratio of CO_2 has been promoted as a tool to differentiate terrestrial ecosystem photosynthesis and respiration for their effects on the composition of atmospheric CO_2 (Farquhar *et al.*, 1993). The independent information recorded by carbon and oxygen isotopes of atmospheric CO_2 reflect their different mechanisms of isotope fractionation during photosynthesis and respiration (see Chapter 14) and this is illustrated by the contrasting patterns of seasonal change for these two isotopic species (Fig. 10.1).

One important pattern that has been recorded in data collected by the Climate Monitoring and Diagnostics Laboratory (National Oceanic and

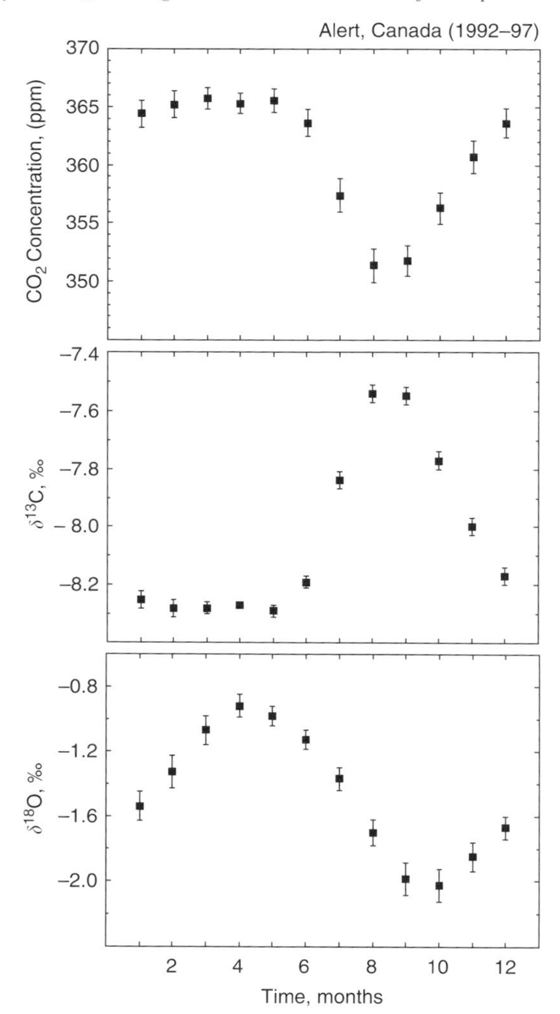

Figure 10.1 Average seasonal changes in the concentration and stable isotope composition of atmospheric carbon dioxide at Alert, Canada during 1992–1997. The data were obtained from the Climate Monitoring and Diagnostics Laboratory, National Oceanic and Atmospheric Administration, United States Department of Commerce (available at: http://www.cmdl.noaa.gov/index.html).

Atmospheric Administration, United States Department of Commerce) flask network is the systematic decline of 0.08‰ yr^{-1} in the $\delta^{18}O$ value of atmospheric CO_2, at least during 1993–1997 (Fig. 10.2). A number of mechanisms can contribute to this pattern. First, the combustion of biomass and fossil fuel releases carbon dioxide into the atmosphere with

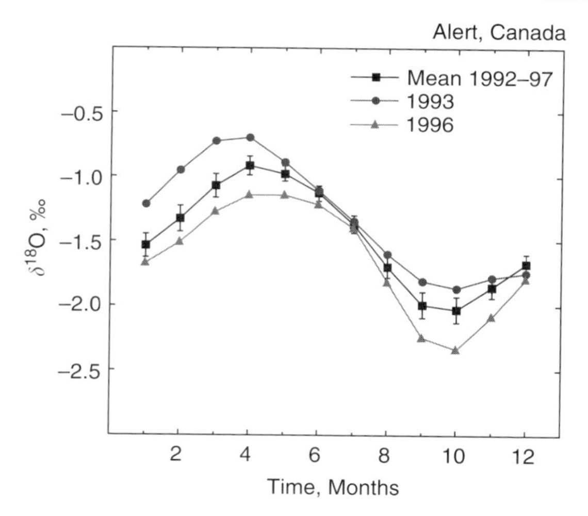

Figure 10.2 Decline in the oxygen isotope composition ($\delta^{18}O$‰, PDB) of atmospheric CO_2 at Alert, Canada during 1993–1996. The monthly values for 1993 and 1996 are compared to the average seasonal pattern during 1992 through 1997. The data were obtained from the Climate Monitoring and Diagnostics Laboratory, National Oceanic and Atmospheric Administration, United States Department of Commerce (available at: http://www.cmdl.noaa.gov/index.html).

a $\delta^{18}O$ (PDB) value of −17‰, and with a total atmospheric forcing of approximately −0.05‰ yr^{-1} (Ciais *et al.*, 1997a; Stern et al., 2001). Second, Gillon and Yakir (2001) suggested that increased conversion of tropical forests (C_3 vegetation) to pasture grasses (C_4 vegetation) could cause a decline of 0.02‰ yr^{-1} in the $\delta^{18}O$ of atmospheric CO_2. This change is a result of the low carbonic anhydrase activity in C_4 plants that limits the equilibration of CO_2 and water in chloroplasts and, therefore, reduces the apparent discrimination against CO_2 molecules containing ^{18}O during photosynthetic gas exchange (Williams *et al.*, 1996; Gillon and Yakir, 2001; Ometto *et al.*, 2004). Stern *et al.* (2001) proposed that other land use changes associated with the expansion of agriculture have increased CO_2 fluxes from the soil to the atmosphere and that this might account for the remaining negative forcing on the atmosphere of approximately 0.01‰ yr^{-1}. There is, however, a great deal of uncertainty in these estimates of the effect of land-use change on the $\delta^{18}O$ of atmospheric CO_2.

An alternative approach to the problem was taken by Ishizawa *et al.* (2002), who used model calculations to illustrate that an increase in both photosynthesis and total ecosystem respiration in the northern hemisphere could cause the observed decline in $\delta^{18}O$ value of atmospheric CO_2, if this change was accompanied by lower $C^{18}OO$ discrimination during photosynthesis.

A decline in photosynthetic discrimination can occur because of larger stomatal limitation of photosynthesis and the associated reduction in the ratio of CO_2 concentration in the chloroplast and that in the ambient atmosphere. Alternatively, a reduction in photosynthetic discrimination could be caused by lower leaf water $\delta^{18}O$ values. While it is possible to conceive of a set of environmental conditions that might result in higher photosynthesis and respiration in northern hemisphere land ecosystems and result in lower photosynthetic discrimination, a limitation of the Ishizawa *et al.* (2002) analysis is that they do not relate their model calculations to actual shifts in environmental conditions that were apparent during the period 1993 to 1997. In this sense there is no mechanistic connection between the current understanding of controls on isotopic fractionation and ecosystem CO_2 fluxes and the observed trend in the $\delta^{18}O$ of atmospheric CO_2.

In this chapter I describe a case study of ecosystem CO_2 exchange and associated isotope effects in a northern, temperate grassland that makes direct mechanistic links between interannual environmental changes and their consequences for variation in the oxygen isotope composition of atmospheric CO_2. This study is complementary to that of Ishizawa *et al.* (2002) and consistent with their general conclusions, but includes a more detailed mechanistic analysis of the interactions involved at the ecosystem level.

Oxygen Isotope Effects during Ecosystem CO_2 Exchange

Several recent studies have documented that terrestrial ecosystems exert the dominant influence on the oxygen isotope composition of atmospheric CO_2 (Ciais *et al.*, 1997a,b; Peylin *et al.*, 1999; Cuntz *et al.*, 2003a,b). The previous chapter of this book described the relevant isotope effects involved. The $^{18}O/^{16}O$ composition of atmospheric CO_2 is dependent on these isotopic fractionation processes, as well as the magnitude of the one-way CO_2 fluxes, gross photosynthesis, and total ecosystem respiration that occur during ecosystem–atmosphere CO_2 exchange. In this study I made use of net ecosystem CO_2 exchange (NEE) data that were collected using the eddy covariance technique (Flanagan *et al.*, 2002). Using the micrometeorological sign convention, gross primary production (GPP) photosynthesis is negative and represents removal of CO_2 from the atmosphere, while total ecosystem respiration (TER) is positive, and represents a gain of carbon dioxide by the atmosphere.

$$\mathrm{NEE} = \mathrm{GPP} + \mathrm{TER} \tag{10.1}$$

Both of the gross fluxes (GPP and TER) have associated isotope compositions that normally represent opposite effects on the $^{18}O/^{16}O$ composition of atmospheric CO_2. The net isotope effect on the atmosphere is termed a net isoflux and was calculated as:

$$\text{Net Isoflux} = \text{GPP}(\delta_a - \Delta) + \text{TER}\,\delta_R \qquad (10.2)$$

where δ_a is the $\delta^{18}O$ value of atmospheric CO_2 (this parameter is held constant at 0‰ [PDB] for this analysis); Δ is discrimination against $C^{18}OO$ that occurs during photosynthetic gas exchange; and δ_R is the $\delta^{18}O$ value of CO_2 respired by plants and soil in the ecosystem. In order to estimate seasonal and interannual changes in the isofluxes for $C^{18}OO$, the continuous eddy covariance measurements of NEE were used to calculate daily-integrated values of GPP and TER, as described in detail by Flanagan *et al.* (2002). The daily average isotope effect associated with GPP and TER was estimated as described below. The apparent discrimination against $C^{18}OO$ during photosynthesis (Δ) was calculated as:

$$\Delta = a + \varepsilon[\theta_{eq}(\delta_e - \delta_a) - (1 - \theta_{eq})a/(\varepsilon + 1)] \qquad (10.3)$$

where a is the average fractionation during diffusion of $C^{18}OO$ from the air into the chloroplast (assumed constant at 7.4‰); $\varepsilon = c_i/(c_a - c_i)$, c_a being the concentration of atmospheric CO_2 (assumed constant at 370 μmol mol^{-1}) and c_i the concentration of CO_2 inside the chloroplast; θ_{eq} is the proportional extent of isotopic equilibrium between oxygen in CO_2 and oxygen in chloroplast H_2O (assumed constant at 0.7); and δ_e is the $\delta^{18}O$ value of CO_2 in the chloroplast. Measurements of the $\delta^{13}C$ of plant biomass samples were used to estimate seasonal changes in the integrated value of c_i by using Eqs 6 and 8 in Farquhar *et al.* (1989) and assuming that discrimination by Rubisco was 27‰, fractionation during diffusion of CO_2 into the leaf was 4.4‰, and the $\delta^{13}C$ of source CO_2 was −8‰. Linear interpolation was used between measurements of plant biomass $\delta^{13}C$ made at regular intervals during each growing season in order to estimate a daily-integrated value for c_i. The Craig and Gordon (1965) model of evaporative enrichment was used to estimate the value of chloroplast water as described by Flanagan *et al.* (1991, 1997). The leaf water model required input values of stem water $\delta^{18}O$, which were estimated from measurements of the $\delta^{18}O$ of soil CO_2 (see below), and the $\delta^{18}O$ of atmospheric water vapor, which was assumed to be constant at −28.3‰ (Standard and Mean Ocean Water, SMOW). The leaf water $\delta^{18}O$ values were calculated for 1-hour intervals during the growing season, and weighted by gross photosynthesis values to obtain daytime averaged leaf water $\delta^{18}O$ values. The $\delta^{18}O$ value of CO_2 in equilibrium with chloroplast water was calculated using the temperature-dependent fractionation factor for CO_2–H_2O exchange as described by Flanagan *et al.* (1997). The daily noon air temperatures and the daytime average (gross photosynthesis weighted) $\delta^{18}O$ value of leaf water were used to calculate the daytime average $\delta^{18}O$ of chloroplast CO_2.

The oxygen isotopic composition of CO_2 released during ecosystem respiration (δ_R) was calculated as the $\delta^{18}O$ value of soil CO_2 minus 7.3‰ (Miller *et al.*, 1999), to account for fractionation that occurs during diffusion of $C^{18}OO$ out of the soil. We made regular measurements of the $\delta^{18}O$ value of soil CO_2 at 10 cm depth during each growing season. This approach assumes that during the day when photosynthesis is active, respiratory CO_2 is only released from the soil. In addition, it assumes that in the dark when photosynthesis is not active, CO_2 released from plant tissues has the same oxygen isotopic composition as that released from the soil. This would require that leaf water (and other plant water) would have the same $\delta^{18}O$ value as stem and soil water at night.

Variation in Environmental Conditions and Associated Changes in Ecosystem Isofluxes

Comparisons are made here among three years with progressively lower amounts of precipitation. Ecosystem CO_2 fluxes and productivity in grassland ecosystems are strongly dependent on growing season precipitation, with the growing season defined as April through August in this system (Flanagan *et al.*, 2002). During our study, precipitation received during 1999 (268 mm) was quite close to the 30-year average for the site (236 ± 86, mean $\pm$ SD, $n = 30$). However, the subsequent 2 years received summer rainfall that was significantly below normal (123 mm in 2000 and 107 mm during 2001). Drought has the potential to influence ecosystem isofluxes in several ways. First, during water stress the magnitude of both ecosystem photosynthesis and respiration is reduced during the growing season. Second, increased stomatal limitation of photosynthetic gas exchange during water stress can result in lower CO_2 concentrations in the chloroplast and consequently reduced discrimination against $C^{18}OO$ during photosynthesis. Potentially counteracting this mechanism for drought-induced decline in $C^{18}OO$ discrimination is an increase in the enrichment of ^{18}O in leaf water caused by higher leaf–air vapor pressure deficits during warmer and drier environmental conditions. Apparent discrimination against $C^{18}OO$ during photosynthesis will increase in association with the ^{18}O content of leaf water. Analyses in our grassland ecosystem suggest that higher leaf water $\delta^{18}O$ values, apparent under drought conditions, have a stronger positive effect on photosynthetic discrimination than the reduction in discrimination caused by lower chloroplast CO_2 concentrations (Fig. 10.3). So there was a progressive increase in ^{18}O discrimination during ecosystem photosynthesis from 1999 through the 2001 growing seasons.

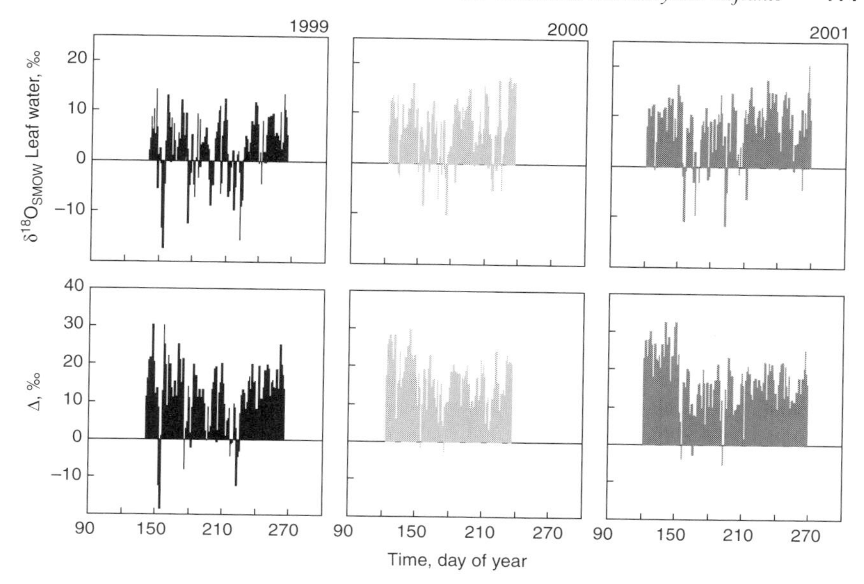

Figure 10.3 (Top) Comparison of daily-average oxygen isotope composition ($\delta^{18}O$‰, SMOW) of leaf water. (Bottom) Comparison of daily-average discrimination against $C^{18}OO$ molecules during photosynthetic gas exchange (Δ‰). Measurements were made in a grassland ecosystem near Lethbridge, Canada during 1999 through 2001.

Isotope effects during ecosystem respiration can also be altered by drought conditions. Evaporation of water from the soil can be a larger fraction of total ecosystem evapo-transpiration and as a consequence soil water in the shallow depths (0–10 cm) can become enriched in ^{18}O. As a result the isotope ratio of soil respired CO_2 could have higher $\delta^{18}O$ values relative to times when soil moisture is abundant. Our measurements illustrated higher $\delta^{18}O$ values for soil CO_2 during 2000 and 2001 compared to 1999, consistent with this mechanism (Fig. 10.4). The evaporative enrichment of ^{18}O in soil water would also alter the source water taken up by plants and contribute to higher leaf water $\delta^{18}O$ values observed under the drier conditions of 2000 and 2001 (Fig. 10.3).

Comparison of the net isofluxes, and the associated photosynthesis and respiration isofluxes, revealed significant differences among the three study years (Fig. 10.5). During the period of active plant growth when ecosystem photosynthetic CO_2 uptake exceeded the loss of CO_2 by respiration, the ecosystem net isoflux had a positive forcing that increased the $\delta^{18}O$ of atmospheric CO_2 in this ecosystem. The magnitude of the positive forcing was increased as water availability declined during 2000 and 2001 because of the higher leaf water $\delta^{18}O$ values. In addition, the ecosystem respiration isoflux was higher in drought years despite the lower respiration rates

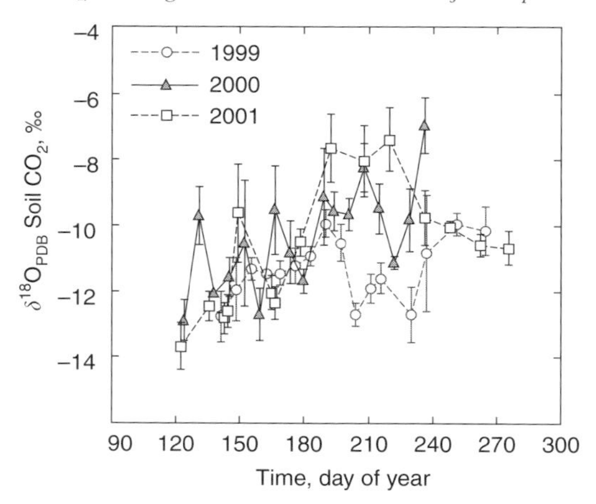

Figure 10.4 Comparison of seasonal changes in the oxygen isotope composition ($\delta^{18}O$‰, PDB) of soil CO_2 (at 10 cm soil depth) in a grassland ecosystem near Lethbridge, Canada during 1999 through 2001.

because of the increase in $\delta^{18}O$ of soil water. So the higher positive net forcing on the atmosphere under dry environmental conditions was caused by an increased positive photosynthetic isoflux and a less negative respiration isoflux. The less-negative respiration isoflux was a consequence of the reduced respiration rates under drought and the associated higher $\delta^{18}O$ values of ecosystem respired CO_2. The cumulative effect of these changes in net isoflux illustrates that drought should cause a positive forcing and increase the $\delta^{18}O$ value of atmosphere during ecosystem CO_2 exchange (Fig. 10.6).

Implications for the Declining Trend in $\delta^{18}O$ of Atmospheric CO_2

While it is unwise to generalize the conclusions of this case study in a single grassland ecosystem to a global pattern in atmospheric CO_2, there are useful mechanistic insights obtained from this study. This analysis showed that reduced ecosystem CO_2 exchange, coupled with a range of associated isotope effects, resulted in an increased positive forcing on the $\delta^{18}O$ value of atmospheric CO_2 as water stress increased (Fig. 10.6). As a consequence it is reasonable to predict that an increase in water availability would result in higher ecosystem productivity and a decline in the positive forcing, which in turn could contribute to a reduction in the $\delta^{18}O$ value of atmospheric CO_2. This is consistent with the global model analysis developed by Ishizawa *et al.* (2002), who suggested that higher ecosystem productivity, higher ecosystem respiration, and lower photosynthetic discrimination in

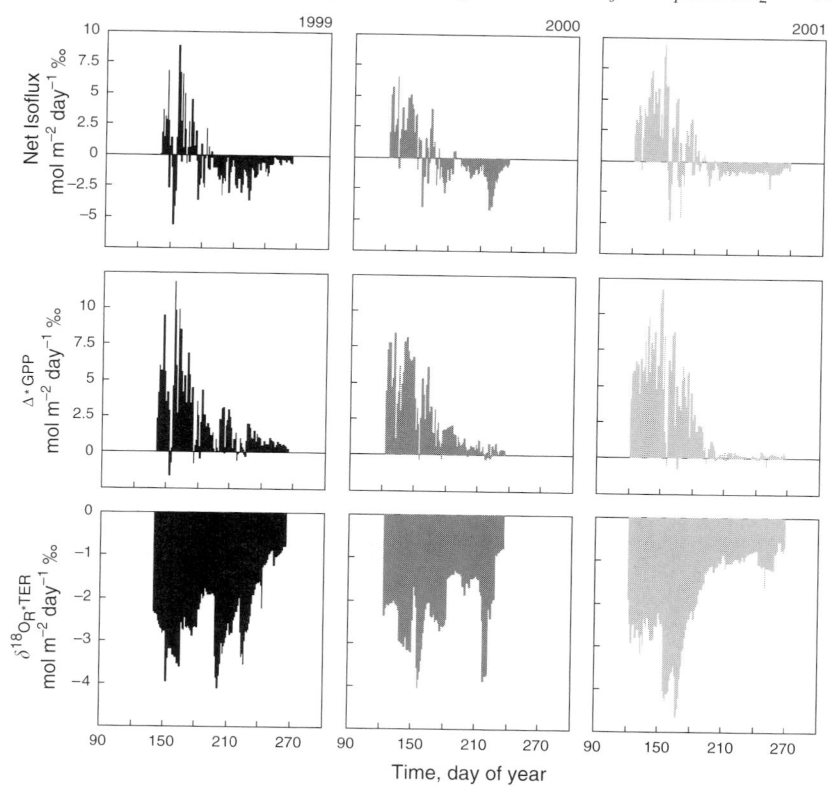

Figure 10.5 Comparison of seasonal variation in the gross one-way isofluxes (for photosynthesis and respiration) and the net isofluxes for ^{18}O in CO_2. Measurements were made in a grassland ecosystem near Lethbridge, Canada during 1999 through 2001.

the northern hemisphere could explain the approximately 0.5‰ reduction in $\delta^{18}O$ value of atmospheric CO_2 observed during 1993–1997. However, Ishizawa *et al.* (2002) suggested that the reduced photosynthetic discrimination was caused by lower chloroplast CO_2 concentrations, while our analysis for the grassland ecosystem indicated that change in the oxygen isotope ratio of leaf water was primarily responsible for the lower photosynthetic discrimination. Normally shifts in environmental factors that cause an increase in photosynthesis also cause an associated rise in stomatal conductance and higher chloroplast CO_2 concentration, although nitrogen fertilization may be an exception to this generalization. The difference in the mechanism proposed here relative to that suggested by Ishizawa *et al.* (2002) is significant for the application of atmospheric measurements as a tool in earth system science. Further analysis of the problem is warranted and can be productively approached using global-scale models that include all the

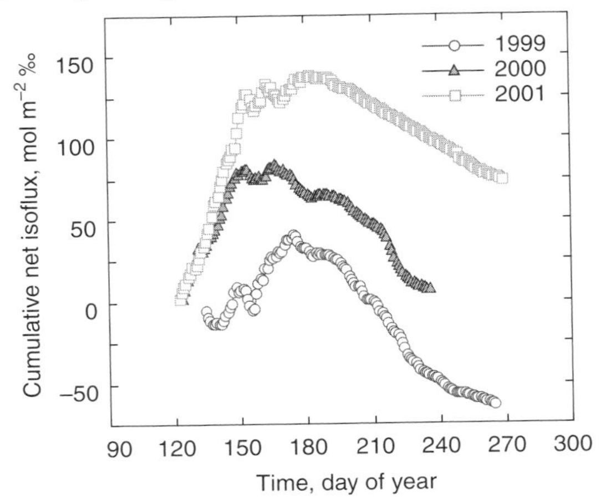

Figure 10.6 Comparison of the cumulative net isofluxes for ^{18}O in CO_2. Measurements were made in a grassland ecosystem near Lethbridge, Canada during 1999 through 2001.

important mechanistic details revealed from theoretical and experimental studies in physiological ecology (Riley *et al.*, 2002; Chapter 9).

References

Canadell J., Mooney H. A., Baldocchi D. D., Berry J. A., Ehleringer J. R., Field C. B., Gower S. T., Hollinger D. V., Hunt J. E., Jackson R. B., Running S. W., Shaver G. R., Steffen W., Trumbore S. E., Valentini R. and Bond B. V. (2000) Carbon metabolism of the terrestrial biosphere: A multitechnique approach for improved understanding. *Ecosystems* **3**: 115–130.

Ciais P., Denning A. S., Tans P. P., Berry J. A., Randall D. A., Collatz G. J., Sellers P. J., White J. W. C., Trolier M., Meijer H. A. J., Francey R. J., Monfray P. and Heimann M. (1997a) A three-dimensional synthesis study of $\delta^{18}O$ in atmospheric CO_2 1. Surface fluxes. *J Geophys Res* **102**: 5857–5872.

Ciais P., Tans P. P., Denning A. S., Francey R. J., Trolier M., Meijer H. A. J., White J. W. C., Berry J. A., Randall D. A., Collatz G. J., Sellers P. J., Monfray P. and Heimann M. (1997b) A three-dimensional synthesis study of $\delta^{18}O$ in atmospheric CO_2 2. Simulations with the TM2 transport model. *J Geophys Res* **102**: 5873–5883.

Craig H. and Gordon L. I. (1965) Deuterium and oxygen-18 variations in the ocean and the marine atmosphere. In *Proceedings of a Conference on Stable Isotopes in Oceanographic Studies and Paleotemperatures* (E. Tongiorgi, ed.) pp. 9–130. Spoleto, Italy. Lischi & Figli, Pisa.

Cuntz M., Ciais P., Hoffmann G. and Knorr W. (2003a) A comprehensive global 3D model of $\delta^{18}O$ in atmospheric CO_2: Validation of surface processes. *J Geophys Res* **108**: doi: 10.1029/2002JD003153.

Cuntz M., Ciais P., Hoffmann G., Allison C. E., Francey R. J., Knorr W., Tans P. P., White J. W. C. and Levin I. (2003b) A comprehensive global 3D model of $\delta^{18}O$ in atmospheric CO_2: 2. Mapping the atmospheric signal. *J Geophys Res* **108**: doi: 10.1029/2002JD003154.

Drake B. G., Gonzalez-Meler M. A. and Long S. P. (1997) More efficient plants: A consequence of rising atmospheric CO_2? *Annu Rev Plant Physiol Plant Mol Biol* **48**: 609–639.

Farquhar G. D., Ehleringer J. R. and Hubick K. T. (1989) Carbon isotope discrimination and photosynthesis. *Annu Rev Plant Physiol Plant Mol Biol* **40**: 503–537.

Farquhar G. D., Lloyd J., Taylor J. A., Flanagan L. B., Syvertsen J. P., Hubick K. T., Wong S. C. and Ehleringer J. R. (1993) Vegetation effects on the isotope composition of oxygen in atmospheric CO_2. *Nature* **363**: 439–443.

Flanagan L. B., Comstock J. P. and Ehleringer J. R. (1991) Comparison of modeled and observed environmental influences on the stable oxygen and hydrogen isotope composition of leaf water in *Phaseolus vulgaris* L. *Plant Physiol* **96**: 588–596.

Flanagan L. B. and Ehleringer J. R. (1998) Ecosystem-atmosphere CO_2 exchange: Interpreting signals of change using stable isotope ratios. *Trends Ecol Evol* **13**: 10–14.

Flanagan L. B., Brooks J. R., Varney G. T. and Ehleringer J. R. (1997) Discrimination against $C^{18}O^{16}O$ during photosynthesis and the oxygen isotope ratio of respired CO_2 in boreal forest ecosystems. *Global Biogeochem Cycles* **11**: 83–98.

Flanagan L. B., Wever L. A. and Carlson P. J. (2002) Seasonal and interannual variation in carbon dioxide exchange and carbon balance in a northern temperate grassland. *Global Change Biol* **8**: 599–615.

Fung I. (2000) Climate change: Variable carbon sinks. *Science* **290**: 1313.

Gillon J. and Yakir D. (2001) Influence of carbonic anhydrase activity in terrestrial vegetation on the ^{18}O content of atmospheric CO_2. *Science* **291**: 2584–2587.

Ishizawa M., Nakazawa T. and Higuchi K. (2002) A multi-box model study of the role of the biosphere metabolism in the recent decline of $\delta^{18}O$ in atmospheric CO_2. *Tellus* **54**B: 307–324.

Miller J. B., Yakir D., White J. W. C. and Tans P. P. (1999) Measurement of $^{18}O/^{16}O$ in the soil–atmosphere CO_2 flux. *Global Biogeochem Cycles* **13**: 761–774.

Ometto J. P. H. B., Flanagan L. B., Martinelli L. A. and Ehleringer J. R. (2004) Oxygen isotope ratios of waters and respired CO_2 in Amazonian forest and pasture ecosystems. *Ecol Appl* **14**: in press.

Peylin P., Ciais P., Denning A. S., Tans P. P., Berry J. A. and White J. W. C. (1999) A 3-dimensional study of the $\delta^{18}O$ in atmospheric CO_2: Contribution of different land ecosystems. *Tellus* **51**B: 642–667.

Riley W. J., Still C. J., Torn M. S. and Berry J. A. (2002) A mechanistic model of $H_2{}^{18}O$ and $C^{18}OO$ fluxes between ecosystems and the atmosphere: Model description and sensitivity analyses. *Global Biogeochem Cycles* **16**: doi: 10.1029/2002GB001878.

Stern L. A., Amundson R. and Baisden W. T. (2001) Influence of soils on oxygen isotope ratio of atmospheric CO_2. *Global Biogeochem Cycles* **15**: 753–759.

Tans P. P. and White J. W. C. (1998) In balance, with a little help from the plants. *Science* **281**: 183–184.

Williams T. G., Flanagan L. B. and Coleman J. R. (1996) Photosynthetic gas exchange and discrimination against $^{13}CO_2$ and $C^{18}OO$ in tobacco plants modified by an antisense construct to have low chloroplastic carbonic anhydrase. *Plant Physiol* **112**: 319–326.

11

Stable Isotope Constraints on Net Ecosystem Production Under Elevated CO_2

Elise Pendall, Jennifer Y. King, Arvin R. Moser, Jack Morgan, Daniel Milchunas

Introduction: Approaches for Estimating Net Ecosystem Production

Net ecosystem production (NEP) is the total carbon accumulated in an ecosystem over a time interval, usually a year. The C is stored above- and belowground in plants, animals, and soils, and has residence times ranging from less than one to several hundred years or more. NEP varies in response to disturbance and during plant succession, and human-induced perturbations to climate and atmospheric constituents also affect NEP at local to global scales. Because NEP determines how much carbon dioxide is taken up from the atmosphere by terrestrial ecosystems, it is critical to evaluate what forcing factors might be regulating variations in NEP at various temporal and spatial scales. In particular, the effects of elevated atmospheric CO_2 may be altering NEP in many ecosystems, and techniques such as stable isotope analysis are improving our ability to evaluate the magnitude of these effects (Pendall, 2002).

NEP has often been defined operationally, depending on the approach used to conduct measurements and the temporal and spatial scales of interest, but this has led to ambiguity and a lack of comparability among studies (Randerson *et al.*, 2002). NEP was first estimated from changes in ecosystem pool sizes, after accounting for lateral fluxes (Olsen, 1963; Bormann *et al.*, 1974; Aber and Melillo, 1991; Chapin *et al.*, 2002):

$$NEP = (\Delta plant + \Delta animal + \Delta SOM)/\Delta t \pm F_{lateral} \qquad (11.1)$$

It has been recognized that generally the changes in these inventories are extremely small relative to the entire pool size (particularly for SOM), making accurate measurement difficult unless long time intervals

(10 years or more) are allowed between sampling events. This has led to the development of flux-based methods, which allow NEP estimates to be made at more frequent intervals. A complete accounting can be formulated as the sum of fluxes into and out of the ecosystem (Olsen, 1963; Randerson *et al.*, 2002):

$$\begin{aligned} NEP &= F_{in} + F_{out} \\ &\approx F_{GPP} + F_{R_e} + F_{fire} + F_{leaching} + F_{erosion} + F_{hydrocarbons} \\ &\quad + F_{herbivory} + F_{harvest} \end{aligned} \tag{11.2}$$

where GPP is gross primary production and R_e is ecosystem respiration. Net ecosystem exchange (NEE) has been defined as the balance of GPP and R_e, and in the absence of fire or other disturbance, NEE provides an approximation of NEP at limited temporal and spatial scales (Chapin *et al.*, 2002). NEE is measured at the ecosystem scale by eddy covariance techniques (e.g., Wofsy *et al.*, 1993), or at the plot scale in flow-through chambers by monitoring the change in CO_2 concentration going into and coming out of the chamber (Drake *et al.*, 1996). NEP has also been formulated as the difference between net primary production (NPP) and heterotrophic respiration (R_h):

$$NPP = GPP - R_a \tag{11.3}$$

$$R_e = R_h + R_a \tag{11.4}$$

$$NEP = GPP - R_e = NPP - R_h \tag{11.5}$$

GPP and R_a (autotrophic respiration) are not generally measured, but NPP can be measured directly at the plot scale, or estimated from models based on leaf area index at regional to global scales. Empirical algorithms have been developed to estimate R_h; these are generally based on a relationship between microbial respiration and soil temperature (and sometimes moisture; Chapin *et al.*, 2002). This approach also assumes that the smaller fluxes, such as leaching and erosional losses, can be neglected.

All of the methods used for estimating NEP (or NEE) have limitations. It appears that uptake measured by NEE in open-top chambers is often higher than can be accounted for by measurements of above- and belowground biomass (Drake *et al.*, 1996; Niklaus *et al.*, 2000). This has been termed a 'locally missing' carbon sink (Cheng *et al.*, 2000; Niklaus *et al.*, 2000), and may in part be caused by over-pressurization of chambers suppressing respiration, or by leaching or volatilization of C compounds. There are known issues with eddy covariance technology, particularly with respect to estimation of the respiration flux. Nighttime fluxes can be underrepresented if wind conditions are stable, or if air drainage patterns cause CO_2 to be lost laterally from the system. Inventories miss a portion of the biomass to herbivory, senescence during the growing season, and incomplete sampling, especially belowground. Improvements in NEE methodology and in belowground inventory methods are narrowing the gap between inventory and exchange techniques.

It is inherently difficult to track small changes in NEP if C is being stored belowground, in part because many experiments are often not long enough (2–5 years) to produce changes in soil organic carbon (SOC) detectable using standard techniques. In certain situations, stable C isotopes improve the resolution with which C storage belowground can be detected. Recently-fixed C entering the soil pool has an isotopic signature distinct from pre-existing SOC in many experiments on elevated CO_2, as well as in cropping situations where a C_4 crop has been planted into a C_3 soil, or vice-versa (Balesdent *et al.*, 1988; Leavitt *et al.*, 1994). In these cases, 'new' inputs from rhizodeposition (root exudates and turnover of fine roots and mycorrhizal hyphae) can be quantified (Hungate *et al.*, 1997; Van Kessel *et al.*, 2000; Leavitt *et al.*, 2001). Rhizodeposition is often too small to measure without isotopic techniques, and may thus constitute some portion of the 'missing' C accumulating in terrestrial ecosystems (Pendall *et al.*, 2004 in review).

Carbon dioxide derived from decomposition of SOC will have a $\delta^{13}C$ value different from that of rhizosphere respiration in the same situations where there is an isotopic disequilibrium between new inputs and pre-existing SOC (Rochette and Flanagan, 1997; Pendall *et al.*, 2001b). This has allowed partitioning of the total soil respiration flux into its components of root/rhizosphere respiration and decomposition. No method exists to fully separate heterotrophic and autotrophic respiration fluxes in the field; decomposition by rhizosphere microbes is generally included with the root respiration component (Hanson, 2000). We suggest that in the case of estimating NEP, C loss by rapid turnover of recent (current year) photosynthate in the rhizosphere is functionally equivalent to autotrophic respiration, and the main C decomposition loss of interest is that which has a residence time of 1 year or longer.

Here we demonstrate a method that relies on biomass measurements and stable C isotopes for evaluating NEP in an experiment on elevated CO_2. This approach is a modification of Eq. 11.5 in that we explicitly include measurement of rhizodeposition to the SOM pool, which is otherwise not included in the inventory:

$$\mathrm{NEP} = \mathrm{AG} + \mathrm{BG} + \mathrm{NSC} - \mathrm{R_h} \tag{11.6}$$

where AG is annual aboveground C increment, or NPP, BG is annual belowground NPP, NSC is annual 'new' soil C input or rhizodeposition, and R_h is annual C loss by decomposition. We maintain that inventory estimates of NEP should include rigorous estimates of both decomposition and rhizodeposition. This approach would be suitable in any situation where an isotopic disequilibrium exists between recently fixed C and pre-existing soil organic matter, such as C_3–C_4 land-use or land-cover change. Stable C isotopes allow a more complete accounting of small increases in the soil carbon pool that are contributed by rhizodeposition, and also allow us to estimate decomposition losses by partitioning rhizosphere respiration and SOM decomposition.

Experimental Approach

An elevated CO_2 experiment was conducted in the shortgrass steppe region of northeastern Colorado, at the USDA-ARS Central Plains Experimental Range (CPER; latitude 40°40′ N, longitude 104°45′ W), ~55 km northeast of Fort Collins. The most abundant species at the study site were the C_4 grass, *Bouteloua gracilis* (H.B.K) Lag. (blue grama), and the C_3 grasses *Stipa comata* Trin and Rupr. (needle-and-thread grass) and *Pascopyrum smithii* (Rydb.) A. Love (western wheatgrass). Root biomass (including crowns) is responsible for ~70% of net primary production (NPP) in this ecosystem (Milchunas and Lauenroth, 2001). Beginning in 1997, open-top chambers (OTC; 4.5 m diameter) were used to evaluate the effects of CO_2 on the shortgrass steppe (SGS) ecosystem, with three replicate chambers at ambient ($360 \pm 20\ \mu mol\ mol^{-1}$) and elevated ($720 \pm 20\ \mu mol\ mol^{-1}$) CO_2. Three unchambered plots of the same area allowed evaluation of any chamber effects. Chambers were placed on the plots before growth started in late March or early April, and removed at the end of the growing season in late October. Blowers with ambient or elevated CO_2 ran continuously. The experimental and chamber design has been described in detail by Morgan *et al.* (2001).

Figure 11.1 shows the $\delta^{13}C$ values of ecosystem components in elevated and ambient CO_2 chambers at the SGS OTC experiment in 1999. Non-chambered plots had values that were not significantly different from those with the AC treatment. The tank gas used to double the atmospheric CO_2 concentration had an approximate $\delta^{13}C$ value of −40‰ during 1999, which produced air in the elevated chambers of -24.7 ± 1.4‰, which compared with background air $\delta^{13}C$ values of -8.1 ± 0.2‰ (Pendall *et al.*, 2003). The ^{13}C-depleted air in the EC treatment produced depleted plant material, and by 1999 (after 3 years of elevated CO_2), the tracer was evident in soil organic matter. In the AC and NC plots, an isotopic disequilibrium between currently growing plants and SOM of ~4‰ was attributed to a reduction in grazing c. 20 years prior to the start of our experiment (grazing was eliminated completely 2 years prior). Reduction of grazing pressure in the SGS tends to favor C_3 over C_4 grasses (Lauenroth and Milchunas, 1988).

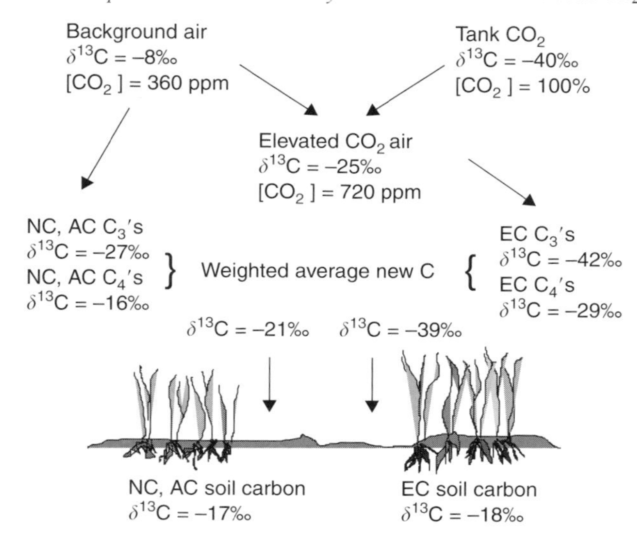

Figure 11.1 Stable C isotope composition of ecosystem components in the shortgrass steppe open-top chamber experiment for 1999.

Biomass Inventory Methods

In grassland ecosystems, aboveground biomass harvests are relatively straightforward. At the end of the growing season in October, aboveground biomass was clipped at the root crown from $1.5\,m^2$ subplots, dried, and weighed. Because plots were clipped to the ground the prior fall, this harvest represents the annual aboveground NPP. We estimate that herbivory losses were less than 5% in all treatments; very occasional pesticide use was needed to reduce aphids. Different subplots were harvested in July and separated by species but this was not practical for plants collected in the fall. Tissue C content was determined by elemental analysis.

Belowground biomass inventories are typically labor-intensive and associated with a wide margin of error. The standing crop of roots in perennial grasslands may have a turnover time of 4–6 years (Milchunas and Lauenroth, 2001), and thus root ingrowth cylinders were used to assess the annual increment of root growth. In the fall prior to the growing season, two 20 cm diameter soil cores were removed from each plot, and replaced with 15 cm diameter PVC pipes to a depth of 40 cm. The outer diameter of each core was lined with plastic mesh (1 mm square openings) and the 5 cm band gap was filled with sieved soil from near the experimental site. At the end of the following growing season, the PVC liners were carefully removed, roots were cut from the mesh liner, and all soil and roots were returned to the laboratory. Roots were separated from soil by flotation and washing over a 1 mm sieve, dried, ground, and analyzed for C content on an elemental analyzer. Subsamples of roots were ashed at 450°C to determine the ash-free C content.

Both above- and belowground NPP inventory methods tend to underestimate NPP, owing to losses such as herbivory and tissue senescence (Chapin *et al.*, 2002). In the case of belowground NPP, our method of accounting for rhizodeposition should at least partly remedy this situation.

Rhizodeposition or 'New' Soil Carbon

Rhizodeposition is soil C derived from turnover of fine roots, secretion of soluble root exudates, and turnover of rhizosphere-associated microbial biomass. This C increment is generally too small to measure on an annual basis using total C analysis because it is within the measurement error, but by using an isotopic tracer, small changes may be detected in 1–2 years (Leavitt *et al.*, 2001). The fraction of C that has been contributed by rhizodeposition (F_{new}) is determined using the two-part mixing model described by Balesdent *et al.* (1988):

$$\delta^{13}C_{SOC} = F_{new}(\delta^{13}C_{new}) + (1 - F_{new})(\delta^{13}C_{old}) \quad (11.7)$$

which can be rearranged to:

$$F_{new} = (\delta^{13}C_{SOC} - \delta^{13}C_{old})/(\delta^{13}C_{new} - \delta^{13}C_{old}) \quad (11.8)$$

In order to use stable isotopes to estimate 'new' C inputs from rhizodeposition to the soil system, it is necessary to define the isotopic 'end-members,' i.e., the $\delta^{13}C$ value of pre-existing SOM and that of the new C being added. Determination of the end-member isotopic values is not a trivial task in many ecosystems, because $\delta^{13}C$ of SOM often increases with depth; therefore, depth-dependent 'old' end-member values must be used. In some ecosystems, the $\delta^{13}C$ of 'new' input C is also challenging to determine. Mixed C_3–C_4 systems require assumptions of the proportion of C allocated belowground as well as the different isotopic values of the different functional groups.

In the case of the SGS OTC experiment, we assumed that the soils on the non-chambered (NC) plots remained at steady-state over the course of the experiment, and that the NC soils represented pre-existing C across the site. (This was confirmed by non-significant changes in $\delta^{13}C$ values in NC soils over time.) Our NC soils therefore provided the old C end-member values. An alternative approach is to use samples collected across the site prior to establishing an experiment (Leavitt *et al.*, 1994). Collecting and analyzing steady-state soils together with treated soils provides analytical control over multi-year experiments. We collected two soil cores (40 cm deep, 15 cm diameter) from each plot at the end of each growing season. Cores were collected volumetrically and divided into five depth increments (0–5, 5–10, 10–20, 20–30, and 30–40 cm). Gravel >2 mm and most roots were removed immediately after sampling, with additional root picking done by hand on dried samples. Soils were then ground to a fine powder for analysis by elemental analyzer-mass spectrometry (EA-MS) for C content and $\delta^{13}C$ value, with precision of 0.1% and 0.2‰, respectively. Soils were carbonate-free to a depth of 40 cm, and we therefore did not acidify the samples.

In the mixed C_3–C_4 shortgrass steppe, the isotopic composition of new inputs can potentially vary spatially and temporally. Aboveground biomass clipping of 1.5 m^2 subplots was done in July 1999, at peak green biomass. Leaves and stems were separated by species, dried, ground, and analyzed for C content and $\delta^{13}C$ value by elemental analyzer-IRMS. Roots were harvested by species at the end of the experiment in 2001, and averaged 1.4‰ heavier than leaves (Pendall *et al.*, 2004). We used the 1999 leaf $\delta^{13}C$ values, weighted them by the 1999 aboveground biomass amounts, and then added 1.4‰ for the 'new' C end-member signatures ($\delta^{13}C_{new}$; Table 11.1). July harvest samples were assumed to represent the proportion of C_3 and C_4 species for the entire growing season; samples harvested at the end of the season (peak total biomass) were not easily separated by species. The three dominant species maintain similar shoot : root ratios at ambient and elevated CO_2, and therefore we assumed that the $C_3 : C_4$ ratio of rhizodeposition was proportional to aboveground biomass (Morgan *et al.*, 1994).

F_{new} was multiplied by the total mass of C in each horizon or depth increment to calculate rhizodeposition going into each horizon on a mass basis. This required measurement of bulk density, which can be a particularly challenging measurement in gravelly or organic-rich soils, but was straightforward in our sandy loam soils. Whole-profile rhizodeposition was calculated by summing the depth increments.

Uncertainties in estimating rhizodeposition were determined by accounting for variability in all of the components, including $\delta^{13}C_{SOC}$, $\delta^{13}C_{old}$, and $\delta^{13}C_{new}$, following a first-order Taylor series approach (Phillips and Gregg,

Table 11.1 $\delta^{13}C$ Values of C_3 and C_4 Plants, Proportion of C_3 Biomass, and $\delta^{13}C$ Values of New C Inputs (After Adding 1.4‰ to Account for Average Root ^{13}C Enrichment) During the Growing Seasons of 1999 at the Shortgrass Steppe OTC Experiment

	$\delta^{13}C$, C_4 (‰)	$\delta^{13}C$, C_3 (‰)	C_3 biomass	$\delta^{13}C$ new (‰)
AC	−15.4 (0.21)	−25.2 (0.99)	0.72 (0.03)	−21.1 (1.01)
EC	−33.2 (0.81)	−42.4 (1.92)	0.81 (0.04)	−39.3 (2.09)
NC	−15.4 (0.27)	−26.5 (0.49)	0.67 (0.07)	−21.4 (0.56)

Standard errors in parentheses include error on the concentration and isotopic composition of C in end-members.

2001). Uncertainty estimates for $\delta^{13}C_{new}$ included standard deviations of % biomass and $\delta^{13}C$ of the dominant grass species. We further corrected for covariance of C content and $\delta^{13}C$ values among soil depth intervals when calculating the standard error for the whole profile (Donald L. Phillips, personal communication, 2002).

Decomposition (R_h)

Determination of the in-situ decomposition flux requires first assessment of the total soil respiration flux. We measured soil respiration in this grassland ecosystem by measuring the CO_2 gradient with depth in the soil, and multiplying that by soil diffusivity to obtain the flux (Pendall *et al.*, 2003). This allowed us to obtain a measurement of belowground respiration without disturbing the aboveground vegetation by clipping. In the fall before the growing season, stainless steel tubes (1/8″ OD) were inserted horizontally 15 cm into a pit face at five depths (3, 5, 10, 15, and 25 cm). The pit was then backfilled. Beginning in May, glass syringes, greased with Apiezon M and fitted with gas-tight valves, were used to collect and store the soil gas samples. Volumes of 10 mL were collected at all depths, except for the 3 cm depth, where 6 mL were collected, to ensure that no atmospheric air was pulled into the syringe. A small amount of magnesium perchlorate was used in-line to absorb moisture; this allowed us to analyze the oxygen isotopes as well as the carbon isotopes (Ferretti *et al.*, 2003). Atmosphere samples were also collected on each sampling date, from each chamber or plot into 0.5 L flasks at ~1.5 m above the ground, after flushing ~10 flask volumes. Diffusivity was calculated from soil temperature and moisture using an approach that accounts for inter- and intra-aggregate differences in diffusion rates (Potter *et al.*, 1996).

Soil gas samples were analyzed within 24 hours of collection for CO_2 concentration using an infra-red gas analyzer (Model LI-6251, LICOR, Inc., Lincoln, NE), with a precision of $\pm 3\,\mu mol\,mol^{-1}$ over the concentration range 360–8000 $\mu mol\,mol^{-1}$; most soil gas samples were about 1000 $\mu mol\,mol^{-1}$. Soil gas was analyzed for stable isotopes of CO_2 using gas chromatography-isotope ratio mass spectrometry (GC-IRMS; Isoprime model, Micromass, UK) (Miller *et al.*, 1999). Tests indicated that standards stored in these greased syringes kept for up to one week without significant leakage or isotopic exchange. To ensure a linear response of the mass spectrometer, sample sizes varying from ~7 to ~ 250 μL of soil gas were injected into a carrier gas stream, which was further split before being introduced into the mass spectrometer. This allowed the peak height of the sample to be within ~10% of the peak height of the standard, and precision was better than ±0.1‰. Flasks of atmospheric air were analyzed for CO_2 mixing ratio by infra-red absorption (Conway *et al.*, 1994) and for $\delta^{13}C$ by dual-inlet mass spectrometry (Trolier *et al.*, 1996).

We used the 'Keeling' plot approach to estimate the $\delta^{13}C$ value of soil respiration ($\delta^{13}C_{SR}$) (Keeling, 1958, 1961; Pendall *et al.*, 2001b, 2003), in order to eliminate the influence of variable amounts of atmospheric CO_2 that diffuse down into the soil profile. When $\delta^{13}C$ is plotted against the inverse of the CO_2 concentration, the y-intercept reflects the flux-weighted average $\delta^{13}C$ value of the biological source of CO_2, resulting from root/rhizosphere respiration and microbial decomposition. The intercepts were calculated using geometric means to account for variability in both independent (CO_2) and dependent (^{13}C) variables (Sokal and Rohlf, 1995). We subtracted 4.4‰ from the y-intercept values to account for kinetic fractionation during diffusion because the soil CO_2 was sampled from within the soil rather than from the soil surface (Cerling, 1984; Amundson *et al.*, 1998). This approach assumes that the biological end-member $\delta^{13}C$ value is constant with depth, reflecting a constant proportion of rhizosphere ('new') and microbial ('old') components.

In order to estimate R_h and ultimately NEP from Eq. 11.6, the soil respiration flux (SR) was partitioned into new and old components using a two-component mixing model similar to Eq. 11.8 (Pendall *et al.*, 2001b, 2003):

$$F_{R_h} = (\delta^{13}C_{SR} - \delta^{13}C_{new})/(\delta^{13}C_{R_h} - \delta^{13}C_{new}) \tag{11.9}$$

where F_{R_h} is the proportion of CO_2 generated by microbial decomposition of soil C. The main difference from Eq. 11.8 is that old end-member values ($\delta^{13}C_{R_h}$) were not taken from bulk soil C analyses, because a portion of bulk C is 'passive,' or unavailable to microbial decomposition. We instead used a biophysical fractionation approach to reflect decomposition of pre-existing SOC that was available to microbes. Soil samples collected at the end of the 1999 growing season were subjected to long-term (300 days) laboratory incubations (Paul *et al.*, 2001). During the 'slow-pool' phase, after day 100, an approximate steady-state $\delta^{13}C$ value was achieved. Average $\delta^{13}C$ values ($n = 4$ for each of two depths) of CO_2 evolved during decomposition of slow-pool C were used as the old C end-member signatures ($\delta^{13}C_{R_h}$). For AC, $\delta^{13}C_{R_h}$ was -17.0 ± 0.47; for EC it was -21.9 ± 1.52; for NC it was -17.8 ± 0.48 (Pendall *et al.*, 2001a).

The new C end-member signatures ($\delta^{13}C_{new}$) were determined as described above for rhizodeposition (Table 11.1). An offset of at least 4‰ between old and new end-members allowed partitioning on all treatments. In pure C_3 ecosystems, it is possible that photosynthate delivered to roots varies at daily to weekly timescales, depending on atmospheric vapor pressure deficit, leading to variability in the isotopic composition of soil-respired CO_2 (Ekblad and Högberg, 2001).

The decomposition flux was estimated by multiplying F_{R_h} by the total soil respiration flux. Uncertainties in estimating the proportion of soil respiration from decomposition were determined by accounting for variability in all of the components, including $\delta^{13}C_{SR}$, $\delta^{13}C_{R_h}$, and $\delta^{13}C_{new}$, following a first-order Taylor series approach (Phillips and Gregg, 2001). Uncertainty in $\delta^{13}C_{SR}$ was based on least squares standard errors of the Keeling intercepts (which were determined by geometric means; Pataki *et al.*, 2003). Uncertainty for $\delta^{13}C_{R_h}$ was estimated as the standard error of the average of 4 incubation subsamples over 7 sampling dates between 100 and 250 days after the start of the incubation experiment.

Constructing the Ecosystem Carbon Balance: NEP

We combined our isotopic measurements with the more standard inventory measurements of above- and belowground net primary production to estimate NEP; rhizodeposition was added to NPP, and decomposition was subtracted (Eq. 11.6). Bulk soil carbon content was not significantly different among treatments in 1999, but $\delta^{13}C$ values were (Fig. 11.2). By assuming that NC SOC was at steady state, we assessed changes in AC and EC SOC that resulted from the experimental perturbation (chambers as well as elevated CO_2). The new C increment (F_{new} from Eq. 11.8) in individual horizons in AC soils was 2–3%, or 20–35 g m^{-2}, compared with 3–7%, or 30–45 g m^{-2} in EC soils (Table 11.2). At some depths in the AC treatment, F_{new} was a negative value, because $\delta^{13}C$ values of SOC were more enriched than the NC soil 'old' end-member value. Natural variability sometimes causes the isotopic composition of a mixture to fall outside the range of the end-members, in which case the mixing model is invalidated.

Rhizodeposition was similar in magnitude to belowground NPP, and was significantly higher in the EC treatment than in the AC treatment (Table 11.2). When summed over the top 40 cm of soil, and the 3 years of the experiment, the EC soils had accumulated 197 g C m^{-2} compared with 78 g C m^{-2} in AC soils. We simply divided these values by 3 to obtain the amount accumulated in 1999 (Table 11.2). When propagating the errors through the individual horizons to the whole profile value, we accounted for the variability in isotopic end-members and the mixture (Phillips and Gregg, 2001), and also for covariance among soil depths. This correction had the effect of increasing the standard error on the whole-profile value from that which would be calculated from the square root of the sum of the variances.

Soil respiration rates were significantly higher under elevated CO_2 during dry periods (Table 11.3; Pendall *et al.*, 2003). Soil-respired CO_2 was depleted in ^{13}C in EC plots relative to AC and NC plots, reflecting metabolism

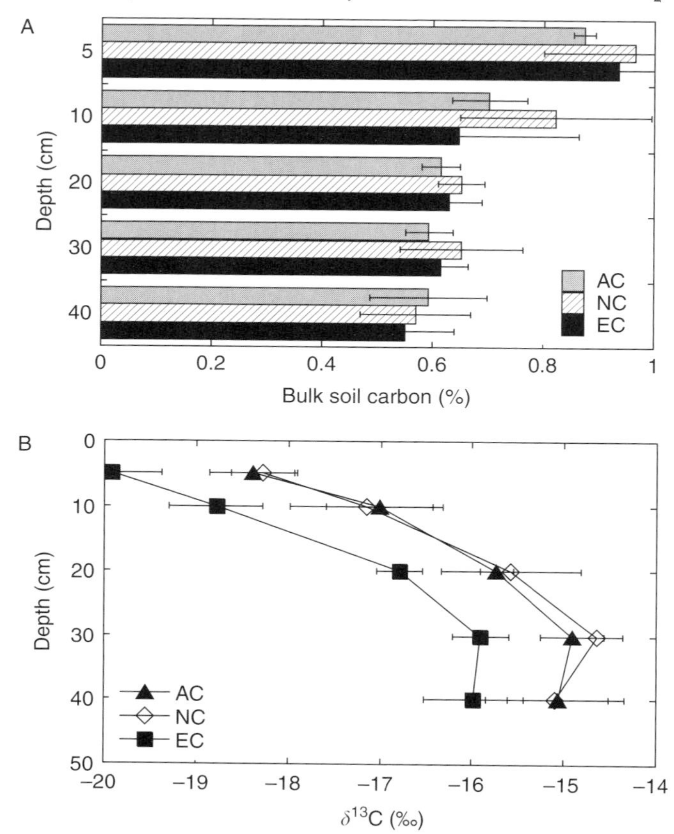

Figure 11.2 (A) Soil C content and (B) $\delta^{13}C$ values in bulk soil collected in October, 1999, after three experimental growing seasons. Ambient CO_2 (AC) and elevated CO_2 (EC) chambers, and non-chambered control plots (NC). Standard errors (bars) for n = 3.

of recently assimilated C (Table 11.3). $\delta^{13}C_{SR}$ values in NC and AC plots were not significantly different from one another. The components of soil respiration, rhizosphere respiration, and decomposition had distinct $\delta^{13}C$ end-member values in all three treatments, which was attributed to reduction of grazing favoring growth and reproduction of C_3 grasses (Pendall *et al.*, 2003). Decomposable organic matter reflected a greater proportion of C_4 carbon, as shown by the relatively enriched microbial incubation $\delta^{13}C$ values.

Elevated CO_2 clearly enhanced decomposition rates throughout the growing season compared with ambient chambers (Fig. 11.3). Annual decomposition losses were estimated by integrating the areas under the curves in Fig. 11.3, assuming the winter fluxes to be negligible (Table 11.4).

Table 11.2 Rhizodeposition (New C) in Ambient CO_2 and Elevated CO_2 Chambers as Estimated From $\delta^{13}C$ Values

Treatment	Depth (cm)	Fraction new C	New C ($g\,m^{-2}$)
AC	0–5	0.02 (.25)	18 (4)
	5–10	−0.03 (.22)	–
	10–20	0.02 (.16)	25 (3)
	20–30	0.03 (.13)	35 (3)
	30–40	−0.004 (.15)	–
EC	0–5	0.07 (.09)	41 (4)
	5–10	0.07 (.09)	35 (3)
	10–20	0.05 (.09)	46 (4)
	20–30	0.05 (.08)	46 (4)
	30–40	0.03 (.08)	29 (3)
	Whole Profile New C ($g\,m^{-2}$)		
	3 years	*1999*	
AC	78 (9)	26 (5)	
EC	197 (9)	66 (5)	

AC, ambient CO_2; EC, elevated CO_2.
Standard errors, calculated following Phillips and Gregg (2001), are in parentheses.
Standard error on Profile New C accumulation corrects for autocorrelation among depth increments in addition to including errors on stable isotopic composition of end-members.

Table 11.3 Soil Respiration Rates and $\delta^{13}C_{SR}$ Values for 1999

		Respiration rate ($mg\,C\,m^{-2}\,h^{-1}$)			$\delta^{13}C_{SR}$ (‰)		
Date	DOY	AC (SE)	NC (SE)	EC (SE)	AC (SE)	NC (SE)	EC (SE)
5/26/99	146	103 (8.6)	126 (15.5)	94 (4.4)	−21.2 (.58)	−21.3 (.54)	−30.7 (.33)
6/16/99	166	90 (32.4)	52 (22.9)	126 (36.3)	−19.8 (.58)	−20.4 (.61)	−32.2 (.98)
6/30/99	181	57 (8.4)	83 (13.8)	76 (21.0)	−20.3 (.68)	−19.5 (.29)	−30.7 (.44)
7/15/99	196	51 (1.5)	76 (7.2)	74 (9.7)	−18.7 (.64)	−18.9 (.26)	−30.6 (.48)
8/3/99	215	90 (11.0)	200 (36.1)	122 (13.3)	−20.3 (.4)	−20.3 (.41)	−33.0 (.61)
9/1/99	244	79 (11.5)	182 (22.3)	93 (17.8)	−20.0 (.55)	−21.5 (.5)	−31.8 (.62)
10/13/99	286	40 (4.7)	47 (5.9)	48 (10.4)	−20.2 (.64)	−20.4 (.64)	−30.7 (.93)
11/17/99	321	17 (2.2)	19 (3.6)	21 (1.4)	−20.8 (.49)	−20.5 (.31)	−29.0 (.84)

Respiration rate was estimated by a flux-gradient approach; $\delta^{13}C_{SR}$ estimated from Keeling plot end-members; standard errors in parentheses. DOY, day of year; AC, ambient chambers; NC, non-chambered plots; EC, elevated CO_2 chambers.

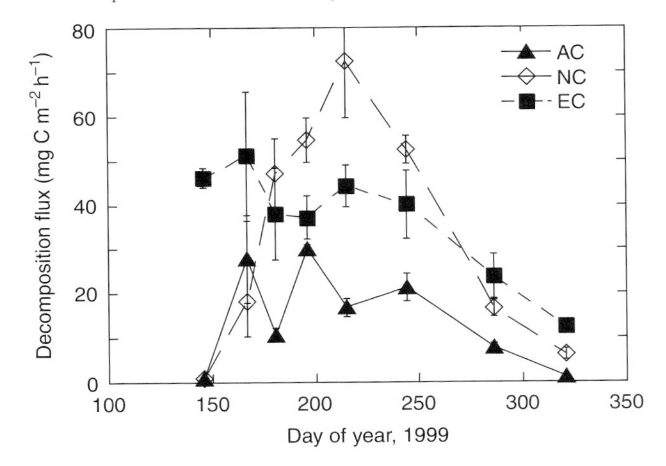

Figure 11.3 Decomposition flux from ambient (AC) and elevated CO_2 (EC) chambers and non-chambered control plots (NC) during 1999. Total soil respiration was multiplied by the fraction from decomposition of old C as estimated from Eq. 11.1. Standard error (bars) included all sources of error: variability of $\delta^{13}C$ values in end-members and mixtures, as well as variability of C_3 and C_4 biomass.

Table 11.4 Components of Net Ecosystem Production (NEP) at the Shortgrass Steppe Elevated CO_2 Experiment in 1999

	EC	AC
AG NPP	74 (9.0)	49 (5.5)
BG NPP	55 (7.3)	40 (6.1)
New soil C	66 (5.2)	26 (4.9)
Decomposition	151 (22.4)	64 (10.2)
NEP	49 (25.7)	51 (14.0)

Values given are g C m^{-2} y^{-1}. EC, elevated CO_2 chambers; AC, ambient CO_2 chambers; AG NPP, aboveground net primary production; BG, belowground; New soil C, also referred to as rhizodeposition.

Errors (in parentheses) were propagated for NEP by taking the square root of the sum of the variance terms for each component (aboveground and belowground biomass, decomposition, and whole-profile rhizodeposition).

The non-chambered plots showed the highest soil respiration and decomposition fluxes, possibly because the soils were slightly drier inside the chambers (Table 11.3; Fig. 11.3). Low decomposition rates early and late in the growing season reflect higher rates of rhizosphere respiration during these moister parts of the growing season. Rhizosphere respiration was positively correlated with soil moisture for all treatments (Pendall *et al.*, 2003).

Elevated CO_2 also altered the temperature sensitivity of decomposition. Significant exponential relationships between soil T and decomposition rate were found for all treatments (Pendall *et al.*, 2003). Interestingly, Q_{10} values for EC soils were lower than for AC soils, with NC soils intermediate. This suggests that decomposition processes under elevated CO_2 are being altered qualitatively in addition to quantitatively, which may relate to an increasing importance of fungal decomposers. Soil xylanase analysis suggests that EC soils have greater fungal activity (E. Kandeler, personal communication), and fungi are known to be generally less temperature sensitive than bacteria (E. Paul, personal communication).

We estimated NEP by summing above- and belowground biomass and rhizodeposition, and subtracting the annual decomposition loss (Fig. 11.4; Table 11.4). During the 1999 growing season, which had 50% greater than average precipitation, NEP was the same on EC and AC treatments, roughly $50\,g\,C\,m^{-2}\,y^{-1}$. Apparently, stimulation of decomposition under elevated CO_2 during a moist year can negate the increased production of biomass in the shortgrass steppe. Carbon cycling rates on this semi-arid grassland were stimulated by elevated CO_2, but no net C accumulated during 1999, as shown by Hungate *et al.* (1997).

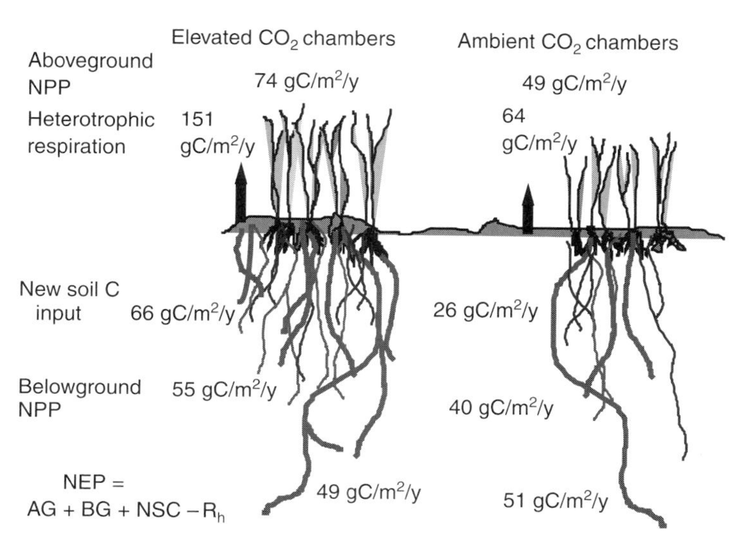

Figure 11.4 Net ecosystem production (NEP) for 1999 (all units are $g\,C\,m^{-2}\,y^{-1}$). Above- and belowground NPP were determined by standard methods. Heterotrophic respiration and new soil C input were determined using stable C isotope tracers.

We defined our experimental plots as the ecosystem boundaries, including soil down to a depth of 40 cm. We have not accounted for leaching losses, but in our semi-arid region this is not likely to be a significant portion of the C balance. Volatilization of organic compounds is also expected to be small; methane is always taken up by the shortgrass steppe ecosystem (Mosier *et al.*, 1991). These factors contribute to unquantified errors in our NEP estimate (Randerson *et al.*, 2002).

Conclusions

Stable C isotopes provided a unique method of tracing C movement through individual components of this shortgrass steppe ecosystem. These analyses enabled us to assign estimates for C pools and fluxes which were otherwise difficult to measure directly, and therefore allowed us to constrain estimates of NEP. Based on the measurements made in our study on the shortgrass steppe, we found that elevated CO_2 stimulated not only plant productivity but rhizodeposition and decomposition fluxes, thus resulting in no overall difference in NEP between ambient and elevated CO_2 during a moist growing season.

Acknowledgments

Generous laboratory support was provided by J. W. C. White, P. P. Tans, the CU-INSTAAR Stable Isotope Lab, and NOAA-CMDL Carbon Cycle Group. The manuscript was improved by the comments of an anonymous reviewer. This research was funded by the Biological and Environmental Research Program (BER), US Department of Energy, through the Great Plains Regional Center of the National Institute for Global Environmental Change (NIGEC) under Cooperative Agreement No. DE-FC03-90ER61010; by the Terrestrial Ecology and Climate Change Initiative, grant NSF-IBN-9524068; by USDA/NRICGP-98-134; and by NSF DEB-9708596; with base support from USDA/ARS.

References

Aber J. and Melillo J. (1991) *Terrestrial Ecosystems.* Saunders College Publishing, Orlando.

Amundson R., Stern L., Baisden T. and Wang Y. (1998) The isotopic composition of soil and soil-respired CO_2. *Geoderma* **82**: 83–114.

Balesdent J., Wagner G. and Mariotti A. (1988) Soil organic-matter turnover in long-term field experiments as revealed by C-13 natural abundance. *Soil Sci Soc Am J* **52**: 118–124.

Bormann F., Likens G., Siccama T., Pierce R. and Eaton J. (1974) The export of nutrients and recovery of stable conditions following deforestation at Hubbard Brook. *Ecol Monogr* **44**: 255–277.

Cerling T. E. (1984) The stable isotopic composition of modern soil carbonate and its relationship to climate. *Earth Planet Sci Lett* **71**: 229–240.

Chapin F. I., Matson P. and Mooney H. (2002) *Principles of Terrestrial Ecosystem Ecology*. Springer-Verlag, New York.

Cheng W., Sims D., Luo Y., Johnson D., Ball J. and Coleman J. (2000) Carbon budgeting in plant-soil mesocosms under elevated CO_2: Locally missing carbon? *Global Change Biol* **6**: 99–109.

Conway T. J., Tans P. P., Waterman L. S., Thoning K. W., Kitzis D. A., Masarie K. A. and Zhang N. (1994) Evidence of interannual variability of the carbon cycle from the NOAA/CMDL global air sampling network. *J Geophys Res* **99**: 22831–22855.

Drake B., Muehe M., Peresta G., Gonzalez-Meler M. and Matamala R. (1996) Acclimation of photosynthesis, respiration and ecosystem carbon flux of a wetland on Chesapeake Bay, Maryland to elevated atmospheric CO_2 concentration. *Plant Soil* **187**: 111–118.

Ekblad A. and Högberg P. (2001) Natural abundance of ^{13}C in CO_2 respired from forest soils reveals speed of link between photosynthesis and root respiration. *Oecologia* **127**: 305–308.

Ferretti D., Pendall E., Morgan J. A., Nelson J., LeCain D. and Mosier A. R. (2003) Partitioning evapotranspiration fluxes from a Colorado grassland using stable isotopes: Seasonal variations and ecosystem implications of elevated atmospheric CO_2. *Plant Soil* **254**: 291–303.

Hanson P., Edwards N., Garten C. and Andrews J. (2000) Separating root and soil microbial contributions to soil respiration: A review of methods and observations. *Biogeochemistry* **48**: 115–146.

Hungate B. A., Holland E. A., Jackson R. B., Chapin F. S. I., Mooney H. A. and Field C. B. (1997) The fate of carbon in grasslands under carbon dioxide enrichment. *Nature* **388**: 576–579.

Keeling C. D. (1958) The concentration and isotopic abundance of atmospheric carbon dioxide in rural areas. *Geochim Cosmochim Acta* **13**: 322–334.

Keeling C. D. (1961) The concentration and isotopic abundance of carbon dioxide in rural and marine air. *Geochim Cosmochim Acta* **24**: 277–298.

Lauenroth W. and Milchunas D. (1988) Shortgrass steppe. In *Natural Grasslands* (R. Coupland, ed.) pp. 183–226. Elsevier, Amsterdam.

Leavitt S. W., Paul E. A., Kimball B. A., Hendrey G. R., MauneyJ. R., Rauschkolb R., Rogers H. J., Lewin K. F., Pinter P. J. and Johnson H. B. (1994) Carbon isotope dynamics of CO_2-enriched FACE cotton and soils. *Agric For Meteorol* **70**: 87–101.

Leavitt S., Pendall E., Paul E., Brooks T., Kimball B., Pinter P., Johnson H. and Wall G. (2001) Stable-carbon isotopes and soil organic carbon in the 1996 and 1997 FACE wheat experiments. *New Phytol* **150**: 305–314.

Milchunas D., Sala O. and Lauenroth W. (1988) A generalized model of the effects of grazing by large herbivores on grassland community structure. *Am Naturalist* **132**: 87–106.

Milchunas D. and Lauenroth W. (2001) Belowground primary production by carbon isotope decay and long-term root biomass dynamics. *Ecosystems* **4**: 139–150.

Miller J., Yakir D., White J. and Tans P. (1999) Measurement of $^{18}O/^{16}O$ in the soil–atmosphere CO_2 flux. *Global Biogeochem Cycles* **13**: 761–774.

Morgan J., Hunt H., Monz C. and LeCain D. R. (1994) Consequences of growth at two carbon dioxide concentrations and two temperatures for leaf gas exchange in *Pascopyrum smithii* (C3) and *Bouteloua gracilis* (C4). *Plant Cell Environ* **17**: 1023–1033.

Morgan J., LeCain D., Mosier A. and Milchunas D. (2001) Elevated CO_2 enhances water relations and productivity and affects gas exchange in C3 and C4 grasses of the Colorado shortgrass steppe. *Global Change Biol* **7**: 451–466.

Mosier A. R., Schimel D. S., Valentine D., Bronson K. and Parton W. J. (1991) Methane and nitrous oxide fluxes in native, fertilized and cultivated grasslands. *Nature* **350**: 330–332.

Niklaus, P., Stocker R., Korner C. and Leadley P. (2000) CO_2 flux estimates tend to overestimate ecosystem C sequestration at elevated CO_2. *Funct Ecol* **14**: 546–559.

Olsen J. (1963) Energy storage and the balance of producers and decomposers in ecologial systems. *Ecology* **44**: 322–331.

Pataki, D., Ehleringer J., Flanagan L., Yakir D., Bowling D., Still C., Buchmann N., Kaplan J. and Berry J. (2003) The application and interpretation of Keeling plots in terrestrial carbon cycle research. *Global Biogeochem Cycles* **17**: 1022, doi: 10.1029/2001/gb001850.

Paul E. A., Morris S. J. and Bohm S. (2001) The determination of soil C pool sized and turnover rates: Biophysical fractionation and tracers. In *Assessment Methods for Soil Carbon* (R. Lal, ed.) pp. 193–206. Lewis Publishers, Boca Raton.

Pendall E. (2002) Where does all the carbon go? The missing sink. *New Phytol* **153**: 207–210.

Pendall E., King J., Mosier A., Morgan J. and White J. (2001a) Changes in turnover rates and pool sizes of soil organic matter under elevated CO_2 on the shortgrass steppe. *EOS Trans Suppl* **82**: F155.

Pendall E., Leavitt S. W., Brooks T., Kimball B. A., Pinter P. J. Jr., Wall G. W., LaMorte R., Wechsung G., Wechsung F., Adamsen F., Matthias A. D. and Thompson T. L. (2001b) Elevated CO_2 stimulates soil respiration in a FACE wheat field. *Basic Appl Ecol* **2**: 193–201.

Pendall E., Del Grosso S., King J., LeCain D., Milchunas D., Morgan J., Mosier A., Ojima D., Parton W., Tans P. and White J. (2003) Elevated atmospheric CO_2 effects and soil water feedbacks on soil respiration components in a colorado grassland. *Global Biogeochem Cycles* **17**: doi: 10.1029/2001GB001821.

Pendall E., Morgan J. and Mosier A. (2004) Rhizodeposition stimulated by elevated CO_2 in a semi-arid grassland. *New Phytol* **162**: 447–458.

Phillips D. L. and Gregg J. W. (2001) Uncertainty in source partitioning using stable isotopes. *Oecologia* **128**: 304, doi: 10.1007/s004420100723.

Potter C., Davidson E. and Verchot L. (1996) Estimation of global biogeochemical controls and seasonality in soil methane consumption. *Chemosphere* **32**: 2219–2245.

Randerson J., Chapin F. I., Harden J., Neff J. and Harmon M. (2002) Net ecosystem production: A comprehensive measure of net carbon accumulation by ecosystems. *Ecol Appl* **12**: 937–947.

Rochette P. and Flanagan L. (1997) Quantifying rhizosphere respiration in a corn crop under field conditions. *Soil Sci Soc Am J* **61**: 466–474.

Sokal R. R. and Rohlf F. J. (1995) *Biometry*. W. H. Freeman and Co., New York.

Trolier M., White J. W. C., Tans P. P., Masarie K. A. and Gemery P. A. (1996) Monitoring the isotopic composition of atmospheric CO_2: Measurements from the NOAA global air sampling network. *J Geophys Res* **101**: 25897–25916.

Van Kessel C., Horwath W., Hartwig U., Harris D. and Luscher A. (2000) Net soil carbon input under ambient and elevated CO_2 concentrations: Isotopic evidence after 4 years. *Global Change Biol* **6**: 435–444.

Wofsy S., Goulden M. and Munger J. (1993) Net exchange of CO_2 in a mid-latitude forest. *Science* **260**: 1314–1317.

12

Stable Isotopes as a Tool in Urban Ecology

Diane E. Pataki, Susan E. Bush, James R. Ehleringer

Introduction

Urban areas currently constitute one of the least understood terrestrial surfaces with regard to land–atmosphere exchange. Despite their small areal extent (2% of the land surface), cities and urban regions have an enormous impact on the atmosphere, as well as on regional and global biogeochemical cycles (Grimm *et al.*, 2000). Fossil fuel emissions, water consumption, food consumption, and waste production are often concentrated in urban areas, which are characterized by high rates of CO_2 emissions, transpiration, and evaporation (Douglas, 1983; Decker *et al.*, 2000). The impacts of urban land–atmosphere exchange on the global atmosphere and biogeochemistry will become increasingly important, as human population growth is expected to be concentrated largely in and around cities (United Nations, 2000). Improving our understanding of the dynamics of material flow through urban regions and the interactions between urbanization and land–atmosphere exchange is critical if we are to consider current and future human alterations to the global carbon and hydrologic cycles.

Isotopes are a natural fit to integrative, atmospheric studies over cities because of the differences in isotopic composition between trace gases resulting from human activities and plant and soil processes. These signatures can be used to identify individual components of trace gas sources and assess the influence of urban pollutants on plants and soils. Identifying the sources and influences of urban trace gases is a critical first step toward understanding urban ecosystem function. Unlike land–atmosphere exchange in natural ecosystems, urban land–atmosphere exchange is strongly influenced by human activities and behavior as well as natural processes such as plant and soil gas exchange (Grimm *et al.*, 2000; Pickett *et al.*, 2001). In order to gain a complete understanding of factors influencing emissions of greenhouse gases and pollutants that have both an anthropogenic and a plant/soil component, we must separate these contributions and evaluate the social, institutional, and environmental factors that influence their temporal and spatial distribution.

Here we discuss previous applications of isotopic sampling of urban air and urban plant biomass for partitioning trace gas emissions and improving our understanding of urban land–atmosphere interactions. We will focus on carbon dioxide and the carbon isotope composition of plant material, although similar methods may be applied to other isotopes and constituents of urban ecosystems.

The Isotopic Composition of Urban CO_2 Sources

The isotopic composition of fossil fuel emissions is of great interest in atmospheric studies of the carbon cycle as well as local studies of urban CO_2 sources (Andres *et al.*, 2000). Emissions of CO_2 from fossil fuel combustion are naturally more depleted in ^{13}C than is atmospheric CO_2, and cause ^{13}C dilution of the atmosphere (Keeling *et al.*, 1979). In order to utilize isotopic tracers to estimate the role of biological uptake of atmospheric CO_2 at the global scale, the contributions of fossil fuels must be removed. Two major sources of fossil-derived CO_2 are gasoline and natural gas combustion, which are generally isotopically distinct. In addition, plants and soils in cities comprise an 'urban forest,' which also contributes to local carbon cycling. Distinguishing between these CO_2 sources can improve estimates of the seasonal and spatial distribution of fossil fuel isotopic composition for regional and global atmospheric studies, as well as provide insight into local urban ecosystem function.

Both coal and petroleum deposits are derived from biological sources and have $\delta^{13}C$ values of approximately -21 to $-32‰$ (Deines, 1980). The carbon isotope composition of various fossil fuels has been estimated globally by Tans (1981) and Andres *et al.* (2000), who considered sources from combustion of coal, natural gas, and petroleum, as well as gas flaring by oil refineries and CO_2 emissions in cement production (Table 12.1). They report that while $\delta^{13}C$ of coal is fairly constant at $-24.1‰$ regardless of origin, the $\delta^{13}C$ of petroleum exhibits variation depending on the location of its source. Andres *et al.* (2000) estimated that petroleum utilized by the world's leading oil-producing nations may be as heavy as -26.1 or as light as $-30.0‰$.

Approximately 80% of commercial natural gas originates from thermogenic (non-biological) processes in buried organic matter (Rice and Claypool, 1981). The largest constituent of natural gas is methane, which exhibits a wide natural range of $\delta^{13}C$ values, but is generally more depleted in ^{13}C than the products of photosynthesis. Methane of thermogenic origin has $\delta^{13}C$ values ranging from about -60 to $-20‰$, while bacterial-derived methane may range from -100 to $-40‰$ (Schoell, 1988). Natural gas

Table 12.1 Globally Averaged Values (‰, V-PDB) for the Carbon Isotope Composition ($\delta^{13}C$) of Fossil Fuel

Fossil fuel emission source	Tans (1981)	Andres *et al.* (2000)	Observed range
Coal	−24.1	−24.1	−20 to −27[1]
Petroleum	−26.5	−26.5	−19 to −35[2,4]
Natural gas	−41	−44	−20 to −100[3]
Gas flaring	−41	−40	−20 to −60[3*]
Cement production	0	0	–

From Tans (1981) and Andres *et al.* (2000).
*The range of values reported for methane of thermogenic, non-biological origin.
Observed ranges are reported from Jeffery *et al.* (1955)[1], Yeh and Epstein (1981)[2], Tans (1981), Schoell (1988)[3], and Andres *et al.* (2000)[4].

also contains small quantities of ethane, butane, propane, and other gases which are isotopically heavier than methane (Deines, 1980; Tans, 1981). Tans (1981) and Andres *et al.* (2000) applied this range of values to obtain weighted global estimates of −41 and −44‰, respectively, for the $\delta^{13}C$ of global natural gas combustion.

Urban regions with natural gas-fired electrical power plants and/or that contain a large proportion of natural gas furnaces for residential heating may be associated with relatively ^{13}C depleted atmospheric CO_2. Oil refineries that distill petroleum into gasoline also emit ^{13}C-depleted CO_2 during gas flaring–the combustion of low molecular weight compounds (Deines, 1980; Andres *et al.*, 2000). There is a great potential to distinguish between combustion of gasoline/coal and natural gas over urban areas using carbon isotopes, although local studies of the isotopic composition of these fossil fuels are required in each case due to the potential for geographic variability, particularly in natural gas. Data from Salt Lake City, USA indicated that emissions from local gasoline and natural gas combustion were isotopically distinct by approximately 10‰ (Table 12.2). Gasoline and diesel exhaust were significantly different, but the difference was small in absolute terms (0.7‰). These results were within 2‰ of measurements made in Paris, France (Widory and Javoy, 2003). In contrast, a study in Dallas, USA showed that gasoline and natural gas emissions differed by almost 15‰ at that location (Clark-Thorne and Yapp, 2003; Table 12.2).

The $\delta^{13}C$ of ecosystem respiration ($\delta^{13}C_R$) has been well studied in natural ecosystems, particularly in temperate forests. Pataki *et al.* (2003b) reviewed 137 measurements of $\delta^{13}C_R$ made in more than 40 C_3 ecosystems (primarily forests) and found that values ranged from −21.4 to −28.9‰ with an average of −26.2‰. Analogous results for urban plant and soil respiration are not available. Two opposing factors may influence $\delta^{13}C_R$ in urban areas: physiological stress caused by atmospheric and soil pollutants may reduce

Table 12.2 The Carbon Isotope Composition ($\delta^{13}C$, V-PDB) of Automobile and Residential Natural Gas Furnace Exhaust in Salt Lake City, Utah*, Paris, France†, and Dallas, TX‡

	Salt Lake City, USA		Paris, France		Dallas, USA	
Exhaust source	$\delta^{13}C$ (‰)	n	$\delta^{13}C$ (‰)	n	$\delta^{13}C$ (‰)	n
Gasoline, vehicle	−27.9(0.1)[a]	42	−28.7(0.2)	10	−27.2(0.1)	3
Diesel, vehicle	−28.6(0.1)[b]	39	−28.9(0.1)	7		
Natural gas, furnace	−37.8(0.3)[c]	6	−39.2(0.5)	6		
Gas, laboratory	−37.3(0.7)[c]	6	−38.4	1	−42.0	1

*Bush *et al.*, unpublished observations.
†Widory and Javoy (2003).
‡Clark-Thorne and Yapp (2003).
Standard errors are given in parenthesis. The sample size n refers to the number of vehicles or furnaces sampled, and letters indicate statistical differences by least squares difference for the Salt Lake City data ($\alpha = 0.05$). Wall gas refers to combustion of the natural gas supplied to the laboratory by the local utility.

the ratio of intercellular to ambient CO_2 and increase expected values of $\delta^{13}C_R$, while ^{13}C dilution of the urban atmosphere by fossil fuel combustion may decrease $\delta^{13}C_R$ if plants incorporate depleted source air into biomass (Kiyosu and Kidoguchi, 2000; Dongarrà and Varrica, 2002; Lichtfouse *et al.*, 2003).

The $\delta^{18}O$ of ecosystem respiration ($\delta^{18}O_R$) is strongly influenced by the oxygen isotope composition of soil and leaf water, as respiratory CO_2 equilibrates with water, a process which is accelerated by the presence of carbonic anhydrase in leaves and microbes in soils (Gillon and Yakir, 2001). While soil water is generally more ^{18}O-enriched at the soil surface than at depth due to evaporation (Barnes and Allison, 1983), Miller *et al.* (1999) found that the effective depth for isotopic exchange between soil CO_2 and soil water was between 5 and 15 cm. Soil water at this depth is less ^{18}O-enriched than leaf water, which is enriched relative to soil water because of the evaporative effects of transpiration (Dongmann *et al.*, 1974; Flanagan and Ehleringer, 1991). Therefore, $\delta^{18}O_R$ is affected by the relative balance of contributions from leaf and soil respiration, which can be offset by as much as 40‰ (Fig. 12.1). Estimates of the proportional contributions of leaf and soil respiration to the total biological signal can be difficult to obtain, and are currently not available at all for urban ecosystems. For the example of Salt Lake City, USA, which has soil water values close to −15‰, applying an estimate of equal contributions of leaf and soil respiration to total respired CO_2 results in values of $\delta^{18}O_R$ that are distinct from the expected value of combustion (Fig. 12.1).

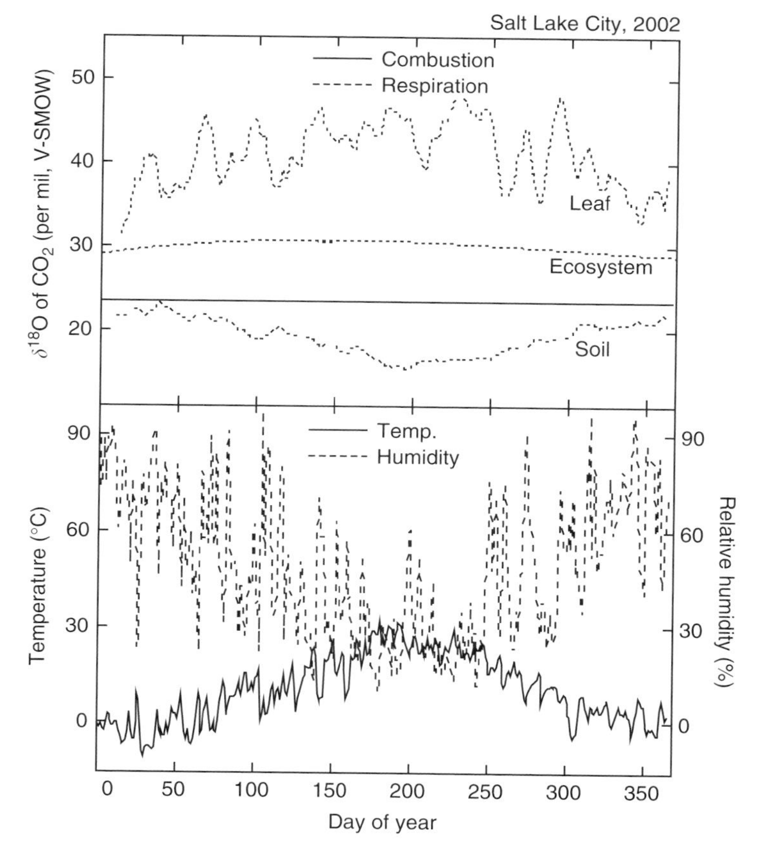

Figure 12.1 *Upper panel*—The oxygen isotope ratio ($\delta^{18}O$) of urban CO_2 sources in Salt Lake City, USA. Plant and soil respiration were modeled hourly according to (Flanagan *et al.*, 1997) with climate data, a source water value of −15‰, and a water vapor value of −21‰; the 10-day running average of nighttime means are shown. Also shown is the annual trend for total nighttime biological respiration if leaf and soil respiration contribute in equal proportions. The solid line shows the expected $\delta^{18}O$ of combustion, equal to $\delta^{18}O$ of diatomic oxygen in the atmosphere (Kroopnick and Craig, 1972). Modified from Pataki *et al.* (2003a). *Lower panel*—Average nighttime temperature and relative humidity for the same period.

Partitioning Urban CO_2 Sources

Isotopes of CO_2 have been used to distinguish between a number of components of ecosystem carbon cycles, including photosynthesis vs respiration (Yakir and Wang, 1996; Bowling *et al.*, 2001) and respiration of plants utilizing the C_3 vs C_4 photosynthetic pathways (Still *et al.*, 2003). In urban ecosystems, ^{14}C is a useful tracer for separating fossil fuel-derived CO_2

from more recent carbon released in plant respiration, as fossil sources contain no ^{14}C. Studies utilizing a combination of the ^{13}C and ^{14}C tracers in urban areas to quantify the relative proportions of respiration and fossil fuel CO_2 sources have found significant contributions from plant and soil respiration (Zondervan and Meijer, 1996; Meijer *et al.*, 1997; Takahashi *et al.*, 2001, 2002). Although these results do not provide direct information about the net carbon balance of urban forests, they indicate that plant and soil processes can be detected in the urban atmosphere, despite the large influence of fossil fuel-derived CO_2 emissions and the reduction in vegetative cover relative to natural forests in some regions. Therefore, isotope partitioning studies may provide the basis for quantifying the role of vegetation in the functioning of urban ecosystems.

The previous sections indicated that the ^{13}C and ^{18}O contents of fossil fuel combustion and plant/soil respiration are often distinct. Florkowski *et al.* (1998) utilized measurements of the carbon and oxygen isotope ratio of CO_2 measured over Krakow, Poland to partition the total CO_2 mixing ratio by simple mass balance:

$$c_\tau = c_A + c_R + c_F \tag{12.1}$$

$$\delta^{13}C_\tau c_\tau = \delta^{13}C_A c_A + \delta^{13}C_R c_R + \delta^{13}C_F c_F \tag{12.2}$$

$$\delta^{18}O_\tau c_\tau = \delta^{18}O_A c_A + \delta^{18}O_R c_R + \delta^{18}O_F c_F \tag{12.3}$$

where c is the CO_2 concentration and the subscript τ is the total, subscript A is the background atmosphere, subscript R is the plant/soil respiration source and subscript F is the fossil fuel source. In this formulation, c_τ, its isotope ratio, and the isotope ratio of all components is specified, while c_A, c_R, and c_F are unknowns.

For the purposes of studying urban ecological function, which encompasses plant physiological processes, biogeochemistry, and socio-economic factors influencing human activity, it may be useful to further separate the fossil fuel CO_2 source. In regions where natural gas combustion may be a significant component of total CO_2 emissions, i.e., areas with a pronounced cold season, the environmental and social influences on the magnitude of emissions from transportation, industrial activities, and residential heating may be vastly different. Isotopic measurements in the urban atmosphere may complement efforts to assess local energy consumption with bottom-up economic inventories. In Krakow, Poland, Kuc and Zimnoch (1998) estimated C_F with radiocarbon measurements and solved Eq. 12.4 for $\delta^{13}C_F$. They then used local end-members for $\delta^{13}C$ of natural gas and coal to estimate that 59% of fossil fuel combustion originated from natural gas in the winter of 1994. This value exceeded estimates obtained from bottom-up inventories, perhaps due to under-estimation of the contribution of industrial activities.

Pataki *et al.* (2003a) further expanded Eqs 12.1–12.3 to include separate gasoline (c_G) and natural gas (c_N) combustion terms, and rearranged these equations to solve for the fractional contributions of each component (f) rather than their absolute mixing ratio:

$$f_R + f_G + f_N = 1 \tag{12.4}$$

$$\delta^{13}C_R f_R + \delta^{13}C_G f_G + \delta^{13}C_N f_N = \delta^{13}C_S \tag{12.5}$$

$$\delta^{18}O_R f_R + \delta^{18}O_G f_G + \delta^{18}O_N f_N = \delta^{18}O_S \tag{12.6}$$

where the subscripts R, G, N, and S denote respiration, gasoline, natural gas, and the total source, respectively, $f_R = c_R/c_S$, $f_G = c_G/c_S$, $f_N = c_N/c_S$, and $c_S = c_R + c_G + c_N$. This approach is convenient in that it utilizes the 'Keeling plot' approach to derive $\delta^{13}C_S$ and $\delta^{18}O_S$ rather than specify c_A, $\delta^{13}C_A$, and $\delta^{18}O_A$.

Keeling (1958, 1961) first showed that total atmospheric CO_2 concentration and isotope measurements could be used to derive the isotopic composition of source CO_2 by applying a linear mixing relationship between the source and background values. This approach has been used extensively in natural ecosystems to determine the isotopic composition of ecosystem respiration. This Keeling plot intercept calculation is based on the assumption that atmospheric CO_2 contains two isotopically distinct components: a background value and a new source added to that background that raises the overall CO_2 concentration:

$$c_\tau = c_A + c_S \tag{12.7}$$

$$\delta^{13}C_\tau c_\tau = \delta^{13}C_A c_A + \delta^{13}C_S c_S \tag{12.8}$$

In this approach, c_τ and $\delta^{13}C_\tau$ are the only measured parameters. Re-arranging Eqs 12.1–12.2 to solve for $\delta^{13}C_\tau$, we can derive the equation for a straight line with an intercept of $\delta^{13}C_S$, the unknown of interest. c_S, $\delta^{13}C_A$, and c_A need not be estimated as they are integrated into the slope of the relationship between $\delta^{13}C_\tau$ and $1/c_\tau$ (Keeling, 1958, 1961). A similar set of equations can be written for $\delta^{18}O$. A constraint is the assumption that the isotopic composition of c_S and c_A does not change during the sample period; for this reason sampling should be limited to the shortest periods possible within one night (Pataki *et al.*, 2003b).

Studies have shown that urban areas are characterized by a high degree of both spatial and temporal variability in atmospheric CO_2 concentrations. Transect studies along rural to urban gradients have shown increased concentrations near urban centers (Berry and Colls, 1990; Idso *et al.*, 1998, 2001). Continuous monitoring of CO_2 concentrations has shown large diurnal changes (Tanaka *et al.*, 1985; Aikawa *et al.*, 1995; Reid and Steyn, 1997; Idso *et al.*, 2002) that vary seasonally (Tanaka *et al.*, 1985; Aikawa *et al.*, 1995; Idso *et al.* 2002) due to variability in both CO_2 emissions and atmospheric mixing. Large, short-term variations in CO_2 concentration may be well suited to utilizing the Keeling plot method to derive the isotopic composition of source CO_2, as the errors in this estimate are often reduced when sampling is conducted with large ranges (Pataki *et al.*, 2003b).

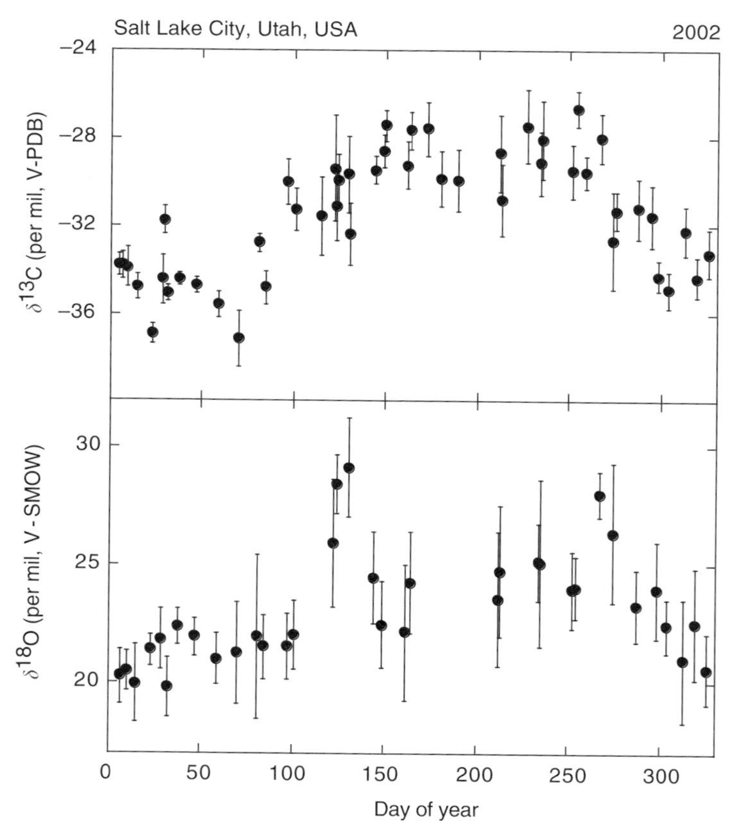

Figure 12.2 Time series of the carbon and oxygen isotope composition of the integrated urban CO_2 source ($\delta^{13}C_S$, $\delta^{18}O_S$) determined by the Keeling plot method for a 1-year period in Salt Lake City, Utah, USA. Error bars show the standard error of the Keeling plot intercept. Measured values of $\delta^{13}C_S$ of local fossil fuel CO_2 sources are given in Table 12.2, and modeled $\delta^{18}O_S$ of local CO_2 sources are shown in Fig. 12.1. Modified from Pataki *et al.* (2003a).

Pataki *et al.* (2003a) sampled urban air weekly to derive Keeling plot intercepts for ^{13}C and ^{18}O for a 1-year period (Fig. 12.2). The values of $\delta^{13}C_S$ in Fig. 12.2 show a distinct seasonal cycle with more depleted values in the winter caused by natural gas combustion, and more enriched values in the spring and summer caused by greater proportions of gasoline combustion and plant/soil respiration. In contrast, $\delta^{18}O_S$ showed relatively constant values during the winter due to the large influence of combustion, and enriched values in the spring and summer (Fig. 12.2). The mean wintertime (Jan 1–Apr 15) value of $\delta^{18}O_S$ was 21.3 ± 0.1‰, or 2.2‰ lighter than the expected value of 23.5‰. This offset could be due to contributions from soil respiration to the CO_2 source, or to fractionation associated with combustion (Pataki *et al.*, 2003a). During the growing season (Apr 16–Nov 15), $\delta^{18}O_S$ became enriched by up to 8‰ relative to depleted winter values of 21‰ (Fig. 12.2), as would be expected by increasing contributions of plant respiration to the total CO_2 source (Fig. 12.1).

Equations 12.4–12.6 were applied to solving for f_R, f_G, and f_N using measured values of $\delta^{13}C_G$ and $\delta^{13}C_N$, the theoretical combustion value for $\delta^{18}O_G$ and $\delta^{18}O_N$, the mean literature value of $\delta^{13}C_R$, and modeled values of $\delta^{18}O_R$ shown in Fig. 12.1. $\delta^{13}C_S$ and $\delta^{18}O_S$ were derived from a smoothed time-series of the data in Fig. 12.2. A seasonal cycle was apparent in the proportion of CO_2 derived from natural gas combustion, which originates largely from heating sources at this location and therefore was a large relative source of CO_2 in the winter (Fig. 12.3). Gasoline combustion and plant/soil respiration had largely opposing effects, with large proportional contributions of respiration in the spring and fall, and small to negligible contributions in the winter and mid-summer (Pataki *et al.*, 2003a; Fig. 12.3).

It is worth emphasizing that this method is most straightforward during periods when all sources of CO_2 are positive, and hence these measurements have been largely conducted in the wintertime or at night. The influence of biogenic respiration on the total CO_2 concentration is not an indication of the total carbon balance of a given urban ecosystem, as these measurements do not account for the uptake of CO_2 during daytime, growing season conditions. However, these measurements illustrate that biological activity is easily detected in many urban ecosystems despite the large influence of fossil-derived CO_2 on the local atmosphere. Isotopic sampling has shown that plant and soil respiration play a significant role in urban land–atmosphere exchange.

The Isotopic Composition of Urban Plant Biomass

We have discussed the influence of plant and soil processes on the urban atmosphere, but the reverse effects are also an essential part of urban ecology. The highly modified urban atmosphere may influence urban forest physiology and growth, and the services that these provide to urban residents (carbon storage, transpirational cooling effects, removal of pollutants, etc.). Stable isotopes also provide a useful tool in this regard. The carbon isotope composition of C_3 plant biomass is a function of the intercellular to ambient ratio of CO_2 partial pressures (c_i/c_A) and the isotopic composition of the atmosphere (Farquhar *et al.*, 1989). Carbon isotopes have been used to infer historical changes in atmospheric CO_2 concentrations (Marino and McElroy, 1991) as well as CO_2 exposure of plants grown in elevated CO_2 experiments (Pepin and Körner, 2002; Pataki *et al.*, 2003c). In the urban environment, the extent to which elevated levels of CO_2 and other products of combustion affect plant physiological processes and biogeochemistry has not been well investigated, such that stable isotope analyses can provide important new information.

C_4 plants provide a convenient tool for assessing CO_2 exposure, as their isotopic composition is relatively insensitive to changes in c_i/c_A (Marino and McElroy, 1991) if bundle sheath 'leakiness' can be neglected (Farquhar, 1983; Buchmann *et al.*, 1996). However, C_4 plants may not be common in all areas where an assessment of urban plant exposure to high CO_2 is desirable. If the variety of environmental conditions that affect c_i/c_A are similar at a given urban location and a rural area that serves as a control, then analysis of C_3 biomass may also be informative. Lichtfouse *et al.* (2003) compared $\delta^{13}C$ of grasses growing near a busy roadway in Paris, France to $\delta^{13}C$ of grasses growing in a rural area. Urban grasses were depleted in ^{13}C

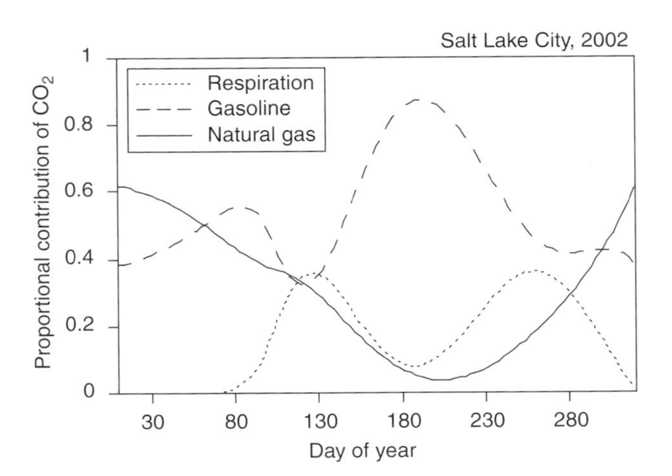

Figure 12.3 Proportional contributions of gasoline combustion, natural gas combustion, and biogenic respiration to total CO_2 source in Salt Lake City, USA determined from Eqs 12.6–12.8. Modified from Pataki *et al.* (2003a).

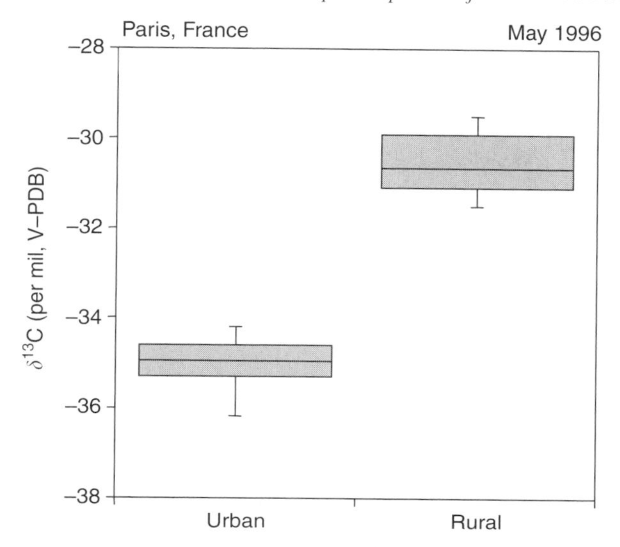

Figure 12.4 The carbon isotope composition ($\delta^{13}C$) of grasses growing near an urban Paris roadway and a rural area. The shaded area encompasses 50% of the data points, and the line shows the median value. The bars show the upper and lower 25% of the data. Modified from Lichtfouse *et al.* (2003).

by 4.5‰ on average (Fig. 12.4). These data were used to calculate the proportion of fossil-fuel derived CO_2 incorporated into urban grass biomass, which ranged from 20.8 to 29.1% (Lichtfouse *et al.*, 2003). Such calculations assume that elevated CO_2 does not in and of itself affect c_i/c_A, which has been supported by many instantaneous measurements (Morison, 1985; Drake *et al.*, 1997) but not all isotope measurements (Pataki *et al.*, 2003c).

Tree rings provide a longer-term record of plant exposure to increasing carbon dioxide. Dongarrà and Varrica (2002) measured $\delta^{13}C$ in tree rings of *Platanus hybrida* individuals in Palermo, Italy and found progressive depletion of 3.6‰ on average during the period 1880–1998 (Fig. 12.5). This is considerably larger than the global background depletion of approximately 1.5‰ during the same period, and indicates a large local effect of urban, ^{13}C depleted CO_2 emissions on plant biomass. These data may be used to reconstruct the historical record of CO_2 concentrations during urban development. Even taking into account long-term changes in c_i/c_A during the tree ring record, Dongarrà and Varrica (2002) inferred that local CO_2 concentrations in Palermo had increased by about 90 ppm on average since the 1950s, or 30 ppm above the global trend. Elevated atmospheric CO_2 has been shown experimentally to affect water use, growth, allocation, and other processes, regardless of changes in c_i/c_A (Drake *et al.*, 1997). Isotopic evidence suggests that elevated CO_2 may play a role in urban plant and soil processes.

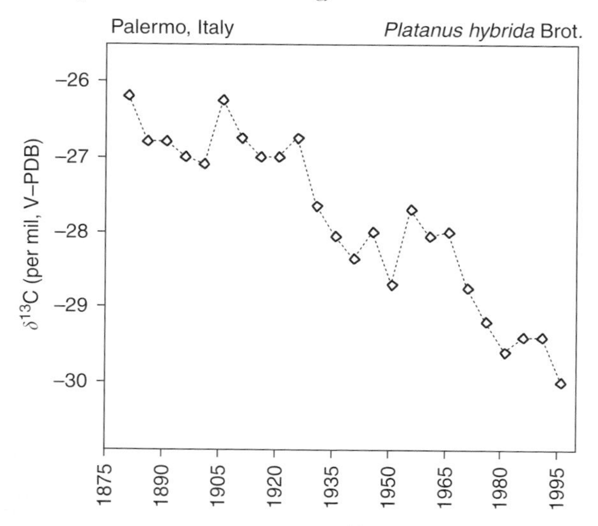

Figure 12.5 The carbon isotope composition ($\delta^{13}C$) of tree rings of *Platanus hybrida* Brot. grown in Palermo, Italy. Modified from Dongarrà and Varrica (2002).

Elevated concentrations of CO_2 in urban areas are often associated with other products of combustion such as SO_2, NO_x, and ozone (Douglas, 1983). While elevated CO_2 tends to increase plant growth, high concentrations of atmospheric pollutants such as sulfur dioxide and ozone may cause tissue damage and stomatal closure in many plants (Beckerson and Hofstra, 1979; Darrall, 1989; Skärby *et al.*, 1998). These effects may reduce or offset the effects of elevated CO_2 on growth (Fiscus *et al.*, 1977; Karnosky *et al.*, 1999; Olszyk *et al.*, 2001; Karnosky *et al.*, 2003). Decreases in c_i/c_A resulting from stomatal closure causes carbon isotope enrichment in C_3 plants, an effect which has been documented in both controlled environment studies of plant growth in response to elevated SO_2 and ozone (Douglas, 1983; Martin *et al.*, 1988), and in field studies comparing tree ring carbon isotope composition near point sources of pollution to unpolluted controls (Martin and Sutherland, 1990; Savard *et al.*, 2002). Isotopes have also been used to document plant uptake of nitrogenous products of pollution such as NH_3/NH_4^+ and NO_2 (Ammann *et al.*, 1999; Stewart *et al.*, 2002) that may act as a fertilizer at low concentrations or may be toxic at high concentrations (Krupa, 2003). Because there are a variety of possible opposing effects of atmospheric pollutants that may enhance or reduce plant gas exchange and growth, the application of multiple isotope tracers to identifying the role of individual trace gases is an important direction of future research.

Conclusions

Land–atmosphere exchange in urban areas is relevant to several areas of global change research, including alterations to global carbon and water cycles and the effect of urbanization on regional landscapes. At present, the dynamics of land–atmosphere exchange in cities are poorly understood with regard to trace gas fluxes. As several major components of urban CO_2 emissions contain distinct combinations of stable isotopes and radioisotopes, isotope measurements are a promising method for understanding the dynamics of urban CO_2 sources, the influence of cities on the isotopic composition of atmospheric CO_2, and the effects of local urban atmospheres on plant function.

Previous studies utilizing radio- and stable isotopes have shown that urban CO_2 emissions are not solely derived from fossil fuel emissions, but contain varying and significant proportions of plant/soil respiration (Zondervan and Meijer, 1996; Meijer *et al.*, 1997; Florkowski *et al.*, 1998; Takahashi *et al.*, 2001, 2002; Pataki *et al.*, 2003a). Conversely, isotopic analyses of plant biomass have confirmed plant exposure to pollutants in urban areas (Ammann *et al.*, 1999; Dongarrà and Varrica, 2002; Lichtfouse *et al.*, 2003), with implications for ecophysiology, biogeochemical cycling, and the influence of urban forests on their local environment.

While we have focused on the utility of measurements of the isotopic composition of urban CO_2 and organic material, stable isotopes offer other opportunities for urban ecological research in measurements of water, methane, hydrocarbons, and other constituents of urban ecosystems. Because urban 'forests' provide important benefits for quality of life (e.g., McPherson, 1995), urban plant and soil dynamics are integral to both social and biogeochemical processes in cities. As urban-atmosphere research progresses, stable isotope integration of these processes may play an important role in the emerging paradigm of urban system science.

References

Aikawa M., Yoshikawa K., Tomida M., Aotsuka F. and Haraguchi H. (1995) Continuous monitoring of carbon dioxide concentration in the urban atmosphere of Nagoya, 1991–1993. *Anal Sci* **11**: 357–362.

Ammann M., Siegwolf R., Pichlmayer F., Suter M., Saurer M. and Brunold C. (1999) Estimating the uptake of traffic-derived NO_2 from ^{15}N abundance in Norway spruce needles. *Oecologia* **118**: 124–131.

Andres R. J., Marland G., Boden T. and Bischof S. (2000) Carbon dioxide emissions from fossil fuel consumption and cement manufacture, 1751–1991, and an estimate of their isotopic composition and latitudinal distribution. In *The Carbon Cycle* (T. M. L. Wigley and D. S. Schimel, eds) pp. 53–62. Cambridge University Press, Cambridge.

Barnes C. J. and Allison G. B. (1983) The distribution of deuterium and ^{18}O in dry soils. *J Hydrol* **60**: 141–156.

Beckerson D. W. and Hofstra G. (1979) Stomatal responses of white bean to O_3 and SO_2 singly or in combination. *Atmos Environ* **13**: 33–35.

Berry R. D. and Colls J. J. (1990) Atmospheric carbon dioxide and sulphur dioxide on an urban/rural transect-II. Measurements along the transect. *Atmos Environ* **24**A: 2689–2694.

Bowling D. R., Tans P. P. and Monson R. K. (2001) Partitioning net ecosystem carbon exchange with isotopic fluxes of CO_2. *Global Change Biol* **7**: 127–145.

Buchmann N., Brooks J. R., Rapp K. D. and Ehleringer J. R. (1996) Carbon isotope composition of C_4 grasses is influenced by light and water supply. *Plant Cell Environ* **19**: 392–402.

Clark-Thorne S. T. and Yapp C. J. (2003) Stable carbon isotope constraints on mixing and mass balance of CO_2 in an urban atmosphere: Dallas metropolitan area, Texas, USA. *Appl Geochem* **18**: 75–95.

Darrall N. M. (1989) The effect of air pollutants on physiological processes in plants. *Plant Cell Environ* **12**: 1–30.

Decker E. H., Elliot S., Smith F. A., Blake D. R. and Rowland F. S. (2000) Energy and material flow through the urban ecosystem. *Annu Rev Energy Environ* **25**: 685–740.

Deines P. (1980) The isotopic composition of reduced organic carbon. In *Handbook of Environmental Isotope Geochemistry* (P. Fritz and J. C. Fontes, eds) pp. 329–406. Elsevier, Amsterdam.

Dongarrà G. and Varrica D. (2002) $\delta^{13}C$ variations in tree rings as an indication of severe changes in the urban air quality. *Atmos Environ* **36**: 5887–5896.

Dongmann G., Nurnberg H. W., Forstel H. and Wagener K. (1974) On the enrichment of H_2O^{18} in the leaves of transpiring plants. *Radiat Environ Biophys* **11**: 41–52.

Douglas I. (1983) *The Urban Environment.* Edward Arnold, London.

Drake B. G., Gonzalez-Meler M. A. and Long S. P. (1997) More efficient plants: A consequence of rising atmospheric CO_2? *Annu Rev Plant Physiol* **48**: 608–637.

Farquhar G. D. (1983) On the nature of carbon isotope discrimination in C4 species. *Aust J Plant Physiol* **10**: 205–226.

Farquhar G. D., Ehleringer J. R. and Hubick K. T. (1989) Carbon isotope discrimination and photosynthesis. *Annu Rev Plant Physiol Plant Mol Biol* **40**: 503–537.

Fiscus E., Reid C., Miller J. and Heagle A. (1997) Elevated CO_2 reduces O_3 flux and O_3-induced yield losses in soybeans: Possible implications for elevated CO_2 studies. *J Exp Bot* **48**: 307–313.

Flanagan L. B. and Ehleringer J. R. (1991) Stable isotope composition of stem and leaf water: Applications to the study of plant water use. *Funct Ecol* **5**: 270–277.

Flanagan L. B., Brooks J. R., Varney G. T. and Ehleringer J. R. (1997) Discrimination against $C^{18}O^{16}O$ during photosynthesis and the oxygen isotope ratio of respired CO_2 in boreal forest ecosystems. *Global Biogeochem Cycles* **11**: 83–98.

Florkowski T., Korus A., Miroslaw J., Necki J., Neubert R., Schimdt M. and Zimnoch M. (1998) Isotopic composition of CO_2 and CH_4 in a heavily polluted urban atmosphere and in a remote mountain area (Southern Poland). In *Isotope Techniques in the Study of Environmental Change*, pp. 37–48. IAEA, Vienna.

Gillon J. and Yakir D. (2001) Influence of carbonic anhydrase activity in terrestrial vegetation on the ^{18}O content of atmospheric CO_2. *Science* **291**: 2584–2587.

Grimm N. B., Grove J. M., Pickett S. T. A. and Redman C. L. (2000) Integrated approaches to long-term studies of urban ecological systems. *Bioscience* **50**: 571–584.

Idso C. D., Idso S. B. and Balling R. C. (1998) The urban CO_2 dome of Phoenix, Arizona. *Phys Geog* **19**: 95–108.

Idso C. D., Idso S. B. and Balling R. C. (2001) An intensive two-week study of an urban CO_2 dome in Phoenix, Arizona, USA. *Atmos Environ* **35**: 995–1000.

Idso S. B., Idso C. D. and Balling R. C. (2002) Seasonal and diurnal variations of near-surface atmospheric CO_2 concentration within a residential sector of the urban CO_2 dome of Phoenix, AZ, USA. *Atmos Environ* **36**: 1655–1660.

Jeffery P. M., Compston W., Greenhalgh D. and De Laeter J. (1955) On the carbon-13 abundance of limstones and coals. *Geochim Cosmochim Acta* **7**: 255–286.

Karnosky D. F., Mankovska B., Percy K., Dickson R. E., Podila G. K., Sober J., Noormets A., *et al.* (1999) Effects of tropospheric O_3 on trembling aspen and interaction with CO_2: Results from an O_3-gradient and a FACE experiment. *Water Air Soil Pollut* **116**: 311–322.

Karnosky D. F., Zak D. R., Pregizter K. S., Awmack C. S., Bockheim J. G., Dickson R. E., Hendrey G. R., *et al.* (2003) Tropospheric O_3 moderates responses of temperate hardwood forests to elevated CO_2: A synthesis of molecular to ecosystem results from the Aspen FACE project. *Funct Ecol* **17**: 289–304.

Keeling C. D. (1958) The concentration and isotopic abundances of atmospheric carbon dioxide in rural areas. *Geochim Cosmochim Acta* **13**: 322–334.

Keeling C. D. (1961) The concentration and isotopic abundance of carbon dioxide in rural and marine air. *Geochim Cosmochim Acta* **24**: 277–298.

Keeling C. D., Mook W. G. and Tans P. P. (1979) Recent trends in the $^{13}C/^{12}C$ ratio of atmospheric carbon dioxide. *Nature* **277**: 121–123.

Kiyosu Y. and Kidoguchi M. (2000) Variations in the stable carbon isotope ratios of *Zelkova serrata* leaves from roadside trees in Toyama City, Japan. *Geochem J* **34**: 379–382.

Kroopnick P. and Craig H. (1972) Atmospheric oxygen: Isotopic composition and solubility fractionation. *Science* **175**: 54–55.

Krupa S. V. (2003) Effects of atmospheric ammonia (NH_3) on terrestrial vegetation: A review. *Environ Pollut* **124**: 179–221.

Kuc T. and Zimnoch M. (1998) Changes of the CO_2 sources and sinks in a polluted urban area (Southern Poland) over the last decade, derived from the carbon isotope composition. *Radiocarbon* **40**: 417–423.

Lichtfouse E., Lichtfouse M. and Jaffrezic A. (2003) $\delta^{13}C$ values of grasses as a novel indicator of pollution by fossil-fuel-derived greenhouse gas CO_2 in urban areas. *Environ Sci Techol* **37**: 87–89.

Marino B. D. and McElroy M. B. (1991) Isotopic composition of atmospheric CO_2 inferred from carbon in C_4 plant cellulose. *Nature* **349**: 127–131.

Martin B., Bytnerowicz A. and Thorstenson Y. R. (1988) Effects of air pollutants on the composition of stable carbon isotopes, 13C, of leaves and wood, and on leaf injury. *Plant Physiol* **88**: 218–223.

Martin B. and Sutherland E. K. (1990) Air pollution in the past recorded in width and stable carbon isotope composition of annual growth rings of Douglas fir. *Plant Cell Environ* **13**: 839–844.

McPherson E. G. (1995) Net benefits of healthy and productive urban forests. In *Urban Forest Landscapes: Integrating Multidisciplinary Perspectives* (G. A. Bradley, ed.) pp. 180–194. University of Washington Press, Seattle.

Meijer H., Smid H., Perez E. and Keizer M. (1997) Isotopic characterisation of anthropogenic CO_2 emissions using isotopic and radiocarbon analysis. *Phys Chem Earth* **21**: 483–487.

Miller J. B., Yakir D., White J. W. C. and Tans P. P. (1999) Measurement of $^{18}O/^{16}O$ in the soil–atmosphere CO_2 flux. *Global Biogeochem Cycles* **13**: 761–774.

Morison J. (1985) Sensitivity of stomata and water use efficiency to high CO_2. *Plant Cell Environ* **8**: 467–474.

Olszyk D. M., Johnson M. G., Phillips D. L., Seidler R. J., Tingey D. T. and Watrud L. S. (2001) Interactive effects of CO_2 and O-3 on a ponderosa pine plant/litter/soil mesocosm. *Environ Pollut* **115**: 447–462.

Pataki D. E., Bowling D. R. and Ehleringer J. R. (2003a) The seasonal cycle of carbon dioxide and its isotopic composition in an urban atmosphere: Anthropogenic and biogenic effects. *J Geophys Res* **108**: 4735.

Pataki D. E., Ehleringer J. R., Flanagan L. B., Yakir D., Bowling D., Still C. J., Buchmann N., Kaplan J. O. and Berry J. A. (2003b) The application and interpretation of Keeling plots in terrestrial carbon cycle research. *Global Biogeochem Cycles* **17**, doi: 10.1029/200/850.

Pataki D. E., Ellsworth D. S., Evans R. D., Gonzalez-Meler, M. A., King J. S., Leavitt S. W., Lin G., Matamala R., Pendall E., Siegwolf R., van Kessel C. and Ehleringer J. R. (2003c) Tracing changes in ecosystem function under elevated carbon dioxide conditions. *Bioscience* **53**: 805–818.

Pepin S. and Körner C. (2002) Web-FACE: A new canopy free-air CO_2 enrichment system for tall trees in mature forests. *Oecologia* **133**: 1–9.

Pickett S. T. A., Cadenasso M. L., Grove J. M., Nilon C. H., Pouyat R. V., Zipperer W. C. and Costanza R. (2001) Urban ecological systems: Linking terrestrial ecological, physical, and socioeconomic components of metropolitan areas. *Annu Rev Ecol Systematics* **32**: 127–157.

Reid K. and Steyn D. (1997) Diurnal variations of boundary-layer carbon dioxide in a coastal city—observations and comparison with model results. *Atmos Environ* **31**: 3101–3114.

Rice D. D. and Claypool G. E. (1981) Generation, accumulation and resource potential of biogenic gas. *Am Assoc Petr Geol Bull* **65**: 5–25.

Savard M. M., Begin C. and Parent M. (2002) Are industrial SO_2 emissions reducing CO_2 uptake by the boreal forest? *Geology* **30**: 403–406.

Schoell M. (1988) Multiple origins of methane in the earth. *Chem Geol* **71**: 1–10.

Skärby L., Ro-Poulsen H., Wellburn F. A. M. and Sheppard L. J. (1998) Impacts of ozone on forests: A European perspective. *New Phytol* **139**: 109–122.

Stewart G. R., Aidar M. P. M., Joly C. A. and Schmidt S. (2002) Impact of point source pollution on nitrogen isotope signatures ($\delta^{15}N$). *Oecologia* **131**: 468–472.

Still C. J., Berry J. A., Ribas-Carbo M. and Helliker B. R. (2003) The contribution of C-3 and C-4 plants to the carbon cycle of a tallgrass prairie: An isotope approach. *Oecologia* **136**: 347–359.

Takahashi H. A., Hiyama T., Konohira E., Takahashi A., Yoshida N. and Nakamura T. (2001) Balance and behavior of carbon dioxide at an urban forest inferred from the isotopic and meteorological approaches. *Radiocarbon* **43**: 659–669.

Takahashi H. A., Konohira E., Hiyama T., Minami M., Nakamura T. and Yoshida N. (2002) Diurnal variation of CO_2 concentration, $\Delta^{14}C$ and $\delta^{13}C$ in an urban forest: Estimate of the anthropogenic and biogenic CO_2 contributions. *Tellus* **54**B: 97–109.

Tanaka M., Nakazawa T. and Aoki S. (1985) Atmospheric carbon dioxide variations in the suburbs of Sendai, Japan. *Tellus* **37**B: 28–34.

Tans P. P. (1981) $^{13}C/^{12}C$ of industrial CO_2. In *Carbon Cycle Modelling* (B. Bolin, ed.) pp. 127–129. Wiley, Chichester.

United Nations (2000) *World Urbanization Prospects: The 1999 Revision*. United Nations, New York.

Widory D. and Javoy M. (2003) The carbon isotope composition of atmospheric CO_2 in Paris. *Earth Planetary Sci Lett* **215**: 289–298.

Yakir D. and Wang X. F. (1996) Fluxes of CO_2 and water between terrestrial vegetation and the atmosphere estimated from isotope measurements. *Nature* **380**: 515–517.

Yeh H.-W. and Epstein S. (1981) Hydrogen and carbon isotopes of petroleum and related organic matter. *Geochim Cosmochim Acta* **45**: 753–762.

Zondervan A. and Meijer H. (1996) Isotopic characterisation of CO_2 sources during regional pollution events using isotopic and radiocarbon analysis. *Tellus* **48**B: 601–612.

Part III

Global Scale Processes

13

Terrestrial Ecosystems and Interannual Variability in the Global Atmospheric Budgets of $^{13}CO_2$ and $^{12}CO_2$

James T. Randerson

Introduction

Several recent ecosystem studies provide evidence that plant discrimination against ^{13}C during photosynthesis is highly variable on synoptic to interannual timescales. Critical axes of variation include the response to drought stress in temperate and tropical forests (Ekblad and Högberg, 2001; Bowling *et al.*, 2002; Mortazavi and Chanton, 2002; Ometto *et al.*, 2002; Fessenden and Ehleringer, 2003) and shifts in the abundance of C_3 and C_4 vegetation both in response to land use (Vandam *et al.*, 1997; Townsend *et al.*, 2002) and climate (Tieszen *et al.*, 1997; Still *et al.*, 2003b).

The purpose of this chapter is to place this variability in the context of the global carbon cycle. First, I present equations describing the global atmospheric budget for $^{13}CO_2$ and $^{12}CO_2$ in graphical form, using 'Robin Hood' diagrams. Robin Hood diagrams provide an intuitive means for comparing isotopic fluxes from terrestrial ecosystems with those from the oceans and fossil fuels, and for partitioning ocean and land carbon sinks using atmospheric $\delta^{13}C$ and CO_2 observations (Enting *et al.*, 1994).

Second, I critically review the mechanisms by which terrestrial ecosystems are thought to impart isotopic anomalies to the atmosphere on interannual timescales. A question that has generated considerable debate in the last few years is what fraction of the interannual variation of $\delta^{13}C$ in the global atmosphere should be attributed to changes in land and ocean carbon sinks, and what fraction should be attributed to changes in plant discrimination and other processes. During the 1990s, the growth rate of $\delta^{13}C$ was highly variable, with the most rapid declines occurring during the middle to latter stage of the 1997–1998 El Nino event (Fig. 13.1; Battle *et al.*, 2000).

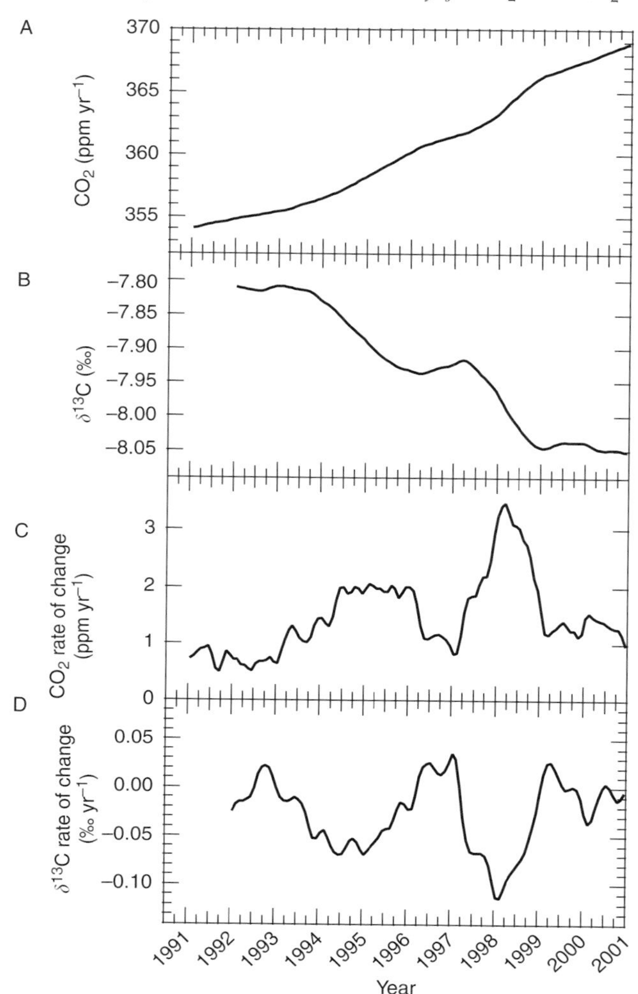

Figure 13.1 A. Global CO_2 concentrations derived from observations from the NOAA Carbon Monitoring Diagnostics Laboratory flask measurements [http://www.cmdl.noaa.gov/ccg/] (Battle *et al.*, 2000). The global mean was constructed by removing a mean seasonal cycle, averaging all stations within 6 latitude zones (0°–30°, 30°–60°, and 60°–90° in each hemisphere), weighting each latitude zone by surface area, and smoothing with a 1-year box car running mean filter. B. Same as A. but for $\delta^{13}C$ (Battle *et al.*, 2000). C. The atmospheric growth rate of CO_2 was highly variable during the 1990s and reached a maximum during late 1997 and early 1998, during the strong El Nino event observed at that time. D. Changes in atmospheric $\delta^{13}C$ mirror changes in CO_2 (shown in C) but have an opposite sign. A flux with a $\delta^{13}C$ of−25‰ explains most of the variability whereas a flux with a $\delta^{13}C$ of −9‰ explains only a small fraction. This suggests that terrestrial ecosystems contributed to most of the variability during this period.

In this context, several recent modeling analyses have highlighted a number of independent mechanisms that may (or may not) work in parallel to explain the atmospheric anomalies, apart from sources and sinks. At least three of the proposed mechanisms are connected to the shifting patterns of precipitation that accompany El Nino events. During El Nino, less rain falls on land, particularly in tropical ecosystems (Ropelewski and Halpert, 1996). This decrease causes global scale drought stress, lowers leaf internal CO_2 concentrations, and causes a temporary decrease in global plant discrimination (Randerson *et al.*, 2002; Scholze *et al.*, 2003). It is also likely that precipitation shifts between C_3 and C_4 ecosystems, causing a change in the relative contribution of these plant functional types to global gross primary production (GPP) and thus global discrimination (Scholze *et al.*, 2003). Finally, global drought stress also changes the distribution and abundance of fires (Langenfelds *et al.*, 2002; Van der Werf *et al.*, 2004), including their isotopic composition, which is not well known.

No contemporary modeling framework has simultaneously considered all of the mechanisms that have been proposed, and it is unlikely that any single mechanism is exclusive. In this context, I conclude by offering several directions for experimental inquiry that could guide model development, and ultimately seem necessary for a predictive understanding of interannual variability in discrimination at large spatial scales.

A Graphical Means to Understanding the Global Atmospheric Budget: Robin Hood Diagrams

Overview

Here I provide a step-by step description of each vector in a Robin Hood diagram of the global atmospheric carbon budget (Fig. 13.2). The budget and figure is constructed arbitrarily for the year 1990. This vector approach is referred to as a Robin Hood diagram because of the abundance of arrows (Inez Fung, personal communication, 2000); and the overall process of solving for land and ocean carbon sinks using $^{13}CO_2$ and $^{12}CO_2$ is known as a 'double deconvolution' (Heimann and Keeling, 1989). In the conventional representation of the global budget, observational changes (labeled with a *1* in Fig. 13.2) represent some combination of fossil fuel emissions (*6*), land disequilibria forcing (*5*), a terrestrial carbon sink (*4*), ocean disequilibria forcing (*3*), and an ocean carbon sink (*2*). In a more general representation of the budget (Fig. 13.2B), both gross primary production (*7*) and a return flux from the biosphere to the atmosphere (*8*) are separately considered (Randerson *et al.*, 2002). In this form, it can be seen that anomalies in discrimination ($\dot{\Delta}_{ab}$) affect GPP, and ensuing disequilibria (defined below) have the potential to affect interannual variability in

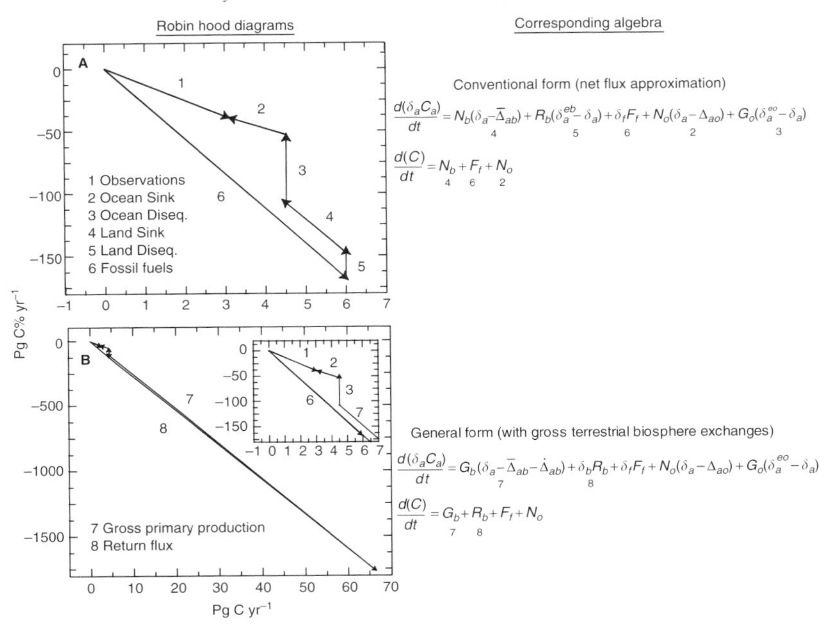

Figure 13.2 A Robin Hood diagram is comprised of multiple vectors, each representing a different carbon cycle process. These vectors are linked to the equations describing the atmospheric budget using numbers. Observations (labeled with a *1*) are given by the left hand side of the equations. A. The conventional double deconvolution representation of the terrestrial budget splits the terrestrial biosphere into a net sink (*4*) and an isotopic disequilibrium term (*5*). B. A more general representation of terrestrial carbon fluxes considers GPP (*7*) and the return flux (respiration, fires, etc., labeled with an *8*). In this representation, it can be seen that small anomalies in discrimination can affect the interpretation of the atmospheric budget because they operate on a very large vector (GPP) and are not associated with a carbon source or sink. Note the change in scale. The isotopic disequilibrium forcing of the biosphere is now explicitly included as a part of the return flux vector (R_b). In this figure, only the fraction of GPP that remains in the biosphere for ~1 year or longer is shown (~30 Pg C yr^{-1}). Adapted from Randerson *et al.* (2002). Modified by permission of the American Geophysical Union.

atmospheric δ^{13}C. The x-axis represents the carbon flux from each process or the observed atmospheric change (with units of Pg C yr^{-1}). The y-axis represents the ^{13}C mass flux, expressed in terms of an isoflux (with units of Pg C‰ yr^{-1}).

Considering only the x-axis, it can be seen in Fig. 13.2A that the observed atmospheric growth rate of carbon (3 Pg C yr^{-1}) is only ~1/2 of fossil fuel emissions (6 Pg C yr^{-1}). Land and ocean sinks must account for this difference, yet partitioning is impossible without additional information. There is one known quantity (the difference between fossil fuel emissions and the atmospheric growth rate) and two unknowns (land and ocean sinks).

Information from the atmospheric rate of change of $\delta^{13}C$ and differences in discrimination against ^{13}C associated with land and ocean sinks allows for a land/ocean sink partitioning. A carbon sink associated with C_3 terrestrial photosynthesis and ecosystem respiration strongly discriminates against ^{13}C (Δ_{ab} is equal to $\sim$19‰), acts to enrich atmospheric $\delta^{13}C$, and is represented in a Robin Hood diagram by a vector with a relatively steep slope (but initially with an unknown length). In contrast, a carbon sink associated with air–sea gas exchange (Δ_{oa} is equal to $\sim$2‰) discriminates weakly against ^{13}C, has a relatively small effect on atmospheric $\delta^{13}C$, and is represented in a Robin Hood diagram by a vector with a relatively shallow slope (and again, initially with an unknown length). Only one combination of land and ocean vectors, with their slopes determined by their respective levels of discrimination, can close the atmospheric budget for both total carbon and ^{13}C, given known values for fossil fuel emissions and isotopic disequilibria forcing terms described below.

With the added dimension that ^{13}C provides, there is a cost. Land and ocean net carbon sinks are not the only processes affecting the rate of change of atmospheric $\delta^{13}C$; ^{13}C fluxes associated with the one-way ecosystem respiration and gross primary production fluxes on land and the gross one-way sea-to-air and air-to-sea fluxes in the ocean also regulate atmospheric $\delta^{13}C$ and thus must be estimated prior to the double deconvolution. These fluxes, known as isotopic disequilibria forcing terms, are not well characterized and contribute to substantial uncertainties in land/ocean partitioning (Fung *et al.*, 1997). Isotopic disequilibria arise from the time delay between entry (loss from the atmosphere) and exit (return to the atmosphere) for carbon in terrestrial ecosystems and ocean reservoirs exchanging with the atmosphere, and the relatively sudden negative perturbation in the atmospheric isotopic composition caused by fossil fuels (Quay *et al.*, 1992; Tans *et al.*, 1993; Fung *et al.*, 1997; Wittenberg and Esser, 1997; Gruber *et al.*, 1999). Other processes besides fossil fuel emissions may induce disequilibria, most notably shifts in photosynthetic discrimination over a period of decades caused by a shift in C_4 vegetation (Ciais *et al.*, 1999; Townsend *et al.*, 2002) or changes in discrimination caused by interannual variability in climate (Randerson *et al.*, 2002).

Background

A primary simplification that is widely employed in ecosystem, regional, and global scale budget calculations is that the product of the total carbon flux and its isotopic composition, in δ notation is a mostly conservative tracer of the ^{13}C flux (and is known as the isoflux). For example, with the Keeling plot approach it is possible to derive the form of the linear regression relating air mass changes to the isotopic source composition of the flux using this notation or more formally by separately considering the ^{12}C and ^{13}C fluxes (Pataki *et al.*, 2003). Biosphere–atmosphere trace gas exchange is quantitatively described by the following two equations:

$$\frac{d(^{12}C_b)}{dt} = {}^{12}F_{ab} + {}^{12}F_{ba} \tag{13.1}$$

and

$$\frac{d(^{12}C_b R_b)}{dt} = \alpha_{ab} R_a {}^{12}F_{ab} + \alpha_{ba} R_b {}^{12}F_{ba} \tag{13.2}$$

where $^{12}F_{ab}$ and $^{12}F_{ba}$ at the global scale are the one-way fluxes of gross primary production and a return flux that is a combination of ecosystem respiration, fires, volatile organic carbon emissions, and other losses. Here a flux out of the surface is defined with a positive sign ($^{12}F_{ba}$) and a flux into the surface is defined with a negative sign ($^{12}F_{ab}$). R_a and R_b represent the isotope ratios ($^{13}C/^{12}C$) of the atmosphere and biosphere, respectively, and α_{ba} and α_{ab} are the fractionation factors associated with the one-way fluxes (Criss, 1999).

Together, Eqs 13.1 and 13.2 can be approximated by the combination of the following two equations (Tans *et al.*, 1993; Fung *et al.*, 1997):

$$\frac{d(C_b)}{dt} = F_{ab} + F_{ba} \tag{13.3}$$

and

$$\frac{d(\delta_b C_b)}{dt} = (\delta_a - \Delta_{ab})F_{ab} + \delta_b F_{ba} \tag{13.4}$$

where in Eq. 13.3, the mass of ^{12}C (from Eq. 13.1) is now approximated by that for total carbon. Equation 13.4, which represents the ^{13}C mass flux anomaly, requires the additional approximation that the isotopic composition of the photosynthesis flux in δ notation is equal to the difference between the background atmosphere (δ_a) and discrimination (Δ_{ab}), where Δ_{ab} is defined according to Farquhar *et al.* (1989). Δ_{ab} is a positive number, typically between 16 and 21‰ for C_3 plants and between 4 and 5‰ for C_4 plants. For many regional and global scale analyses, it is also assumed that discrimination against ^{13}C associated with respiration and the return flux is minimal, and so the isotopic composition of the biosphere to atmosphere flux is equal to the isotopic composition of the biosphere (δ_b). This does not mean, however, that δ_b is necessarily equal to $\delta_a - \Delta_{ab}$, because carbon released today as respiration was fixed at an earlier time with a different δ_a (and also with a potentially different Δ_{ab}).

Observations

The observed atmospheric growth rate of carbon was approximately $3\,Pg\,C\,yr^{-1}$ in 1990. Change in the ^{13}C mass of the atmosphere over time is represented by the time derivative of the product of the total mass and its $\delta^{13}C$. Using the Product Rule, this derivative can be split into two components:

$$\frac{d(C_a\delta_a)}{dt} = \frac{d(C_a)}{dt}\bar{\delta}_a + \bar{C}_a\frac{d(\delta_a)}{dt} \tag{13.5}$$

where the first term on the right hand side represents the change of ^{13}C mass caused by the change in the total carbon inventory of the atmosphere, and the second term represents the change of ^{13}C mass caused by the change in atmospheric $\delta^{13}C$. In 1990, both terms contributed substantially to the ^{13}C mass anomaly ($3\,\text{Pg C yr}^{-1} \times -7.8‰ = -23.4\,\text{Pg C‰ yr}^{-1}$ and $760\,\text{Pg C yr}^{-1} \times -0.02‰\,\text{yr}^{-1} = -15.0\,\text{Pg C‰ yr}^{-1}$).

Fossil Fuels

Fossil fuels represent the single largest perturbation to the atmospheric budget on an annual basis, for both total carbon and its isotopic composition ($\delta^{13}C$). Without CO_2 exchange with ocean or land carbon reservoirs, the observed decrease in atmospheric $\delta^{13}C$ would have been $-0.16‰\,\text{yr}^{-1}$ in 1990, about 8 times greater than that observed ($-0.02‰\,\text{yr}^{-1}$). Cumulatively, if all of the ~220 Pg C of fossil carbon emitted prior to 1990 had remained in the atmosphere (again without ocean or land exchange) the atmosphere in 1990 would have had a $\delta^{13}C$ of $-12.5‰$, about 4 times greater than the observed perturbation ($-7.8‰$) as compared with the preindustrial atmosphere ($-6.3‰$).

In Fig. 13.2, I assumed that fossil fuel emissions were $6\,\text{Pg C yr}^{-1}$ in 1990, had a global $\delta^{13}C$ signature of $-28‰$, and thus an isoflux of $-168\,\text{Pg C‰ yr}^{-1}$ (Andres *et al.*, 2000). A quantitative analysis of the uncertainties in the $\delta^{13}C$ content of fossil fuels remains to be done, and it is unlikely that this term is known to within $\pm 1‰$. At higher resolution temporal and spatial scales, even greater uncertainty exists, primarily from variability in the use of natural gas products that have a $\delta^{13}C$ of $-40‰$ to $-60‰$ (Andres *et al.*, 2000).

Terrestrial Isotopic Disequilibria

Even without a net carbon sink, gross exchange with the terrestrial biosphere enriches the contemporary atmosphere in $\delta^{13}C$. Before the industrial revolution, δ_a was relatively constant for many hundreds of years (Francey *et al.*, 1999). During this time, ^{13}C fluxes into and out of the terrestrial biosphere were roughly at steady state: enrichment of the atmosphere by photosynthesis was balanced by depletion of the atmosphere by a return flux on annual and decadal timescales (Fig. 13.3). Since the eighteenth century, however, the $\delta^{13}C$ of flux entering the biosphere has steadily decreased

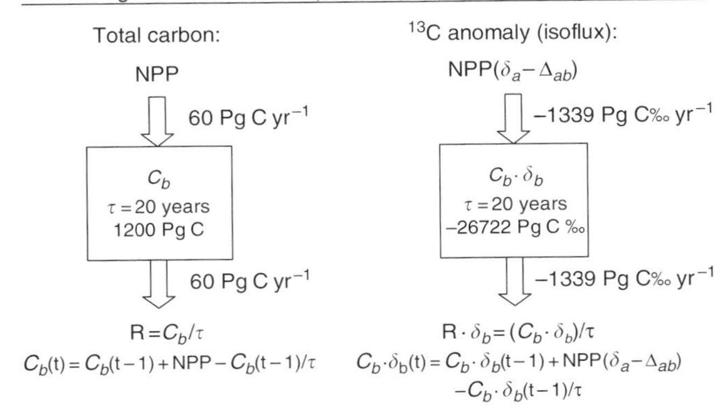

Figure 13.3 Steady-state fluxes for the preindustrial biosphere for both total carbon and the ^{13}C mass flux anomaly (isoflux). Preindustrial δ_a was assumed to be −6.3‰ and global Δ_{ab} was assumed to be −16‰, reflecting a combination of C_3 and C_4 ecosystems. Equations are also provided for describing the time rate of change of total carbon and the ^{13}C mass anomaly under a changing δ_a.

because of the fossil fuel perturbation to the atmosphere (Keeling *et al.*, 1979; Francey *et al.*, 1999). The return flux is also decreasing, but with an additional delay that reflects the residence time of carbon in plant, litter, and soil pools. The difference between the δ^{13}C of the return flux (that was fixed by photosynthesis in previous years) and a hypothetical flux that would be in equilibrium with the contemporary atmospheric state (i.e., that had a value of $\delta_a - \Delta_{ab}$), is known as the isotopic disequilibrium.

In general the longer the carbon residence time, the larger the isotopic disequilibrium (Randerson *et al.*, 1999). The maximum possible value of the disequilibrium is the difference between the preindustrial atmosphere and the contemporary atmosphere. In 1990 this maximum was ~1.6‰. For this maximum to occur, all of the carbon in the return flux would have to have an age greater than 2–3 centuries. Given the relatively large allocation of net primary production (NPP) to fine roots and leaves, most estimates of the terrestrial disequilibrium are considerably smaller, in the order of 0.2 to 0.5‰ when defined with respect to NPP (Thompson and Randerson, 1999).

From the perspective of modeling or measuring isotopic disequilibria caused by a changing δ_a, it is worth noting several points. First, most of the carbon that enters an ecosystem via photosynthesis leaves quickly via autotrophic respiration and leaf and fine root turnover (Trumbore, 2000). Thus the magnitude of the isotopic disequilibrium for a given year is highly sensitive to changes in δ_a in the 10 or so preceding years. Second, an isotopic disequilibrium induced by a changing δ_a does not depend on Δ_{ab}, but only on the residence time of carbon in the surface reservoir (with the residence time defined with respect to the atmosphere). Finally, any estimate of the isotopic disequilibrium must be made in reference to a flux with a defined magnitude. For example, the isotopic disequilibrium in an ecosystem associated with NPP would be considerably greater than one defined for the same ecosystem relative to GPP. The reason for this is that while GPP is a much greater flux, much of GPP rapidly returns to the atmosphere via plant respiration and thus has a very short residence time.

The product of the isotopic disequilibrium and the gross flux is known as an 'isotopic disequilibrium forcing.' Estimates of the global terrestrial disequilibrium forcing require spatially distributed estimates of both GPP and the age distribution of carbon in ecosystem respiration, volatile organic carbon emissions, and fires. In turn, the age distribution of carbon requires knowledge of the distribution of plant functional types over the land surface, allocation, lifetimes of different plant tissues, decomposition rates, and the disturbance regime. In Fig. 13.2, I assumed that the isotopic disequilibrium forcing from the terrestrial biosphere was 20 Pg C‰ yr^{-1}, following from Randerson *et al.* (2002).

Using the single reservoir model of the terrestrial biosphere described in Fig. 13.3, in Table 13.1 I show the components of the budget required to calculate the isotopic disequilibrium and isotopic disequilibrium forcing arising from the fossil fuel perturbation to δ_a. These simplified estimates are derived for a terrestrial biosphere with an NPP of 60 Pg C yr^{-1} and a single carbon pool with a turnover time of 20 years. This calculation is easy to reproduce in a spreadsheet program, once a continuous time history of δ_a has been constructed from observations, such as those of Francey *et al.* (1999).

A second class of disequilibria arises from changes in Δ_{ab} through time (and not changes in δ_a). On weekly to interannual timescales, changes in drought stress can drive trends in Δ_{ab}, while over a period of decades to centuries long-term shifts in vegetation or climate are likely to be the most important drivers.

Terrestrial Carbon Sink

In the double deconvolution inversion, the magnitude of the terrestrial carbon sink is one of two unknowns. The slope of the vector is set by Δ_{ab} (Fig. 13.2). Only one combination of lengths of the ocean and land sink vectors can match the observed vector, once fossil fuel and isotopic disequilibrium vectors are known.

The slope of the terrestrial carbon sink vector is proportional to Δ_{ab}. If the terrestrial carbon sink is distributed in proportion to productivity in C_3 and C_4 ecosystems, then a GPP-weighted discrimination (one that combined C_3 and C_4 discrimination values) would be appropriate for use in a global analysis (Still *et al.*, 2003a). If, however, most of the terrestrial carbon sink is in forests recovering from natural or human disturbance,

Table 13.1 Changes in the Terrestrial Isotopic Disequilibrium Forcing Caused by Fossil Fuel Emissions

		Total carbon mass and fluxes[1]			^{13}C mass anomaly (isoflux)[2]				Isotopic disequilibrium	Isotopic disequilibrium forcing
Year	δ_a (‰)	NPP ($Pg\,C\,yr^{-1}$)	C_b ($Pg\,C$)	$R_h = C_b/\tau$ ($Pg\,C\,yr^{-1}$)	$NPP \cdot (\delta_a - \Delta_{ab})$ ($Pg\,C$ ‰ yr^{-1})	$C_b \cdot \delta_b$ ($Pg\,C$ ‰)	$R_h \cdot \delta_b = (C_b \cdot \delta_b)/\tau$ ($Pg\,C$ ‰ yr^{-1})	$\delta_b = (R_h \cdot \delta_b)/R_h$ (‰)	$\delta_a^{eb} - \delta_a = \delta_b - (\delta_a - \Delta_{ab})$ (‰)	$NPP \cdot (\delta_a^{eb} - \delta_a)$ ($Pg\,C$ ‰ yr^{-1})
1765	−6.31	60	1200	60	−1339	−26 771	−1339	−22.3	0.00	0.0
1865	−6.57	60	1200	60	−1354	−27 022	−1351	−22.5	0.05	3.3
1965	−7.17	60	1200	60	−1390	−27 490	−1374	−22.9	0.28	16.8
1990	−7.79	60	1200	60	−1427	−28 065	−1402	−23.4	0.42	25.3

[1] The turnover time of carbon in terrestrial biosphere reservoir was assumed to be 20 years ($\tau = 20$ years). Total NPP was assumed to remain at steady state throughout the model run.

[2] Δ_{ab} was assumed to be 16‰, representing a combination of C_3 and C_4 vegetation. This value does not affect the magnitude of the isotopic disequilibrium or the isotopic disequilibrium forcing.

then this vector should have a slope reflecting primarily C_3 vegetation (Still *et al.*, 2003a). At the ecosystem scale, estimates of Δ_{ab} derived from Keeling plot observations of nighttime respiration may also be offset from the Δ_{ab} associated with a net annual carbon sink because long-term accumulation of carbon within an ecosystem occurs primarily within wood, coarse woody debris, and slowly turning-over soil organic matter pools, whereas nighttime respiration consists largely of recently fixed sugars and carbohydrates that have not undergone the additional biochemical synthesis steps required for the formation of cellulose or lignin.

Ocean Disequilibrium Forcing

This term is analogous to the terrestrial biosphere disequilibrium forcing term, with a magnitude approximately 2–3 times larger than that for the terrestrial biosphere. The gross flux from the atmosphere to the ocean is ~80 Pg C yr^{-1}, based on radiocarbon and chlorofluorocarbon tracer constraints (Wanninkhof, 1992). Much of the carbon in the return flux, from the ocean to the atmosphere, has a residence time within the oceans of years to decades, and thus has an isotopic composition that is more enriched than that expected for a flux in isotopic equilibrium with the contemporary atmosphere (Gruber and Keeling, 2000). Estimates of the ocean isotopic disequilibrium forcing for the 1980s and early 1990s range between 37 Pg C‰ yr^{-1} and 77 Pg C‰ yr^{-1} (Tans *et al.*, 1993; Francey *et al.*, 1995; Gruber and Keeling, 2000) and critically depend on the relationship between wind speed and rates of air–sea gas exchange rate. In Fig. 13.2, I assumed an ocean disequilibrium forcing of 54 Pg C‰ yr^{-1} (Gruber and Keeling, 2000).

Ocean Carbon Sink

As with the terrestrial carbon sink, the length of ocean carbon sink vector is one of two unknowns that must account for the difference between the observed atmosphere, and fossil fuel and isotopic disequilibria terms. The slope of the ocean vector is fixed from measurements of discrimination associated with the one-way ocean-to-atmosphere, and atmosphere-to-ocean fluxes (Zhang *et al.*, 1995).

Mechanisms Contributing to Interannual Variability in Atmospheric $\delta^{13}C$

At least three additional terrestrial processes probably contribute to the variability in δ_a visible in Fig. 13.1, apart from global scale imbalances between GPP and ecosystem respiration.

Global Drought Stress

During El Nino events, less rain falls on land, particularly in the tropics, and air temperatures increase. Combined, these shifts in climate increase drought stress, and lead to decreases in both GPP and stomatal conductance. From both leaf and ecosystem level studies, there is strong empirical evidence that stomatal conductance decreases more rapidly than GPP in response to drought stress, causing a decline in internal leaf CO_2 concentrations, and thus a decrease in Δ_{ab} (Farquhar *et al.*, 1989; Ekblad and Högberg, 2001; Bowling *et al.*, 2002; Mortazavi and Chanton, 2002; Fessenden and Ehleringer, 2003). A smaller terrestrial discrimination provides less resistance to the δ_a decline caused by fossil fuels, and so the decrease in δ_a accelerates. While the anomalies in discrimination ($\dot{\Delta}_{ab}$) are likely to be small when integrated globally (much less than 1‰) they operate on GPP ($\sim$120 Pg C yr^{-1}) and so the instantaneous isotopic forcing can be quite large. Even after a year, it is likely that 30–40 Pg C of an original annual cohort of GPP remains in the biosphere, and so it is likely that much of the anomaly in discrimination contributes to interannual changes in the atmosphere (Randerson *et al.*, 2002).

Shifts in C_3 and C_4 Productivity

Even if drought stress has no effect on discrimination within C_3 or C_4 ecosystems, it is still possible that global discrimination can change from year to year from a shift in C_3 and C_4 GPP.

$$\Delta_{ab} = \frac{\Delta_{ab,C_3} \cdot GPP_{C_3} + \Delta_{ab,C_4} \cdot GPP_{C_4}}{GPP_{C_3} + GPP_{C_4}} \tag{13.6}$$

where Δ_{ab,C_3} is equal to $\sim$19‰ and $\Delta_{ab,C4}$ is equal to $\sim$5‰, and GPP_{C_3} and GPP_{C4} vary from year to year in response to variability in climate. The large difference in discrimination between these two plant functional types creates a large sensitivity to shifts in GPP. Field *et al.* (1996) proposed that shifts in C_3 and C_4 GPP may contribute to the observed variations in atmospheric $\delta^{13}C$; therefore smaller shifts in ocean and land carbon sinks were required to explain the observations from Francey *et al.* (1995) and Keeling *et al.* (1995).

More recently, Scholze *et al.* (2003) integrated ^{13}C isotopes into a global biosphere–atmosphere model, and explored interannual variation in Δ_{ab} caused by climate variability. In their model, both C_3 and C_4 photosynthesis and discrimination processes were represented using equations developed by Farquhar *et al.* (1980). The authors find that global Δ_{ab} changes substantially from year to year and that if this variability is not considered within double deconvolution analyses, partitioning of carbon sinks between land and ocean reservoirs could have biases of up to 0.8 Pg C yr^{-1}.

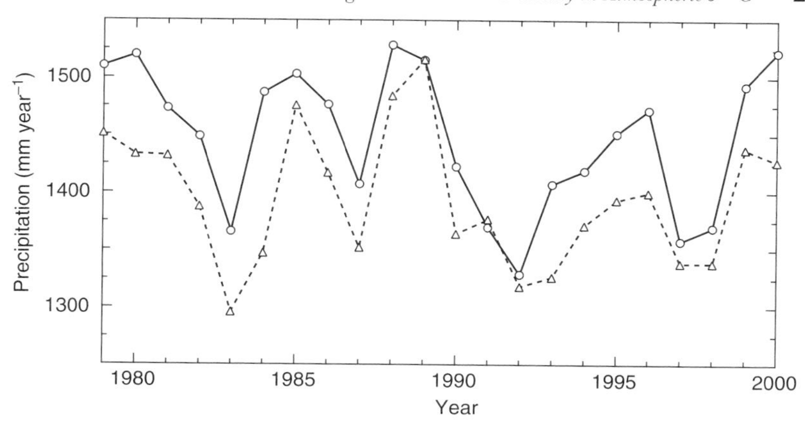

Figure 13.4 Precipitation over C_3 (solid) and C_4 (dashed) regions had considerable interannual variability over the 1979 to 2000 period, but occurred mostly in phase with one another. Precipitation estimates from the GPCP in each grid cell were weighted by mean annual NPP from the CASA model (a climatological mean NPP) and C_3 and C_4 plant cover fractions, obtained from the work of Still *et al.* (2003a).

A key mechanism identified by the authors was shifting productivity between areas dominated by C_3 and C_4 plant functional types.

In Fig. 13.4, I further explore the possibility of shifts in C_3 and C_4 GPP using the Global Precipitation Climatology Project dataset of monthly precipitation (Huffman *et al.*, 1997; Susskind *et al.*, 1997). Precipitation in each year was weighted according to a spatial map of climatologically mean NPP for either C_3 or C_4 plant functional types (Still *et al.*, 2003a). At the global level, precipitation levels over C_3 and C_4 plant functional types were highly correlated over the period from 1979 to 2000, and were both substantially lower during major El Nino events like the ones observed in 1982–1983, 1987, 1992–1994, and 1997–1998. This analysis suggests that at least for one major driver of GPP, the interannual pattern of resource availability is mostly synchronous for the two plant functional types.

In some years, however, the C_3 and C_4 precipitation levels diverged. In 1996, precipitation increased substantially more in C_3 areas. If the added rain increased GPP in this region more than in adjacent C_4 regions, then part of the arrested decline in δ_a observed in 1996 and early 1997 (Fig. 13.1D) may be attributable to this shift in global Δ_{ab} (thus requiring less interannual variability in the land sink during this period to balance the atmospheric budget).

Fires

During El Nino events, drought stress also dries fuel loads and leads to increased fire frequency and intensity in tropical woodland and closed canopy tropical forests (Van der Werf *et al.*, 2004). While the isotopic composition of fires, reflecting some combination of C_3 and C_4 biomass, should cause a decrease in atmospheric δ_a, these fluxes would also increase total carbon emissions. In this sense, fire emissions are a legitimate source that should be resolved by the double deconvolution, in contrast to the other two mechanisms proposed above, which may or may not be accompanied by changes in a net sink. However, in contrast to GPP and ecosystem respiration, fire dynamics are frequently overlooked in terms of explaining atmospheric CO_2 anomalies in many models and their isotopic composition is uncertain, and may also vary from year to year.

Unresolved Questions

Is the decrease in Δ_{ab} following drought stress that is observed in C_3 ecosystems derived from a mostly uniform increase in water-use efficiency across individuals and species or does it reflect a shift in GPP and ecosystem respiration to plants within the community that are more drought tolerant? More specifically, during drought stress, does the relative contribution to gross and net carbon exchange shift from rapidly growing plants with high internal leaf CO_2 concentrations to slowly growing plants with lower internal leaf CO_2 concentrations? If the primary ecosystem level response is mostly uniform across individuals and species, then models that use a single canopy parameterization in each region may successfully capture patterns of ^{13}C exchange at regional scales. In contrast, if the primary response to drought stress arises from a shift in species contributions, then a single canopy parameterization may still work, but its formulation would be sensitive to changes in plant biodiversity and the distribution of plant functional types. The relative importance of these two mechanisms may also shift from biome to biome. For example, in the Pacific Northwest it is clear that stands dominated by a single species often show a large increase in water use efficiency following drought (Goldstein *et al.*, 2000; Bowling *et al.*, 2002). In contrast, in tropical forests, where there are high levels of biodiversity, the relative importance of leaf level and community level responses to drought stress is not well known. Tree ring analyses from multiple species (where tree rings exist) and ongoing drought stress experimental manipulations in the tropics have the potential to provide insight.

What is the impact of diffuse/direct beam changes on the $\delta^{13}C$ of ecosystem fluxes? Roderick *et al.* (2001) provide evidence that increases in diffuse light following the Pinatubo volcano (which erupted in June of 1991) may have increased global photosynthesis and the net terrestrial carbon sink because they increased the light-use efficiency of closed canopy forests. NEE measurements from the Harvard Forest confirm at the ecosystem scale that, under clear sky conditions, diffuse light from Pinatubo aerosols enhanced summertime photosynthesis in 1992 (Gu *et al.*, 2003). Diffuse light penetrates more deeply into the canopy than direct-beam radiation, illuminating more leaves at lower levels in the canopy (Roderick *et al.*, 2001). Shifting the vertical distribution of photosynthesis downward through the canopy may increase Δ_{ab} because humidity tends to be higher within the canopy. Thus, in a way that is similar to the effect of drought stress induced by El Nino, GPP and Δ_{ab} may covary in response to volcanic events. With respect to a double deconvolution inversion, this covariance would require less interannual variability in net carbon sinks in terms of explaining variation in δ_a.

What is the isotopic composition of fires, and how does it vary in response to ENSO? A fraction of biomass burning and tropical fire emissions is associated with net deforestation (Nepstad *et al.*, 1999; Cochrane *et al.*, 1999; Siegert *et al.*, 2001) and another fraction is associated with a background disturbance regime that is balanced by C accumulation in forests in various stages of recovery. Fire associated with the background disturbance regime has both C_3 and C_4 components; in Africa and Australia savanna fires are the primary source of emissions (Scholes *et al.*, 1996; Barbosa *et al.*, 1999). However, it is likely that the increase of fires during an El Nino event (e.g., Van der Werf *et al.*, 2004) primarily occurs in closed forests and that the isotopic signal imparted by the fire emission anomalies is actually more depleted in ^{13}C (more C_3) than what would be expected during the La Nina phase of ENSO.

Conclusions

It is likely that several terrestrial processes that are linked to ENSO contribute to interannual variability in δ_a in addition to net carbon sinks. With modifications to the conventional double deconvolution approach, it is possible to account for these processes when partitioning land and ocean sinks by building on a foundation of past and contemporary physiological and ecosystem studies that link changes in Δ_{ab} with climate and plant functional type.

Acknowledgments

I gratefully acknowledge research support from NASA (NAG5-11200, NAG5-9462), NSF (NSF OPP-0097439), and NOAA (GC02-324). I thank

C. B. Field, I. Y. Fung, and C. J. Still for discussions on biosphere–atmosphere isotope exchange, and J. W. C. White, B. Vaughn, P. P. Tans, and T. J. Conway for making the atmospheric flask observations of $\delta^{13}CO_2$ and CO_2 publicly available.

References

Andres R. J., Marland G., Boden T. and Bischof S. (2000) Carbon dioxide emissions from fossil fuel consumption and cement manufacture, 1751–1991, and an estimate of their isotopic composition and latitudinal distribution. In *The Carbon Cycle* (T. M. L. Wigley and D. S. Schimel, eds) pp. 53–62. Cambridge University Press, Cambridge.

Barbosa P. M., Stroppiana D., Gregoire J. M. and Pereira J. M. C. (1999) An assessment of vegetation fire in Africa (1981–1991): Burned areas, burned biomass, and atmospheric emissions. *Global Biogeochem Cycles* **13**: 933–950.

Battle M., Bender M. L., Tans P. P., White J. W. C., Ellis J. T., Conway T. and Francey. R. J. (2000) Global carbon sinks and their variability inferred from atmospheric O-2 and delta C-13. *Science* **287**: 2467–2470.

Bowling D. R., McDowell N. G., Bond B. J., Law B. E. and Ehleringer. J. R. (2002) C-13 content of ecosystem respiration is linked to precipitation and vapor pressure deficit. *Oecologia* **131**: 113–124.

Ciais P., Friedlingstein P., Schimel D. S. and Tans. P. P. (1999) A global calculation of the delta C-13 of soil respired carbon: Implications for the biospheric uptake of anthropogenic CO_2. *Global Biogeochem Cycles* **13**: 519–530.

Cochrane M. A., Alencar A., Schulze M. D., Souza C. M., Nepstad D. C., Lefebvre P. and Davidson. E. A. (1999) Positive feedbacks in the fire dynamic of closed canopy tropical forests. *Science* **284**: 1832–1835.

Criss R. E. (1999) *Principles of Stable Isotope Distribution.* Oxford University Press, New York.

Ekblad A. and Högberg P. (2001) Natural abundance of C-13 in CO_2 respired from forest soils reveals speed of link between tree photosynthesis and root respiration. *Oecologia* **127**: 305–308.

Enting I. G., Wigley T. M. L. and Heimann M. (1994) In *CSIRO Division of Atmospheric Research Technical Paper No. 31.* CSIRO, Canberra.

Farquhar G. D., von Caemmerer S. and Berry J. A. (1980) A biochemical model of photosynthetic CO_2 assimilation in leaves of C3 plants. *Planta* **149**: 78–90.

Farquhar G. D., Ehleringer J. R. and Hubick K. T. (1989) Carbon isotope discrimination and photosynthesis. *Annu Rev Plant Physiol Plant Mol Biol* **40**: 503–537.

Fessenden J. E. and Ehleringer. J. R. (2003) Temporal variation in delta C-13 of ecosystem respiration in the Pacific Northwest: Links to moisture stress. *Oecologia* **136**: 129–136.

Field C. B., Thompson M. V., Randerson J. T. and Malmstrom C. M. (1996) Deriving land and ocean sinks from atmospheric ^{13}C: The role of C4 photosynthesis. In *Proceedings of the American Geophysical Union Fall Meeting*, San Francisco.

Francey R. J., Tans P. P., Allison C. E., Enting I. G., White J. W. C. and Trolier M. (1995) Changes in oceanic and terrestrial carbon uptake since (1982). *Nature* **373**: 326–330.

Francey R. J., Allison C. E., Etheridge D. M., Trudinger C. M., Enting I. G., Leuenberger M., Langenfelds R. L., Michel E. and Steele L. P. (1999) A 1000-year high precision record of delta C-13 in atmospheric CO_2. *Tellus Series B–Chem Phys Meteorol* **51**: 170–193.

Fung I. Y., Berry J. A., Field C. B., Thompson M. V., Randerson J. T., Malmstrom C. M., Vitousek P. M., Collatz G. J., Sellers P., Randall D., Badeck F. and John J. (1997) Carbon-13 exchanges between the atmosphere and biosphere. *Global Biogeochem Cycles* **11**: 507–533.

Goldstein A. H., Hultman N. E., Fracheboud J. M., Bauer M. R., Panek J. A., Xu M., Qi Y., Guenther A. B. and Baugh W. (2000) Effects of climate variability on the carbon dioxide, water, and sensible heat fluxes above a ponderosa pine plantation in the Sierra Nevada (CA). *Agric Forest Meteorol* **101**: 113–129.

Gruber N., Keeling C. D., Bacastow R. B., Guenther P. R., Lueker T. J., Wahlen M., Meijer H. A. J., Mook W. G. and Stocker T. F. (1999) Spatiotemporal patterns of carbon-13 in the global surface oceans and the oceanic Suess effect. *Global Biogeochem Cycles* **13**: 307–335.

Gruber N. and Keeling C. D. (2000) An improved estimate of the isotopic air-sea disequilibrium of CO_2: Implications for the oceanic uptake of anthropogenic CO_2. *Geophys Res Lett* **8**: 555–558.

Gu L. H., Baldocchi D. D., Wofsy S. C., Munger J. W., Michalsky J. J., Urbanski S. P. and Boden T. A. (2003) Response of a deciduous forest to the Mount Pinatubo eruption: Enhanced photosynthesis. *Science* **299**: 2035–2038.

Heimann M. and Keeling C. D. (1989) A three-dimensional model of atmospheric CO_2 transport based on observed winds: 2. Model description and simulated tracer experiments. In *Aspect of Climate Variability in the Pacific and the Western Americas* (D. H. Peterson, ed.) Vol. 55. pp. 237–275. American Geophysical Union, Washington D.C.

Huffman G. J., Adler R. F., Arkin P., Chang A., Ferraro R., Gruber A., Janowiak J., McNab A., Rudolf B. and Schneider U. (1997) The global precipitation climatology project (GPCP) combined precipitation dataset. *Bull Am Meteorol Soc* **78**: 5–20.

Keeling C. D., Mook W. G. and Tans P. P. (1979) Recent trends in the $^{13}C/^{12}C$ ratio of atmospheric carbon dioxide. *Nature* **277**: 121–123.

Keeling C. D., Whorf T. P., Wahlen M. and van der Plicht J. (1995) Interannual extremes in the rate of rise of atmospheric carbon dioxide since 1980. *Nature* **375**: 666–669.

Langenfelds R. L., Francey R. J., Pak B. C., Steele L. P., Lloyd J., Trudinger C. M. and Allison C. E. (2002) Interannual growth rate variations of atmospheric CO_2 and $\delta^{13}C$, H_2, CH_4, and CO between 1992 and 1999 linked to biomass burning. *Global Biogeochem Cycles* **16**: 1048.

Mortazavi B. and Chanton J. P. (2002) Carbon isotope discrimination and control of night-time canopy $\delta^{18}O$-CO_2 in a pine forest in the southeastern U.S. *Global Biogeochem Cycles* **16**: 1008.

Nepstad D. C., Verissimo A., Alencar A., Nobre C., Lima E., Lefebvre P., Schlesinger P., Potter C., Moutinho P., Mendoza E., Cochrane M. and Brooks V. (1999) Large-scale impoverishment of Amazonian forests by logging and fire. *Nature* **398**: 505–508.

Ometto J., Flanagan L. B., Martinelli L. A., Moreira M. Z., Higuchi N. and Ehleringer J. R. (2002) Carbon isotope discrimination in forest and pasture ecosystems of the Amazon Basin, Brazil. *Global Biogeochem Cycles* **16**: 1109.

Pataki D. E., Ehleringer J. R., Flanagan L. B., Yakir D., Bowling D. R., Still C. J., Buchmann N., Kaplan J. O. and Berry. J. A. (2003) The application and interpretation of Keeling plots in terrestrial carbon cycle research. *Global Biogeochem Cycles* **17**: 1022.

Quay P. D., Tilbrook B. and Wong C. S. (1992) Oceanic uptake of fossil fuel CO_2: Carbon-13 evidence. *Science* **256**: 74–79.

Randerson J. T., Thompson M. V. and Field C. B. (1999) Linking ^{13}C-based estimates of land and ocean sinks with predictions of carbon storage from CO_2 fertilization of plant growth. *Tellus Series B—Chem Phys Meteorol* **51**: 668–678.

Randerson J. T., Collatz G. J., Fessenden J. E., Munoz A. D., Still C. J., Berry J. A., Fung I. Y., Suits N. and Denning. A. S. (2002) A possible global covariance between terrestrial gross primary production and ^{13}C discrimination: Consequences for the atmospheric ^{13}C budget and its response to ENSO. *Global Biogeochem Cycles* **16**: 1136.

Roderick M. L., Farquhar G. D., Berry S. L. and Nobel I. R. (2001) On the direct effect of clouds and atmospheric particles on the productivity and structure of vegetation. *Oecologia* **129**: 21–30.

Ropelewski C. F. and Halpert M. S. (1996) Quantifying Southern Oscillation–precipitation relationships. *J Climate* **9**: 1043–1059.

Scholes R. J., Kendall J. and Justice C. O. (1996) The quantity of biomass burned in southern Africa. *J Geophys Res Atmos* **101**: 23667–23676.

Scholze M., Kaplan J. O., Knorr W. and Heimann M. (2003) Climate and interannual variability of the atmosphere-biosphere (CO_2)-C-13 flux. *Geophys Res Lett* **30**: 1097.

Siegert F., Ruecker G., Hinrichs A. and Hoffmann A. A. (2001) Increased damage from fires in logged forests during droughts caused by El Nino. *Nature* **414**: 437–440.

Still C. J., Berry J. A., Collatz G. J. and DeFries R. S. (2003a) The global distribution of C3 and C4 vegetation: Carbon cycle implications. *Global Biogeochem Cycles* **17**: 1006.

Still C. J., Berry J. A., Ribas-Carbo M. and Helliker B. R. (2003b) The contribution of C-3 and C-4 plants to the carbon cycle of a tallgrass prairie: An isotopic approach. *Oecologia* **136**: 347–359.

Susskind J., Piraino P., Rokke L., Iredell T. and Mehta A. (1997) Characteristics of the TOVS Pathfinder Path A dataset. *Bull Am Meteorol Soc* **78**: 1449–1472.

Tans P. P., Berry J. A. and Keeling R. F. (1993) Oceanic $^{13}C/^{12}C$ observations: A new window on ocean CO_2 uptake. *Global Biogeochem Cycles* **7**: 353–368.

Thompson M. V. and Randerson J. T. (1999) Impulse response functions of terrestrial carbon cycle models: Method and application. *Global Change Biol* **5**: 371–394.

Tieszen L. L., Reed B. C., Bliss N. B., Wylie B. K. and DeJong D. D. (1997) NDVI, C-3 and C-4 production, and distributions in Great Plains grassland land cover classes. *Ecol Appl* **7**: 59–78.

Townsend A. R., Asner G. P., White J. W. C. and Tans P. P. (2002) Land use effects on atmospheric 13C imply a sizable terrestrial CO_2 sink in tropical latitudes. *Geophy Res Lett* **29**: 1426.

Trumbore S. (2000) Age of soil organic matter and soil respiration: Radiocarbon constraints on belowground C dynamics. *Ecol Appl* **10**: 399–411.

Van der Werf G. R., Randerson J. T., Collatz G. J., Giglio L., Kasibhatla P. S., Arellano A. F., Olsen S. C. and Kasischke E. S. (2004) Continental-scale partitioning of fire emissions during the 1997 to 2001 El Nino/La Nina period. *Science* **303**: 73–76.

Vandam D., Veld Kamp E. and Van Breemen N. (1997) Soil organic-carbon dynamics: Variability with depth in forested and deforested soils under pasture in Costa-Rica. *Biogeochemistry* **39**: 343–375.

Wanninkhof R. (1992) Relationship between wind-speed and gas-exchange over the ocean. *J Geophys Res Oceans* **97**: 7373–7382.

Wittenberg U. and Esser G. (1997) Evaluation of the isotopic disequilibrium in the terrestrial biosphere by a global carbon isotope model. *Tellus Series B—Chem Phys Meteorol* **49**: 263–269.

Zhang J., Quay P. D. and Wilbur D. O. (1995) Carbon-isotope fractionation during gas–water exchange and dissolution of CO_2. *Geochim Cosmochim Acta* **59**: 107–114.

14

Remarks on the Use of ^{13}C and ^{18}O Isotopes in Atmospheric CO_2 to Quantify Biospheric Carbon Fluxes

Philippe Ciais, Matthias Cuntz, Mark Scholze, Florent Mouillot, Philippe Peylin, Vincent Gitz

Introduction

In this review chapter, we write the mass-conservation equations for CO_2 and its isotopomers $^{13}CO_2$ and $CO^{18}O$ that can be used to infer globally biospheric and oceanic net fluxes in the case of ^{13}C, and gross terrestrial biospheric fluxes in the case of ^{18}O. The quantitative use of atmospheric measurements of ^{13}C and ^{18}O in CO_2 to better constrain those fluxes requires knowledge of various processes specific to each isotopomer. The chapter is divided into two parts, one on each isotope. For ^{13}C, we review existing work that calculated isofluxes either using global estimates, or derived isofluxes from spatially and temporally explicit models. In addition, we estimate the magnitude of new isofluxes that were not addressed in former studies. These cover the effects of biomass burning, rock weathering and volcanism, and the oxidation of reduced carbon gases into CO_2 within the atmosphere. We also performed a new global calculation of the ^{13}C isoflux caused by the replacement of C_3 vegetation by C_4 plants following changes in land use. Those isotopic disequilibrium terms which roughly correspond to a ^{13}C flux to or from the atmosphere that has no counterpart in ^{12}C, prove to be important in the global apportionment of ocean vs terrestrial carbon fluxes using atmospheric ^{13}C data. Altogether, they have an aliasing effect of up to 1 GtC yr^{-1}. For ^{18}O, which is less understood than ^{13}C, we discuss in detail the different processes and controlling variables that contribute to the atmospheric signal. To do so, we refer to the calculations of spatially explicit global models of the soil–plant–atmosphere system. In the specific case of ^{18}O, the isotopic disequilibrium terms are of secondary importance because ^{18}O directly constrains the gross fluxes (photosynthesis and ecosystem respiration) rather than the net fluxes as for ^{13}C. One exception to that rule is the disequilibrium induced by stratosphere–troposphere mixing of the enriched stratospheric ^{18}O signal.

Formulation of Global Budgets

Globally, the change of CO_2 in time is the sum of all surface fluxes, F_i:

$$\frac{dC_a}{dt} = M_a \sum_i F_i \tag{14.1}$$

where C_a denotes the global mean atmospheric CO_2 mixing ratio and M_a the conversion factor between fluxes in GtC and mixing ratios in ppm. $1/M_a = 2.12$ GtC ppm^{-1}, i.e., about 2 GtC are required to change the atmospheric CO_2 mixing ratio by 1 ppm (M_a is independent of the CO_2 mixing ratio). The same equation can be used for the relationship of the mixing ratio of an isotopomer of CO_2, for example, like $^{13}CO_2$ or $CO^{18}O$ (a plain C stands thereby for ^{12}C and a plain O stands for ^{16}O), with the surface fluxes of the isotopomer. Denoting all isotopomer variables with a prime gives:

$$\frac{dC'_a}{dt} = M_a \sum_i F'_i \tag{14.2}$$

R_x is the ratio of C'_a over C_a or of F'_i over F_i, so Eq. 14.2 can be written as:

$$\frac{d(C_a R_a)}{dt} = M_a \sum_i R_i F_i \tag{14.3}$$

Writing this in delta notation, $\delta = R_x/R_{standard} - 1$, and using Eq. 14.1 gives:

$$\frac{d(C_a \delta_a)}{dt} = M_a \sum_i \delta_i F_i \tag{14.4}$$

with δ_a the global mean atmospheric delta value of CO_2 and δ_i the delta value of the CO_2 flux, i. $C_a\delta_a$ is thereby a mass conservative tracer, just like CO_2. $dC_aR_a = R_a dC_a + C_a dR_a$, so Eq. 14.3 can be written as:

$$C_a \frac{dR_a}{dt} = M_a \sum_i R_i F_i - R_a M_a \sum_i F_i = M_a \sum_i F_i (R_i - R_a) \tag{14.5}$$

Or in delta notation:

$$\frac{d\delta_a}{dt} = \frac{M_a}{C_a} \sum_i F_i(\delta_i - \delta_a) = \frac{M_a}{C_a} \sum_i F_i \Delta_i \tag{14.6}$$

Δ_i is therefore just the difference between the delta value of flux i and the atmospheric delta value. The product of F_i and Δ_i is called isoflux, so the temporal evolution of CO_2 is mainly the sum of all CO_2 fluxes and the evolution of the atmospheric delta value is the sum of all isofluxes. Conceptually, the index i in Eqs 14.1 and 14.2 need not be the same. That means that the net flux of a process for CO_2 can be zero but there can be a flux of the isotopomer. For example, the ocean in steady-state gives a zero net flux of CO_2 but it can absorb $^{14}CO_2$, changing the delta value of ^{14}C in atmospheric CO_2 without changing the atmospheric CO_2 mixing ratio. In this case, one writes the zero net CO_2 flux as two canceling CO_2 gross fluxes in Eq. 14.6. Here 'canceling' means that both $F_{i,in}$ and $F_{i,out} = -F_{i,in}$ would appear as F_i in Eq. 14.1. More generally, if we group all canceling gross fluxes F_j in Eq. 14.6, it changes to:

$$\begin{aligned}\frac{d\delta_a}{dt} &= \frac{M_a}{C_a}\left[\sum_i F_i(\delta_i - \delta_a) + \sum_j F_j(\delta_{j,in} - \delta_{j,out})\right] \\ &= \frac{M_a}{C_a}\left(\sum_i F_i\Delta_i - \sum_j D_j\right)\end{aligned} \tag{14.7}$$

where F_i are resulting net (for ^{13}C) or gross (for ^{18}O) fluxes that appear also in Eq. 14.1, and D_j values are called 'isotopic disequilibria' and are the product of the CO_2 gross flux and the difference between the delta value of the gross flux into and out of the atmosphere. The D_j values stay unchanged if one regards the mass conservative tracer $C_a\delta_a$ and Eq. 14.4 becomes:

$$\frac{d(C_a\delta_a)}{dt} = M_a\left(\sum_i \delta_i F_i + \sum_j D_j\right) \tag{14.8}$$

In the text below, we will suppress M_a and M_a/C_a for simplicity. This is the same as expressing the CO_2 fluxes and isofluxes in the correct units.

Understanding $\delta^{13}C$ in Atmospheric CO_2

Atmospheric ^{13}C is fractionated by C_3 terrestrial photosynthesis ($\sim -17‰$) with only minor effects caused by air–sea gas exchange processes ($\sim -2‰$). On the other hand, C_4 plants fractionate very little ($\sim -4.4‰$), and therefore C_4 plant exchange of carbon with the atmosphere cannot be distinguished from air–sea fluxes. Atmospheric measurements of ^{13}C and CO_2 have been used to apportion ocean and land fluxes globally with their year-to-year fluctuations, using time series from few stations (Francey *et al.*, 1995; Keeling, 1995), and regionally using a network of stations and atmospheric tracer transport models (Heimann and Keeling, 1989; Ciais *et al.*, 1995a,b; Fung, 1997; Bousquet, 1999a,b; Rayner *et al.*, 1999). This method, called 'double deconvolution' has also been applied to ice-core records of CO_2 and $\delta^{13}C$ to reconstruct the temporal evolution of ocean and land net fluxes over the twentieth century (Joos and Bruno, 1998; Trudinger *et al.*, 1999). The double deconvolution is based on the mass balance of CO_2 and $^{13}CO_2$ in the atmosphere. The mass balance of CO_2 can be expressed as:

$$\frac{dC_a}{dt} = F_o + F_b + F_f \tag{14.9}$$

where F_o and F_b are the netalgebraic air–sea fluxes and air–land fluxes, and F_f represents the fossil fuel emissions. The F_b term can be broken down into:

$$F_b = F_{bur} + F_{bur\_regrow} + F_{def\_resp} + F_{def\_assim} + F_{res} \tag{14.10}$$

where F_{bur} is the CO_2 source implied by biomass burning that is partially offset by a carbon sink due to re-growth over burned areas F_{bur_regrow}. F_{def_resp} is the gross loss of carbon to the atmosphere implied by forest conversion to either grasslands or pastures, and F_{def_assim} is the gross carbon uptake due to re-growth of newly established ecosystems following deforestation. F_{res} is the residual terrestrial flux, globally a sink, which is often referred to as the 'missing sink' of anthropogenic CO_2. The mass balance of atmospheric ^{13}C is given by:

$$\frac{d\delta_a}{dt} = F_o \cdot \varepsilon_{ao} + F_b \cdot \Delta_b^{13} + F_f \cdot (\delta_f - \delta_a) + D_o + D_b + D_{bur} + D_{def} \tag{14.11}$$

where Δ_b^{13} is the average ^{13}C discrimination of terrestrial plants and ε_{oa} is the sea-to-air isotopic fractionation factor. The last four terms on the right hand side are called isotopic disequilibria, and their expressions are given by:

$$D_o = F_{oa} \cdot (\delta_o - \delta_o^e) \tag{14.12}$$

$$D_b = F_{HR} \cdot (\delta_b - \delta_b^e) \tag{14.13}$$

$$D_{bur} \approx F_{bur} \cdot (\delta_{bur} - \delta_{bur\_regrow}) \tag{14.14}$$

$$D_{def} = F_{def\_resp} \cdot (\delta^*_{def\_resp} - \delta^*_{def\_assim}) \tag{14.15}$$

where the subscript *HR* represents heterotrophic respiration. The disequilibria have in common that they cause a net flux of δ to or from the atmosphere in Eq. 14.11, with no counterpart in the net CO_2 fluxes in Eq. 14.9. Thus, isotopic disequilibria are very important terms in double deconvolutions, which must be estimated in order to properly infer F_b and F_o separately from atmospheric measurements. Both isoflux terms and isotopic disequilibria terms, expressed globally in Eqs 14.11–14.15, are spatially distributed of course. If we are to use transport models instead of a single-box atmosphere, then the spatial and temporal distribution of the disequilibria flux must be estimated and subsequently transported to be accounted for (for instance pre-subtracted) in the $\delta^{13}C$ signal at atmospheric stations. The isotopic disequilibria are generally proportional to the gross exchange fluxes, with the exception of fossil fuel emissions. In the following sections, we review the magnitude of the isotopic disequilibria, their uncertainty at the global scale, and wherever possible their spatial distribution. Finally in the last paragraph, we focus our attention on spatial and temporal patterns of plant discrimination.

Air–Sea Isotopic Disequilibrium (D_o)

The air–sea ^{13}C disequilibrium is proportional to the sea-to-air gross flux, $F_{oa} = K_{ex}p_o$, and hence its estimation requires knowledge of the CO_2 partial pressure in surface waters (p_o) and the air–sea gas transfer coefficient K_{ex}. The air–sea disequilibrium therefore bears an uncertainty at least as large as that of the air–sea gas exchange transfer coefficient [±30% (Wanninkhof, 2002)]. It is also noted that F_{oa} increases with time along with the secular atmospheric increase in CO_2, as over most ocean gyres, p_o tracks the atmospheric partial pressure signal p_a. Today's average value of F_{oa} is on the order of 85 ± 17 GtC yr^{-1} (Heimann and Maier-Reimer, 1996). It should be pointed out (Joos and Bruno, 1998) that modeling D_o requires an estimation of p_o, and K_{ex} and thus is not independent of inferring the net air–sea flux $F_o = K_{ex}(p_o - p_a)$. The sea-to-air flux has already increased by roughly 30% since preindustrial times and will likely continue to do so in the future, making the air–sea disequilibrium a dominant term in future double deconvolutions. The difference between δ_o, the real world ocean $\delta^{13}C$ in surface waters total carbon and the hypothetical value that would correspond to isotopic equilibrium with today's atmosphere $\delta_o^e = \delta_a - \varepsilon_{ao} + \varepsilon_{oa}$, is positive. Indeed, ocean carbon integrates atmospheric ^{13}C that has invaded the ocean earlier on, when atmospheric $\delta^{13}C$ was higher than today because of the secular decrease in atmospheric $\delta^{13}C$ due to fossil fuel accumulation in the atmosphere. Estimates of the $\delta_o - \delta_o^e$ difference of 0.43‰ made by Tans *et al.* (1993) using earlier ocean data has been revised by Gruber and Keeling (2001) to 0.62 ± 0.08‰, based on more recent oceanographic surveys, which makes the overall air–sea disequilibrium term on the order of 53 ± 13 GtC ‰ yr^{-1}. Other published estimates are 47 GtC ‰ yr^{-1} using a 3-D ocean carbon model (Murname and Sarmiento, 2000), and 60 GtC ‰ yr^{-1} using a reduced ocean model (Joos *et al.*, 1999), and 50 GtC ‰ yr^{-1} using latitudinal averages of ocean $\delta^{13}C$ data and gas transfer coefficient fields (Ciais *et al.*, 1995b; Battle *et al.*, 2000). In comparison, in

Eq. 14.11 a 1 $GtC\,yr^{-1}$ sink in C_3 biomes ($\varepsilon_{ab} = -17‰; \delta_a = -8‰$) would increase δ_a by 25 $GtC\,‰\,yr^{-1}$.

Soil-Respired Isotopic Disequilibrium (D_b)

The soil-respired isotopic disequilibrium is proportional to the return flux of CO_2 from land ecosystems to the atmosphere F_{ba}. Strictly speaking, if F_{ab} equals net carbon assimilation, then F_{ba} equals ecosystem respiration. Generally, averaged over long enough timescales (longer than a few weeks) the difference in isotopic composition between assimilates incorporated by plant photosynthesis and their respiration by maintenance and growth respiration is close to zero. This assumption seems reasonable in light of isotope marking experiments (Ekblad and Högberg, 2001), but measurements over sub-daily periods suggest that assimilation and ecosystem respiration rarely reach isotopic equilibrium (Bowling *et al.*, 1999, 2001; Ogée *et al.*, 2003), except perhaps during the early morning and late afternoon. Yet assuming isotopic equilibrium between CO_2 respired by autotrophic processes and CO_2 fixed by photosynthesis, F_{ab} can be identified to net primary productivity (NPP) and thereof F_{ba} to heterotrophic respiration F_{HR}. As remarked for the air–sea disequilibrium, F_{HR} is not strictly independent of the 'unknown' F_b in Eq. 14.11, since $F_b = -NPP + F_{HR}$ ignoring disturbances, hence introducing another degree of complexity in modeling the soil-respired isotopic disequilibrium, D_b. Since the globally averaged net biospheric flux F_b generally amounts to a small fraction (only 1–2% of the gross flux F_{HR}), one can make for global studies the reasonable assumption that $F_{HR} \approx NPP$. This simplification is not verified regionally at the scale of current atmospheric inversions, or at the stand scale. For instance, one can compare regional NPP estimates from models (Cramer *et al.*, 2001) and NBP from inversions (Gurney *et al.*, 2002) at the scale of continents, or NPP and NEP values at stand level (Valentini *et al.*, 2000; Wirth *et al.*, 2002) to realize that F_b may be regionally or locally a sizeable fraction of NPP. Third, the value of the soil-respired disequilibrium D_b should increase with time as the input of carbon to ecosystems via NPP may increase to yield a net uptake of carbon in ecosystems. If NPP has globally increased as driven by CO_2 fertilization (Thompson *et al.*, 1996) or by other factors (Nemani *et al.*, 2003) over the past decades, then D_b must have paralleled the trend in NPP. Today's estimate of global F_{HR} ranges from 39.9 to 80.5 $GtC\,yr^{-1}$ if based on NPP from 15 global models (Cramer *et al.*, 1999) and is 76.5 $GtC\,yr^{-1}$ based on soil respiration measurements (Raich and Potter, 1995). This latter estimate is 30 to 60% higher than global NPP because it takes into account root autotrophic respiration of recent assimilates, which would make it inappropriate to estimate the global isotopic disequilibrium D_b. The isotopic difference between soil-respired CO_2 issued from organic matter decomposition and recently formed phytomass (δ_b^e) depends on the turnover time of excess carbon in ecosystems, which increases with increasing latitude and decreasing temperatures (Schimel *et al.*, 1994). Note again that turnover times of carbon are not independent from the inferred biospheric uptake F_b in double deconvolutions (Randerson *et al.*, 1998). The difference in isotopic composition $\delta_b - \delta_b^e$ between newly formed phytomass and soil-respired CO_2 from previously formed biomass has been evaluated at 0.5‰ (Heimann and Keeling, 1989) and 0.5‰ (Ciais *et al.*, 1995a) using zonally averaged turnover times, and 0.56‰ (Ciais *et al.*, 1999), 0.33‰ (Fung *et al.*, 1997) and 0.53‰ (Scholze *et al.*, 2003) using spatially explicit global biosphere models. These numbers correspond to a flux-weighted global average. The value of $\delta_b - \delta_b^e$ is generally higher in biosphere models at high northern latitudes where a long residence time of carbon in soils prevails, and in tropical forests where carbon is immobilized in long-lived trees. To our knowledge, few measurements of $\delta_b - \delta_b^e$ are available, but it should be kept in mind that, generally, most newly formed biomass gets respired rapidly, and thus bears only a small isotopic disequilibrium. For instance, using radiocarbon tracing methods, Balesdent and Recous (1997) determined that over a tilled temperate cropland, 75% of the litter is degraded within a year, whereas 14% of soil carbon had a residence time of 40 years. Gaudinski *et al.* (2000) at Harvard Forest suggest that 59% of soil respiration was issued from carbon that resided in the plant/soil pools for less than a year, with an average of CO_2 produced by heterotrophs of 8 years only. It should finally be noted that there must be seasonal variations in $\delta_b - \delta_b^e$, in the range of 0.3‰ as modeled by Fung *et al.* (1997, see their Fig. 7) as different pool sizes are seasonally degraded by respiration processes. Together with estimates of heterotrophic respiration, the global soil-respired isotopic disequilibrium D_b ranges from 19.8 GtC ‰ yr^{-1} (Fung *et al.*, 1997) to 33.6 GtC ‰ yr^{-1} (Ciais *et al.*, 1999) and 37.4 GtC ‰ yr^{-1} as the mean over the 1990s (Scholze *et al.*, 2003).

Biomass Burning Isotopic Disequilibrium (D_{bur})

The isotopic composition δ_{bur} of burned aged biomass is isotopically enriched compared to recently formed plant tissues in recovering ecosystems after fire (δ_{bur_regrow}). Assuming $\delta_{bur} = \delta_{bur\_regrow}$, this would yield to aliasing towards a higher terrestrial sink in double deconvolutions. We used the historical record of atmospheric $\delta^{13}C$ together with estimates of the mean age of burned biomass (from fire return times) and combustion fluxes to estimate the value $\delta_{bur} = -7.55$‰, that is, 0.45‰ heavier than for today's atmosphere. Overall, it is difficult to compute the net imbalance in the terrestrial carbon budget implied by biomass burning sources, partly offset by post-fire recovery sinks. Assuming that $F_{bur} = F_{bur\_regrow}$

(range 2.1–3.4 GtC yr^{-1}) yields a biomass burning disequilibrium isoflux D_{bur} = 1.66 GtC ‰ yr^{-1}. That is a small contribution to the mean trend in $\delta^{13}C$, equivalent to aliasing a terrestrial sink of 0.1 GtC yr^{-1}.

Isotopic Effects Caused by Land Use and C_3/C_4 Shifts in Vegetation

The conversion of forests to croplands and grasslands globally decreases the turnover time of carbon in ecosystems and thus acts to reduce the value of the soil-respired disequilibrium. In addition, in places where forests are converted to C_4 crops (e.g., sugar cane) or C_4 pastures, a difference in isotopic composition between newly formed C_4 biomass and respired CO_2 from former C_3 forest soils, of the order of $\Delta_{b-C_4} - \Delta_{b-C_3} \approx 12.5‰$, will be seen in the atmosphere. This disequilibrium term, D_{def}, unlike the ocean and the soil-respired disequilibria D_o and D_b, acts to deplete the atmosphere of ^{13}C. Ignoring it would be equivalent to inferring too small a terrestrial sink F_b in double deconvolution studies. The land use disequilibrium accounts for a source of CO_2 to the atmosphere right after forest conversion followed by delayed emissions during the release of former C_3 soil carbon, the sum of both being F_{def_resp}, which is counterbalanced by carbon uptake from the newly established vegetation F_{def_assim}. After all former C_3 soil carbon has been oxidized, a new carbon equilibrium state is reached, which is also a carbon isotopic equilibrium state, where inputs compensate for outputs. Typically this takes 20 to 30 years in the tropics (Townsend *et al.*, 1997). The global significance of this effect, outlined by Ciais *et al.* (1999), was quantified by Townsend *et al.* (2002) using the CENTURY soil carbon model. Over all fractions of land that have been converted during a time period beginning τ years ago until today, the land use change-induced isotopic disequilibrium at instant t is:

$$D_{def}(t) = \int_{u=t-\tau}^{u=t} F_{def\_resp}\,\delta_{def\_resp}\,du - \int_{u=t-\tau}^{u=t} F_{def\_assim}\,\delta_{def\_assim}\,du \tag{14.16}$$

This equation is unpleasant to account for because it would mean that F_b alone cannot be solved separately in Eqs 14.9 and 14.11 but rather the sum $F_b - F_{def\_assim} + F_{def\_resp}$, which is not a very useful quantity in global carbon cycle studies. Therefore, Townsend *et al.* (2002) have made implicitly the approximation $F_{def\_assim} \approx F_{def\_resp}$, and therefore:

$$D_{def}(t) = \int_{u=t-\tau}^{u=t} F_{def\_resp}(\delta_{def\_resp} - \delta_{def\_assim})\,du \tag{14.17}$$

It is also possible without that simplification to rewrite Eq. 14.16 as

$$D_{def} = F_{def\_resp}(t) \cdot (\delta^*_{def\_resp} - \delta^*_{def\_assim}) \qquad (14.18)$$

Where:

$$\delta^*_{def\_resp} = \frac{1}{F_{def\_resp}(t)} \int_{t-\tau}^{t} F_{def\_resp}(t') \cdot \delta_{def\_resp}(t') \cdot dt' \qquad (14.19)$$

$$\delta^*_{def\_assim} = \frac{1}{F_{def\_resp}(t)} \int_{t-\tau}^{t} F_{def\_assim}(t') \cdot \delta_{def\_assim}(t') \cdot dt' \qquad (14.20)$$

with those quantities being flux-weighted by the spatial distribution of F_{def_assim}, F_{def_resp} in the case where a spatially explicit biosphere model is used. The value of D_{def} (Townsend *et al.*, 2002) is -15 GtC ‰ yr^{-1}, that is about one-third of the soil-respired disequilibrium, but of opposite sign. With the land use bookkeeping model of Gitz and Ciais (2003), we performed an independent calculation of D_{def} and obtained -18.9 GtC ‰ yr^{-1}. Both calculations made the same assumption that 80% of deforested areas in the Tropics are replaced by C_4 grasses (50% in Asia because of rice cultivation). Land use changes, yielding to the substitution of forests by C_4 grasslands, mostly occur in the tropics, so adding D_{def} in double deconvolutions increases the tropical sink by roughly 1 GtC yr^{-1} (Townsend *et al.*, 2002).

In addition to tropical land use changes, the establishment of C_4 croplands (maize) at mid-northern latitudes during the twentieth century in North America and Europe has had an impact on the atmospheric $\delta^{13}C$ signal (60 Mha were cultivated for corn in 1980–2000 compared to 2–5 Mha in the early twentieth century). This adds a disequilibrium not accounted for by Townsend *et al.* (2002) that we estimated at 0.5 GtC ‰ yr^{-1} using the simplified land use model (forest–croplands) of Gitz and Ciais (2003). Accounting for additional plowing of grasslands into C_4 croplands in North America would increase that estimate up to 2 GtC ‰ yr^{-1} (M. Scholze, LPJ results, unpublished).

Rock Weathering ^{13}C Isotope Effects

Carbonate rock weathering results in small 'background' natural fluxes in the carbon cycle, acting over land as a sink, $-F_{carb\_w}$, of soil CO_2 reacting with $CaCO_3$ capping the Earth's crust to yield dissolved bicarbonates ions:

$$CO_{2(soil)} + CaCO_3 \rightarrow 2HCO_3^- \qquad (14.21)$$

The HCO_3^- ions are transported into freshwater systems, where a small fraction may re-precipitate as carbonates, whereas the rest get transported to the coastal seas by rivers, and a sub-fraction of it to the deep ocean.

In the sea, the inverse of reaction 14.21 precipitates $CaCO_3$ in marine sediments, causing a CO_2 source, F_{carb_w}, to become available to the atmosphere. Regarding the $\delta^{13}C$ transfers in the weathering cycle, the initial sink has a signature, $\delta_{w\_uptake} = \delta_b$. As soil CO_2 is enriched by 4.4‰ with respect to δ_b, and reaction 14.21 consumes one mole of soil CO_2 for one mole of soil carbonate (at +1‰), the two HCO_3^- ions in reaction 14.21 take an intermediate $\delta^{13}C$ approximately equal to $(\delta_b + 4.4 + 1)/2 \approx -11$‰, a value effectively observed in freshwaters carbonate systems. Once transferred by rivers to the sea, the dissolved HCO_3^- get isotopically diluted with the huge marine bicarbonate reservoir and when CO_2 returns to the atmosphere, fractionation occurs to reconstitute the isotopic balance. Yet overall, carbonate rock weathering results in the lateral displacement of isofluxes, with an iso-uptake over land of:

$$-F_{carb\_w} \cdot \Delta_b^{13} \tag{14.22}$$

compensated by an equivalent iso-release over the oceans.

Parallel to the weathering of carbonates, the weathering of silicate rocks proceeds by the reaction:

$$3H_2O + 2CO_2 + MSiO_3 \rightarrow M^{2+} + 2HCO_3^- + H_4SiO_4 \tag{14.23}$$

Once delivered to the ocean via rivers, the two bicarbonates produced by reaction 14.23 contribute to the formation of marine carbonates:

$$2HCO_3^- \rightarrow CaCO_3 + CO_2 \tag{14.24}$$

Unlike for carbonate weathering where the land sink balances an ocean source, for silicate weathering, two moles of atmospheric CO_2 are consumed on land for only one being released over the ocean. The carbon cycle gets balanced over geologic timescales (>1 Myr) since carbonates deposited at the bottom of the sea get subducted in the hot mantle, yielding carbon outgassing by volcanoes. The overall isoflux due to silicate rock weathering can be expressed as:

$$D_{silicate\_weath} = F_{silicate\_weath}(\delta_{w\_release} - 2 \cdot \delta_{w\_uptake}) + F_{volcanoes}\delta_{mantle} \tag{14.25}$$

With $F_{silicate\_weath} = 0.2\,\mathrm{GtC\,yr^{-1}}$, $F_{volcanoes} = 0.02\,\mathrm{GtC\,yr^{-1}}$, $\delta_{mantle} = +6$‰, this yields $D_{silicate\_weath} = 3.6$ GtC ‰ $\mathrm{yr^{-1}}$, equivalent to the aliasing effect of a sink of $0.2\,\mathrm{GtC\,yr^{-1}}$ by land ecosystems. Uncertainties are large on this term and there could also be isotopic equilibrium on long timescales. Overall, both carbonate and silicate rock cause an isoflux of ≈4 GtC % $\mathrm{yr^{-1}}$, mostly a disequilibrium. This would diminish the mean value of the terrestrial sink currently inferred in double deconvolutions by roughly $0.4\,\mathrm{GtC\,yr^{-1}}$. Since 'background' fluxes only evolve on geologic timescales, they cause a systematic offset to the mean fluxes, but do not affect the interannual variability and secular changes inferred in double deconvolutions.

Isotope Effects from the Oxidation of Reactive Carbon Gases in the Atmosphere

Carbon compounds are released from plants to the atmosphere as non-CO_2 gases (isoprene, terpenes, CH_4, CO, acetone, methanol, ...). Non-CO_2 gases react with OH and O_3 to be oxidized into CO and eventually into CO_2 (a small direct oxidation channel by R-O_2 radicals into CO_2 also exists). Atmospheric CO has a lifetime of a few months and it is, on average, depleted in ^{13}C compared to atmospheric CO_2 (average $\delta^{13}C\text{-}CO \approx -27‰$ with a range of -20 to $-40‰$, after Bergamaschi *et al.* [2000b]). Using the oxidation reaction $CO + OH \rightarrow CO_2$ in respect of the kinetic isotope effect (fractionation) determined by Röckmann *et al.* (1998) we estimated that CO_2 produced from CO should be enriched by 3‰ globally with regard to the CO atmospheric reservoir: $\delta^{13}C\text{-}CO_2$ from CO oxidation, $\delta_{CO\_oxidation} = -27 + 3 = -24‰$. With a total source of carbon produced by CO oxidation of 0.92 GtC yr^{-1} in the atmosphere (Novelli *et al.*, 1995), this yields a net isoflux (not a disequilibrium) of -25.6 GtC ‰. This isoflux is almost entirely compensated for by the fact that part of the *NPP* is not returned to the atmosphere as CO_2, but via emissions of non-CO_2 gases. Globally, the annual growth rates of CO_2 and $\delta^{13}C\text{-}CO_2$ in Eqs 14.10 and 14.11 thus already implicitly include the oxidation of non-CO_2 gases ($F_{CO\_oxidation}$). It should, however, be remarked that atmospheric transport models based on space and time gradients in $\delta^{13}C$ and CO_2 to infer regional fluxes may be biased regionally when not accounting for the emission of non-CO_2 carbon gases and the oxidation of CO into CO_2 within the atmosphere.

Atmosphere–Land Fractionation Factor or Discrimination (Δ_b^{13})

Δ_b^{13} is controlled by the photosynthetic pathway (C_3/C_4) and by the ratio of the chloroplast to canopy CO_2 concentration (C_c/C_a). C_3 and C_4 photosynthetic fractionation differences are associated with the biochemical fixation of CO_2; the C_c/C_a ratio depends on the opening of the leaf stomata and therefore is a function of the plant water stress, and of the mesophyll resistance (Lloyd and Farquhar, 1994). The large-scale spatial distribution of Δ_b^{13} reflects mostly the extent of C_4 photosynthesis relative to C_3 (Fig. 14.1) and partly the water limitations for C_3 plants. Figure 14.1 shows two annual maps of Δ_b^{13} as calculated by Scholze *et al.* (2003) using the LPJ terrestrial biosphere model and by Cuntz *et al.* (2003a,b) using the MECBETH photosynthesis–respiration model. Other modelers have produced similar maps of Δ_b^{13} and sometimes calculated the associated

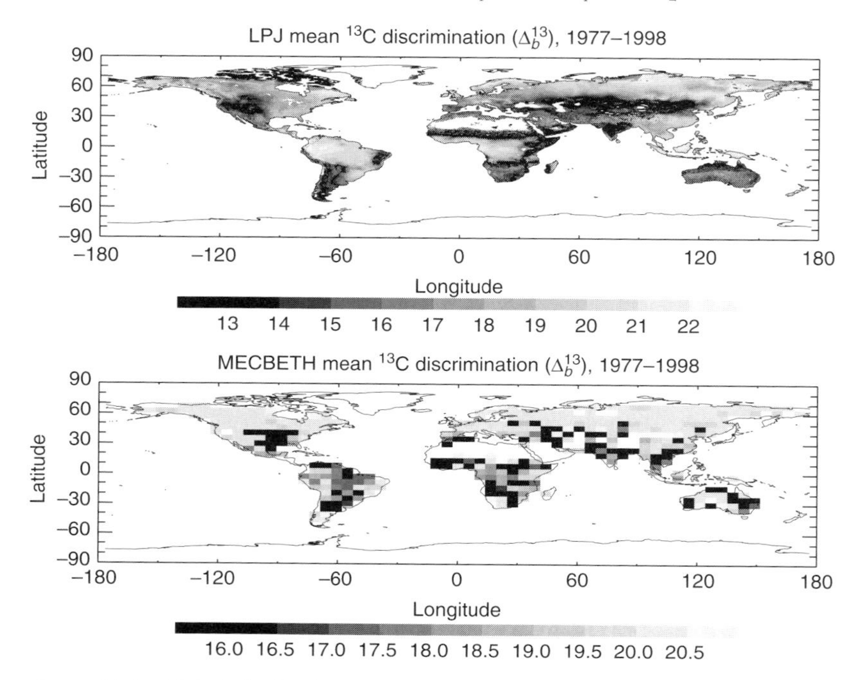

Figure 14.1 Global distribution of photosynthetic discrimination as simulated by two different models; (top) LPJ biogeochemical model at 0.5° spatial resolution forced by climate input data in Scholze *et al.* (2003); (bottom) MECBETH photosynthesis–respiration model coupled to an atmospheric general circulation model at 5° resolution in Cuntz *et al.* (2003a).

$\delta^{13}C$-CO_2 atmospheric distribution using an atmospheric transport model (Fung *et al.*, 1997; Kaplan *et al.*, 2002). Lloyd and Farquhar (1994) mapped Δ_b^{13} by extrapolation of observed physiological relationships. In all these studies, Δ_b^{13} was modeled to vary with latitude across the range from 7 to 10‰ for changes from C_4 to C_3 plants (subtropical to tropical and temperate regions) and from 2 to 4‰ for C_3 plants only (temperate and high latitude regions) reflecting the stomatal response to water stress caused by arid conditions.

It is difficult to validate such Δ_b^{13} maps against observations. Local canopy data are rather scarce and difficult to upscale, although the situation is quickly improving as part of the BASIN and SIBAE isotope networks (see http://basin isotopes.org/ and http://www.esf.org/sibae and references therein). An alternative is to use regional Δ_b^{13} deduced from the relationships between CO_2 and $\delta^{13}C$-CO_2 atmospheric measurements, as did Bakwin *et al.* (1998) and Randerson *et al.* (2002b), but atmospheric measurements are also scarce and difficult to downscale. In Fig. 14.2, we provide such a plot showing different zonally averaged modeled Δ_b^{13} values against atmospheric

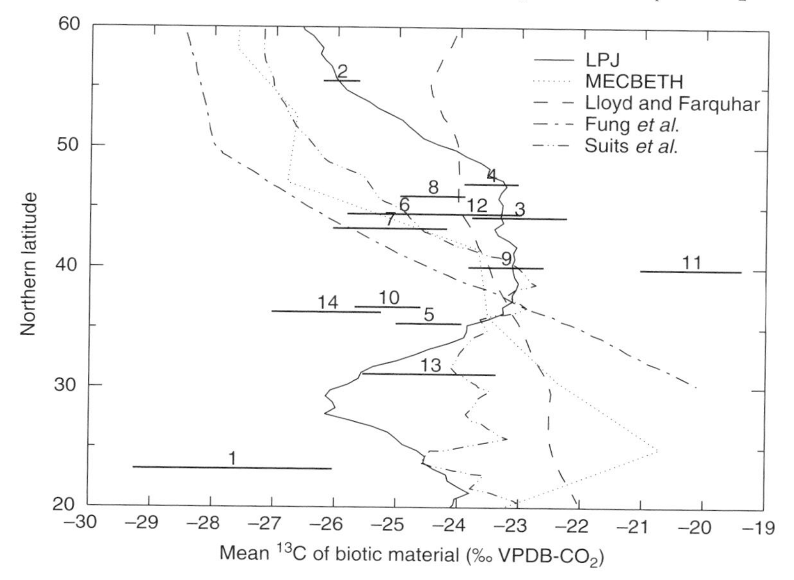

Figure 14.2 Comparison between modeled zonally averaged discrimination north of 20° N (references in the text) and the discriminations inferred from the relationship between CO_2 and $\delta^{13}C$ at atmospheric stations.

station inferred ones. Comparing models with each other yields little information on processes causing differences among them, since the modeled value of Δ_b^{13} is very sensitive to the assumed fractionation parameters (a and b values) as well as to the modeled drop in CO_2 partial pressures between chloroplast (C_c) and leaf stomatal cavity (C_i) corresponding to mesophyll resistance effects.

Temporal variations in Δ_b^{13} relevant to double deconvolution studies occur mainly on an interannual timescale. Year-to-year changes in Δ_b^{13} (as driven for instance by El Niño occurrences) may also possibly covary with photosynthesis to amplify changes in $\delta^{13}C\text{-}CO_2$ in Eq. 14.11 both directly through the product $F_b\Delta_b^{13}$ (Randerson *et al.*, 2002a) and indirectly through the disequilibria D_b and D_{bur}. Randerson *et al.* (submitted) calculated that C_4 grasses contribute 31% of global fire emissions, but only 6% of the interannual emission anomalies. Randerson *et al.* (2002b) modeled the interannual changes in the isotopic discrimination of C_3 plants covarying with GPP. Seasonal variations in Δ_b^{13} with an amplitude of 1.8 to 15‰ exist but accompany variations in gross photosynthesis and are not likely to affect inferences from inverse studies (Fung *et al.*, 1997).

There are indications of a long-term trend in the global mean fractionation factor, i.e., a decrease of 0.4‰ during the twentieth century (Scholze *et al.*, 2003); however on the time-span covered by data for atmospheric inversions (~20 years) such a trend is hardly visible. Two modeling studies found similar amplitudes in the interannual variability of the fractionation factor: less than 1.0‰ (Ito and Oikawa, 2002) and less than 0.3‰ (Scholze *et al.*, 2003). This variability is mainly caused by climate anomalies such as ENSO generally leading to higher plant water stress conditions over tropical forest ecosystems. Although the influence on the partitioning of carbon fluxes is minor compared to the disequilibria term it can lead to biases of 10% in those fluxes.

Understanding $\delta^{18}O$ in Atmospheric CO_2

Carbon dioxide in contact with water can exchange its ^{18}O isotopic signature. If this reaction occurs in nature, in most cases there are several orders of magnitude more water associated than CO_2. That means that the isotopic signature of CO_2 is fully determined by the isotopic signature of the equilibrating water, which is barely changing itself. Due to evaporation, the water in leaves which equilibrates with CO_2 is highly enriched in ^{18}O compared to soil water, which in turn equilibrates with CO_2 leaving the soil from respiration processes. The relevant biosphere CO_2 fluxes are thus the gross fluxes, assimilation and respiration, which determine the $\delta^{18}O$ in atmospheric CO_2. These two gross fluxes are about two orders of magnitude greater than the biospheric net flux. Other CO_2 exchange processes, such as ocean exchange, contribute a small fraction to the total $\delta^{18}O$-CO_2 isoflux. The temporal evolution of $\delta^{18}O$-CO_2 in the atmosphere is therefore mainly the sum of two distinct, opposing isofluxes, from assimilation and respiration respectively. Using this feature, a double deconvolution of $\delta^{18}O$-CO_2 and CO_2 allows determination of the two biospheric CO_2 gross fluxes, provided that the other components can be reasonably estimated. This feature was used locally to determine the gross fluxes over a crop field (Yakir and Wang, 1996) and in a spruce forest (Langendörfer *et al.*, 2002), and to determine the biosphere CO_2 gross fluxes globally (Peylin, 1999). But all these studies pointed out that the estimated gross fluxes are still uncertain and strongly rely on *a priori* information about key parameters that control the computation of the isotopic discriminations. Recent global model studies of $\delta^{18}O$ in atmospheric CO_2 showed that we are not yet able to satisfactorily reproduce the atmospheric $\delta^{18}O$-CO_2 signal with our current knowledge of the $\delta^{18}O$-CO_2 cycle (Peylin, 1999; Cuntz *et al.*, 2003a,b).

For the mass balance of CO_2 (Eq. 14.9), the CO_2 net fluxes of ocean and biosphere can be split into gross fluxes:

$$\frac{dC_a}{dt} = F_o + F_b + F_f = F_{oa} - F_{ao} + F_R - F_A - F_f - F_{bur} \tag{14.26}$$

where F_{oa} is the flux from the ocean to the atmosphere, F_{ao} the flux from the atmosphere to the ocean, F_R the total respiration flux, and F_A the assimilation flux (all four terms defined as positive values). Biomass burning is thereby not included in the respiration term because, unlike for ^{13}C, it has an isotopic signature significantly different from that for ^{18}O respiration. The global $\delta^{18}O$-CO_2 budget is then (Francey and Tans, 1987; Farquhar *et al.*, 1993; Ciais *et al.*, 1997a,b; Peylin *et al.*, 1997; Tans, 1998):

$$\frac{d\delta_a}{dt} = F_{oa}\Delta_{oa} + F_{ao}\Delta_{ao} + F_R\Delta_R + F_A\Delta_A + (F_f + F_{bur})\Delta_{f,bur} + F_{inv}\Delta_{inv} + F_{strat}\Delta_{strat} \tag{14.27}$$

with

$$\Delta_{oa} = \delta_o + \varepsilon_o - \delta_a \tag{14.28}$$

$$\Delta_{ao} = -\varepsilon_o \tag{14.29}$$

$$\Delta_R = \delta_s + \varepsilon_s - \delta_a \tag{14.30}$$

$$\Delta_A = -\varepsilon_l \frac{C_a}{C_a - C_{cs}} + (\delta_l + \varepsilon_l - \delta_a)\frac{C_{cs}}{C_a - C_{cs}} = -\varepsilon_l + \frac{C_{cs}}{C_a - C_{cs}}(\delta_l - \delta_a) \tag{14.31}$$

$$\Delta_{f,bur} = \delta^{18}_{O_2} - \delta_a \tag{14.32}$$

$$\Delta_{inv} = \delta_s - \delta_a \tag{14.33}$$

$$\Delta_{strat} = \delta_{strat} - \delta_a \tag{14.34}$$

The last two terms in Eq. 14.27 are the isotope effects due to 'invasion' of CO_2 into the soil and to the stratosphere–troposphere gross exchange of CO_2. They can be called 'isotopic disequilibria', as for ^{13}C, because they have no CO_2 counterpart. We will discuss in the following each term of Eq. 14.27 in order to determine its importance for the global budget of $\delta^{18}O$-CO_2, or even regional budget to a lesser extent, and describe how well they are currently known.

As mentioned before, CO_2 equilibrating with water is almost always fully determined by the isotopic composition of the water. This equilibration process is temperature dependent and the fractionation between water and CO_2 equilibrated with water follows the relationship (Brenninkmeier *et al.*, 1983):

$$\varepsilon_{eq}(T) = \left(\frac{17604}{T} - 17.93\right) \Big/ 1000 \tag{14.35}$$

such that CO_2 equilibrated with water has the isotopic composition of:

$$\delta_{CO_2} = \delta_{H_2O} + \varepsilon_{eq} \tag{14.36}$$

The ε's in Eqs 14.28–14.34 are kinetic fractionations and will be dealt with in the following sections.

Air–Sea Isoflux of $\delta^{18}O$-CO_2

The first two terms of Eq. 14.27 can be rewritten in order to separate the effect of the net air–sea flux and that of the ocean-to-atmosphere gross flux:

$$F_{oa}\Delta_{oa} + F_{ao}\Delta_{ao} = F_o\varepsilon_o + F_{oa}(\delta_o - \delta_a) \tag{14.37}$$

The net contribution, usually referred as the equilibrium term, is proportional to the net air–sea exchange, which is of the order of $-2\,\mathrm{GtC\,yr^{-1}}$. Given the small kinetic fractionation from diffusion of CO_2 at the air–sea interface, $\varepsilon_o = 0.8‰$ (Vogel *et al.*, 1970), the isoflux is only of the order of $-1.6\,\mathrm{GtC\,‰\,yr^{-1}}$ and thus negligible. The second term, usually referred to as the ocean isotopic disequilibrium, tends to adjust the atmospheric $\delta^{18}O$-CO_2 signal towards the $\delta^{18}O$ of CO_2 dissolved in ocean surface water, δ_o. This latter is dictated by the isotopic composition of ocean surface water δ_o^w and by the surface temperature through Eqs 14.35 and 14.36. Based on ocean measurement campaigns, Craig and Gordon (1965) proposed an empirical regression that relates the $\delta^{18}O$ of H_2O to surface salinity, a relation that has been used by most global studies (Francey and Tans, 1987; Farquhar *et al.*, 1993; Ciais *et al.*, 1997a,b; Peylin *et al.*, 1997; Cuntz *et al.*, 2003a,b): $\delta_o^w = 0.5 \cdot \text{salinity} - 16.75$, with the salinity expressed in ‰ and δ_o^w in ‰ VSMOW. The resulting δ_o^w field decreases towards high latitudes where salinity is lower. Note however that in some particular areas with very low salinity, such as the Baltic Sea, this relation produces too-low values and should thus be corrected (Förstel *et al.*, 1975). The temperature dependence of the fractionation factor ε_{eq} (Eq. 14.37) counterbalances this latitudinal effect and the resulting δ_o field increases from low latitudes toward high latitudes with values around 0 and 6‰ VPDB-CO_2, respectively (Peylin, 1999). Given tropospheric values of δ_a (between -2 and $+2$‰) the overall effect of the air–sea exchange is to enrich the atmosphere in ^{18}O. The amplitude of this effect is directly proportional to the gross sea-to-air flux $F_{oa} = K_{ex}p_o$. Like the case with ^{13}C, the estimation of this isotopic disequilibrium requires the knowledge of the CO_2 partial pressure in surface waters (p_o) and of the air–sea gas exchange transfer coefficient K_{ex}. As mentioned above, F_{oa} increases with time (with an increase of roughly 30% since preindustrial times) and today's average value is of the order of $85\pm17\,\mathrm{GtC\,yr^{-1}}$ (Heimann and Maier-Reimer, 1996). Weighted according to F_{oa}, the global value of $(\delta_o - \delta_a)$ is around $+0.7$ to $+1.6$‰ (Farquhar *et al.*, 1993; Ciais *et al.*, 1997a,b;

Peylin, 1999; Cuntz *et al.*, 2003a), which in turn would increase δ_a by ~140 GtC ‰ yr^{-1}. Such a contribution should be accounted for properly in a double deconvolution of $\delta^{18}O$-CO_2 and CO_2 as it corresponds to around 20 GtC yr^{-1} of terrestrial respiration. This air–sea isoflux bears also a large uncertainty given the uncertainty of F_{oa} (at least 30% from K_{ex}).

$\delta^{18}O$-CO_2 Isoflux of the Terrestrial Respiration

Carbon dioxide in soil equilibrates isotopically with soil water. But the isotopic composition of soil changes with depth. Ciais *et al.* (1997a) estimated that, on average, a CO_2 molecule can travel about 12 cm in soil before it takes the isotopic signature of water (value adjusted for the correct reaction rate). Miller *et al.* (1999) developed from their measurements the rule of thumb that the relevant water isotopic composition is the one at a depth of 15 cm. Riley *et al.* (2002) compared this with their numerical solution of the diffusion equation of CO_2 in soil and found a good agreement with the 15 cm rule of thumb in most cases. So it is common sense to take the water isotopic composition at 15 cm soil depth for CO_2 equilibration. But Riley *et al.* (2002) recognized in addition strong deviations from the 15 cm rule, for example in semi-arid regions, where high soil evaporation leads to steep water isotope profiles. In regions with high water pressure deficits, the water content near the surface can diminish sharply, coupled with very high $\delta^{18}O$-H_2O values (Mathieu and Bariac, 1996a,b; Melayah *et al.*, 1996). There is then the competition of CO_2 fractionation due to diffusion and CO_2 equilibration with soil water to take into account. Respiration isofluxes, for example in savannas, are therefore most probably erroneously calculated using the simple approach of Eq. 14.30. When CO_2 diffuses out of the soil, it becomes fractionated with a maximum fractionation of −8.8‰. In the atmosphere, it will then be transported without fractionation. But CO_2 has to pass an intermediate, laminar air layer just at the ground level where the fractionation is only −5.9‰ (Merlivat and Jouzel, 1979). The effective fractionation will be something in between and the most recent estimate of the global mean is $\varepsilon_s = -7.2‰$ (Miller *et al.*, 1999).

The concentration of CO_2 in soil can rise to very high levels in deep soil (Hesterberg and Siegenthaler, 1991). Therefore, the CO_2 flux coming from soil was determined to be a one-way diffusive flux and the flux from the air into the soil was neglected. Tans (1998) showed that the back-diffusion flux should be considered for $\delta^{18}O$-CO_2. He divided the 'net' flux on the ground into the 'normal' CO_2 respiration flux just leaving the soil and a CO_2 flux which enters the soil from the air (with the isotopic signature of atmospheric CO_2), equilibrates with soil water, and leaves the soil again (with the soil water isotopic signature). This effect is called 'invasion' and the flux from air to soil and back to air is called 'invasion flux' (Tans, 1998) or 'abiotic flux' (Stern *et al.*, 2001). Tans (1998) estimated a global CO_2 invasion flux of roughly 30 GtC yr^{-1} from a back-of-the-envelope calculation and Cuntz *et al.* (2003) calculated a global CO_2 invasion flux of 18.6 GtC yr^{-1} with a global 3-D model of $\delta^{18}O$-CO_2.

Soil water integrates precipitation inputs, and likewise the isotopic signature of soil water integrates incoming meteoric water isotopic composition. Annual mean meteoric water over land is around −7‰ vs VSMOW (Farquhar *et al.*, 1993; Gat, 2000). The soil seems to damp down, almost completely, incoming meteoric inputs (Yurtsever and Gat, 1981; Hesterberg and Siegenthaler, 1991) but measurements of the isotopic composition of nighttime CO_2 fluxes indicate that soil isotopes could change seasonally by about 10‰ (Cuntz *et al.*, 2003b). These observations support a mean respiration weighted value of $\delta_s = -6 \pm 2$‰.

$\delta^{18}O$-CO_2 Isoflux of the Terrestrial Assimilation

To close the $\delta^{18}O$-CO_2 budget, Francey and Tans (1987) speculated that plants had to assimilate about 200 GtC yr^{-1}, which seemed unreasonable. But Farquhar *et al.* (1993) showed that the assimilation of CO_2 is a diffusive process and therefore CO_2 enters and leaves the stomata, resulting in a diffusive net CO_2 flux: assimilation, driven by the CO_2 mixing ratio gradient between chloroplasts and canopy air. Plants use the enzyme carbonic anhydrase to accelerate the fixation of CO_2 and therefore Farquhar *et al.* suggested that all CO_2 leaving the stomata by diffusion is isotopically equilibrated with leaf water. The flux from canopy air to the stomata is thus $g_s C_a$ and the flux from the stomata to canopy air $g_s C_{cs}$ (g_s being roughly the stomatal conductance), resulting in assimilation:

$$F_A = g_s(C_a - C_{cs}) = \frac{C_a}{(C_a - C_{cs})} F_A - \frac{C_{cs}}{(C_a - C_{cs})} F_A \qquad (14.38)$$

The value of C_{cs} is about $0.7 C_a$, so the first ratio gives about $3F_A$ and the second ratio about $2F_A$. The influx only fractionates kinetically and the outflux is isotopically equilibrated with leaf water, leading to Eq. 14.31. Carbon dioxide from the canopy has to diffuse to the leaf surface, through a laminar boundary layer at the leaf surface (fractionation: −5.9‰) and through the stomata (fractionation: −8.8‰). Again, the effective fractionation falls somewhere in between. Farquhar *et al.* (1993) estimated a global mean value of $\varepsilon_l = -7.4$‰. But this came from a simple hand calculation and canopy measurements suggest that the effective fractionation could be closer to –5.9‰ (Langendörfer *et al.*, 2002).

Carbon dioxide inside the stomata is hydrated very quickly due to the enzyme carbonic anhydrase in the mesophyll cell water. It diffuses next to the chloroplasts where it enters the Calvin cycle. This is 'driven' by the gradient between the stomata, C_i, and the chloroplasts, C_c. Gillon and Yakir (2000a) claimed that carbonic anhydrase effectively erases the gradient of CO_2 mixing ratios between the stomata and the chloroplasts, leading to a CO_2 mixing ratio which equilibrates with leaf water, C_{cs}, somewhere between C_i and C_c. They suggested that a reasonable assumption would be halfway between the two CO_2 mixing ratios. The average drawdown between C_c and C_i lies between $0.1C_a$ (Lloyd and Farquhar, 1994) and $0.2C_a$ (Yakir and Sternberg, 2000), so Gillon and Yakir (2000a) suggested taking a value of $C_{cs} - C_i = 0.1C_a$. The actual drawdown between stomata and chloroplasts can be calculated if one knows the stomatal and the mesophyll conductances. However, the latter is very hard to measure and hence not very well known (for details see Gillon and Yakir, 2000a).

There is also a difference between leaf water isotopic composition at the surface of the mesophyll cells, δ_l^w, where water is evaporating, and the supplying vein water isotopes, δ_s^w. Craig and Gordon (1965) developed a formulation for the isotopic composition of evaporative water. When the isotopic composition of water vapor flux becomes stationary, it tends towards the isotopic composition of the supplying water δ_s^w (Bariac *et al.*, 1990), which is an approximately root-density weighted mean of soil water (Bariac *et al.*, 1987; Bariac, 1988). This formulation is called the Craig and Gordon equation and in a simplified form is:

$$\delta_{l-CG}^w = \delta_s^w + \varepsilon_l^w - \varepsilon_k^w + h(\delta_s^w + \varepsilon_k^w - \delta_a^w) \tag{14.39}$$

The superscript w denotes thereby that water isotopes are involved. ε_l^w is the equilibrium fractionation between water and water vapor, ε_k^w the kinetic fractionation from the site of evaporation to canopy air, h the relative humidity normalized to leaf temperature, and δ_a^w the isotopic composition of water vapor in the canopy. But this formulation assumes steady state conditions that are probably not achieved in nature very often (Roden and Ehleringer, 1999; Barbour *et al.*, 2000). Modifying the Craig and Gordon formulation to non-steady state conditions leads to transitional models where one has to make certain assumptions. Assuming a well-mixed leaf water pool of composition δ_s^w which is not changing over time, this results in the transitional model of leaf water at the evaporating site (Förstel *et al.*, 1975):

$$\delta_l^w|_t = \delta_l^w|_{t-1} - (\delta_l^w|_{t-1} - \delta_{l-CG}^w|_t)\exp\{-\Delta t/\tau\} \tag{14.40}$$

with $\delta_l^w|_t$ the leaf water isotopic composition at timestep t and that $\delta_l^w|_{t-1}$ that at timestep $t-1$; τ the turnover time of leaf water corrected for isotope fractionation, $\tau = \zeta V_l/E_v$ with $\zeta = (1-h)(\varepsilon_l^w + 1)(\varepsilon_k^w + 1)$; V_l the leaf water volume; and E_v the transpiration rate. So, leaf water tends to the steady state but only reaches it if evaporation is very high. But this formulation neglects the existence of a gradient in leaf water isotopic composition from incoming source water to the site of evaporation (Péclet effect; Farquhar and Lloyd (1993); and also neglects the fact that the source water can change its isotopic composition during the day. Nevertheless, the transitional model builds a lower limit and the Craig and Gordon model an upper limit for leaf water isotopic composition. Farquhar *et al.* (1993) calculated a global mean Craig and Gordon value of $\delta_l^w = 4.4‰$ (vs VSMOW) and Ciais *et al.* (1997a) a value of around 3‰. Neither used the transitional model but they calculated their estimates on a monthly basis which (almost) averages out the differences between the Craig and Gordon formulation and the transitional model. Cuntz *et al.* (2003b) included a diurnal cycle in the calculation of Craig and Gordon and the transitional model and calculated a difference of 1‰ between the two formulations, as assimilation-weighted values. They estimated a non-assimilation-weighted Craig and Gordon value of 5.6‰, which is already higher than the estimates of Farquhar *et al.* (1993) or Ciais *et al.* (1997a). Using assimilation-weighted values increased the Craig and Gordon result markedly (7.0‰), but the transitional model in contrast diminished the actual δ_l^w to 6.3‰. These estimates seem to be more realistic because they also fit better with estimates of δ_l^w from the global isotope budget of O_2 (Bender *et al.*, 1994).

Recently, it was doubted that CO_2 leaving the stomata by diffusion is always in isotopic equilibrium with leaf water (Gillon and Yakir, 2000b, 2001). This could be due to reduced activity of the enzyme carbonic anhydrase but also due to the diffusive nature inside the stomata. Gillon and Yakir added an extra term to Eq. 14.33, Θ, the fraction of CO_2 molecules which become hydrated in leaf water. This term effectively reduces Δ_A.

$$\Delta_A = -\varepsilon_l + \frac{C_{cs}}{C_a - C_{cs}}(\delta_l - \delta_a) - \frac{C_{cs}}{C_a - C_{cs}}(1 - \Theta)(\delta_l - \delta_a - [1 - C_{cs}/C_a]\varepsilon_l) \tag{14.41}$$

This seems to be mostly important for C_4 plants, C_4 grasses for example having $\Theta = 0.4$, whereas C_3 plants have values of $\Theta = 0.9$ or higher. Gillon and Yakir (2001) estimate a global mean value of 0.78 and Cuntz *et al.* (2003b) estimate 0.8, where differences derive only from the different vegetation distribution used.

$\delta^{18}O$-CO_2 Isoflux of Burning Processes

Burning processes transform carbonaceous material and atmospheric oxygen into CO_2. It is assumed that the newly formed CO_2 bears the isotopic signature of atmospheric oxygen: $\delta_f = -17‰$ (Kroopnick and Craig, 1972). To our knowledge, there are no observations of this value for different burning processes, and it could potentially be different for specific burning processes. As detailed above for ^{13}C, burning occurs over very different areas of the world but mainly involves tropical forest and savannas with a global total of 2 to 3.4 GtC yr^{-1} (Table 14.1). This source of carbon is partially offset by a carbon sink due to re-growth over burned areas (F_{bur_regrow} term in ^{13}C section). Such an offset concerns essentially tropical savannas with values of the order of 0.36 GtC yr^{-1}, 1.04 GtC yr^{-1}, and 0.1 GtC yr^{-1} for South America, Africa, and South Asia, respectively. These estimates are based on the difference between the burning flux of Hao *et al.* (1990) and the net deforestation flux of Houghton (1999). In Africa, 75% of the CO_2 emitted from biomass burning is reincorporated within the same year by re-growth of savannas while the fractions in South America and South Asia are only 35% and 15%, respectively. For these latter regions, it implies a large loss of carbon. The effect of re-growth should be treated as a $\delta^{18}O$-CO_2 isoflux of terrestrial assimilation (see above). However, at a global scale the re-growth concerns only a small amount of carbon compared to the gross assimilation flux F_A, so its isoflux will not significantly impact a double deconvolution of $\delta^{18}O$-CO_2 and CO_2.

Stratosphere to Troposphere $\delta^{18}O$-CO_2 Isoflux

Carbon dioxide in the stratosphere is enriched by 2–3‰ in ^{18}O compared to the troposphere (Gamo *et al.*, 1989) due to isotopic exchange of CO_2 with mass-independently enriched ozone in the stratosphere (Thiemens, 1999). Carbon dioxide enters the stratosphere mainly in the tropics and this flux is mostly counterbalanced by a CO_2 flux from the stratosphere to the troposphere in the extra-tropics (Zahn *et al.*, 1999, 2000). This leads to a contribution of the stratosphere–troposphere exchange (STE) to the north–south gradient in tropospheric $\delta^{18}O$-CO_2 (Peylin *et al.*, 1997). STE is assumed to have a gross CO_2 flux of about 200 GtC yr^{-1} (Hesshaimer, 1997), which leads to an isoflux of about 400 GtC ‰ yr^{-1}. Because the STE is strongest in spring in the northern hemisphere, STE could also lead to seasonal variations in the tropospheric $\delta^{18}O$-CO_2 signal. But until now, this flux has not been estimated reliably and is a major unknown in the global $\delta^{18}O$-CO_2 cycle.

Concluding Remarks

The apportionment of sources and sinks between ocean and land using $\delta^{13}C$ and CO_2 records in the atmosphere is mainly dependent on isotopic disequilibrium terms, which have an impact on the atmospheric trend as large as the net fluxes themselves. At the global scale, isotopic

Table 14.1 Estimates of Carbon Emissions due to Biomass Burning ($TgC\,yr^{-1}$) and their Isotopic $\delta^{13}C$ Signature (‰ PDB) for the Major Ecosystems ($TgC = r^{-3}G+C$)

	Average carbon emission by fires ($TgC\,yr^{-1}$)	Average fire return interval (yr)	References	Average $\delta^{13}C$ of burned biomass (‰ PDB)
Tropical forest	600–1000	250 (100–500)	Levin (1994)	
	(200–1200) (Amazonia alone)		Cochrane and Laurence (2002)	
			Potter *et al.* (2001)	
Mixed forest	38 (11–60) USA	100	Leenhouts (1998)	−7.03
	17 (10–49) China	20–120	Wang *et al.* (2001)	−7.03
	8 Europe (mostly Mediterranean)	10–110	Mouillot *et al.* (2002)	−7.17
	22 Australia	30	Trabaud (1994)	−7.57
			Gill *et al.* (1997)	
Boreal forest	40 (10–100) Canada	200	Chen *et al.* (2000)	−6.76
	50 (10–100) Russia	50 (25–100)	Shvidenko and Nilsson (2000)	−7.32
			Conard and Ivanova (1997)	
Grasslands	1100 (500–1200) Africa	3	Barbosa *et al.* (1999)	−7.91
Savanna	82 India	3	Hao and Liu (1994)	−7.91
	550 South America	5	Seiler and Crutzen (1980)	−7.88
	100 (70–150) Australia	4		−7.90
	66 Central Asia	10		−7.83
Agriculture	910	1	Andreae (1991)	−8
Total	(2081–3449)			−7.46

Today's atmosphere is set to −8‰ and ice core data have been used in this calculation. Averages and minimum/maximum values (in brackets when available) are presented.

disequilibria have an uncertainty range of 30% for the air–sea disequilibrium (range 47–60 GtC ‰ yr^{-1}) and 50% for the soil-respired disequilibrium (range 19.8–33.6 GtC ‰ yr^{-1}). Further, we have quantified other important disequilibrium and isofluxes, which are generally ignored in double deconvolutions:

(1) The replacement of C_3 by C_4 vegetation, which amounts to 50–100% of the soil respiration disequilibrium but has an opposite sign (−15 to −22 GtC ‰ yr^{-1}).
(2) The (small) disequilibrium induced by biomass burning processes (1.7 GtC ‰ yr^{-1}).
(3) The disequilibrium of rock weathering processes (10.5 GtC ‰ yr^{-1}).
(4) The correction to soil-respired disequilibrium due to non-CO_2 gas emissions and oxidation within the atmosphere (17 GtC ‰ yr^{-1}).

At the regional level, some isotopic disequilibrium can become proportionally larger than the isoflux of net CO_2 fluxes, especially at high northern latitudes for the aging of soil-respired CO_2, and in the Tropics for the shifts from C_3 to C_4 vegetation. At even smaller scales, the ecosystems are probably never approaching isotopic equilibrium, neither over long timescales because of disturbances (biomass formed never has the same age as soil-respired carbon), nor on very short timescales (the isotopic composition of photosynthates does not equal that of respiration). Over time, interannual climate-induced fluctuations in the biospheric fractionation factor alter the interference of land and ocean sinks anomalies using atmospheric records of $\delta^{13}C$ and CO_2 by less than 0.1 GtC yr^{-1}. On the other hand, the C_3 shift to C_4 land use-induced disequilibrium should not change strongly from one year to the next; neither should the disequilibrium due to rock weathering. The soil-respired and air–sea isotopic disequilibrium should have a rate of interannual variability similar to one of gross fluxes, that is 10–20% globally, according to terrestrial and ocean model calculations. Priorities in future research should be to include all disequilibria in inverse atmospheric transport models to re-analyze regional fluxes. As recently performed for $CO^{18}O$ by Cuntz *et al.* (2003a,b), it should also be important to incorporate 'on-line' the discrimination of $^{13}CO_2$ by canopy photosynthesis and its subsequent transport in the atmosphere in coupled land-surface–atmosphere models.

For $\delta^{18}O$-CO_2, the atmospheric transport and the biospheric $\delta^{18}O$-CO_2 isofluxes determine almost completely the atmospheric signal (see Table 14.2). $\delta^{18}O$-CO_2 has therefore a high potential to deduce the CO_2 gross fluxes of the terrestrial biosphere, but the ^{18}O isotopic exchange between water and CO_2 makes the $\delta^{18}O$-CO_2 cycle more complex than the $\delta^{13}C$ cycle. On the other hand, $\delta^{18}O$ in atmospheric CO_2 is determined by the gross fluxes of the carbon cycle compared to $\delta^{13}C$ that is determined

Table 14.2 Global $\delta^{18}O$-CO_2 Isofluxes of Different Processes[a]

Process	Isoflux (GtC ‰ yr^{-1})
Assimilation	600–1850
Respiration	−600 – −1450
Ocean net flux	1.5–2
Ocean gross flux	70–160
Fossil fuel	∼−100
Biomass burning	∼−50
Invasion	∼−130
Stratosphere	200–400
Carb. anhydrase	∼−300

[a]Estimated from Peylin (1999) and Cuntz *et al.* (2003a,b).

by the net fluxes. Isotopic disequilibria play therefore an important role in the $\delta^{13}C$ cycle but are of secondary importance for $\delta^{18}O$-CO_2. The main unknown in the $\delta^{18}O$-CO_2 cycle is leaf discrimination (Δ_A) due to large uncertainties in its determining variables: mainly the relevant CO_2 mixing ratio inside the leaf (C_{cs}), the leaf water isotopic composition at the site of evaporation (δ_l^w), and the carbonic anhydrase activity. Global estimates for leaf water isotopic composition range between 3 and 8.8‰ VSMOW (Gillon and Yakir, 2001) and lead to an uncertainty in Δ_A of about 25%. This puts a limit to the accuracy at which the assimilation can currently be retrieved in double ^{18}O deconvolutions. The stratosphere–troposphere exchange is the other important uncertain term in the $\delta^{18}O$-CO_2 cycle. The stratosphere contributes most probably an isoflux of about 400 GtC ‰ yr^{-1} to the troposphere, which is approximately one fourth of the assimilation isoflux. Not taking it into account in double ^{18}O deconvolutions would lead to a similar bias in the derived terrestrial biosphere CO_2 gross fluxes to that caused by the uncertainty in Δ_A. The contribution of the mass-independent enrichment of $\delta^{18}O$-CO_2 in the stratosphere has not yet been incorporated in global models because of the low vertical resolution of the models and it was speculated that this could lead to the observed differences between modeled and measured $\delta^{18}O$-CO_2 at observatories of marine background air (Cuntz *et al.*, 2003b).

Notations

C_a atmospheric CO_2 mixing ratio
C_a' atmospheric $^{13}CO_2$ or $CO^{18}O$ mixing ratio

C_c	CO_2 mixing ratio in chloroplasts
C_{cs}	CO_2 mixing ratio at chloroplast surface
C_i	CO_2 mixing ratio inside stomatal cavity
D	isotopic disequilibria
D_b	soil-respired isotopic disequilibrium
D_{bur}	biomass burning isotopic disequilibrium
$D_{carbonate\_weath}$	carbonate weathering isotopic disequilibrium
D_{def}	land use-induced isotopic disequilibrium
D_o	air–sea isotopic disequilibrium
$D_{silicate\_weath}$	silicate weathering isotopic disequilibrium
E_V	transpiration rate
F	CO_2 flux
F'	$^{13}CO_2$ or $CO^{18}O$ flux
F_A	CO_2 net assimilation
F_{ab}	atmosphere–biosphere CO_2 exchange flux
F_{ao}	atmosphere–ocean CO_2 exchange flux
F_b	biospheric CO_2 net flux
F_{ba}	biosphere–atmosphere CO_2 exchange flux
F_{bur}	biomass burning CO_2 flux
F_{bur_regrow}	CO_2 flux of recovering ecosystem after fire
F_{carb_w}	soil CO_2 uptake with $CaCO_3$
$F_{CO\_oxidation}$	CO_2 produced from CO oxidation
F_{def_assim}	photosynthetic CO_2 uptake of new biomes after deforestation
F_{def_resp}	immediate and delayed CO_2 release from deforestation
F_f	CO_2 release by burning of fossil fuels
F_{HR}	heterotrophic respiration
F_{inv}	invasion or abiotic CO_2 flux
F_o	air–sea CO_2 net flux
F_{oa}	ocean–atmosphere CO_2 exchange flux
F_R	CO_2 net ecosystem respiration
F_{res}	residual terrestrial CO_2 flux, often referred to 'missing sink'
$F_{silicate\_weath}$	CO_2 uptake from silicate rock weathering
F_{strat}	stratosphere–troposphere CO_2 exchange flux
$F_{volcanoes}$	CO_2 outgassed by volcanoes
g_s	stomatal conductance
h	relative humidity
K_{ex}	air–sea gas transfer coefficient
$KIE(\delta^{13}C\text{-}CO)$	kinetic isotope effect (fractionation) of $CO + OH \rightarrow CO_2$

M_a	conversion factor between fluxes in GtC and mixing rations in ppm
NPP	Net Primary Productivity
p	atmospheric pressure
p_a	atmospheric CO_2 partial pressure
p_o	ocean surface CO_2 partial pressure
R	$^{13}C/^{12}C$ or $^{18}O/^{16}O$ isotope ratio
R_a	isotope ratio of atmospheric air
T	absolute temperature
V_l	leaf water volume
δ	relative deviation of isotope ratio from standard material
δ_a	isotopic composition of atmospheric CO_2
δ_a^w	isotopic composition of atmospheric water vapor
δ_b	isotopic composition of soil-respired CO_2 from previously formed biomass
δ_b^e	isotopic composition of soil-respired CO_2 from recently formed biomass
δ_{bur}	isotopic composition of burned material in absence of isotopic fractionation associated with combustion
δ_{bur_regrow}	isotopic composition of recovering ecosystems after fire
$\delta_{CO\_oxidation}$	$\delta^{13}C\text{-}CO_2$ from CO oxidation
δ_{def_assim}	isotopic composition of photosynthetic flux of new biomes after deforestation
$\delta^*_{def\_assim}$	apparent isotopic composition of photosynthetic flux of new biomes after deforestation
δ_{def_resp}	isotopic composition of immediate and delayed CO_2 release from deforestation
$\delta^*_{def\_resp}$	apparent isotopic composition of immediate and delayed CO_2 release from deforestation
δ_f	isotopic composition of CO_2 released from burning fossil fuel
δ_{in}	δ-value of flux to the atmosphere
δ_l	isotopic composition of CO_2 equilibrated with leaf water at the evaporation site
δ_l^w	isotopic composition of leaf water at the evaporation site
δ_{l-CG}^w	steady-state isotopic composition of leaf water at the evaporation site after Craig and Gordon formulation
δ_{mantle}	isotopic composition of CO_2 released from volcanoes

δ_o	isotopic composition of CO_2 equilibrated with ocean surface water
δ_o^e	hypothetical isotopic composition of CO_2 if ocean in isotopic equilibrium with today's atmosphere
δ_o^w	isotopic composition of ocean surface water
δ_{out}	δ-value of flux from the atmosphere
$\delta_{O_2}^{18}$	isotopic composition of atmospheric oxygen
δ_s	isotopic composition of CO_2 equilibrated with soil water
δ_s^w	isotopic composition of soil water
δ_{strat}	isotopic composition of stratospheric CO_2
$\delta_{w\_release}$	isotopic composition of CO_2 released from rock weathering
δ_{w_uptake}	isotopic composition of CO_2 uptake by rock weathering
Δ	discrimination; here difference between flux and atmospheric isotopic composition
Δ_A	^{18}O discrimination of assimilation
Δ_{ao}	^{18}O discrimination of CO_2 entering the ocean from the atmosphere
Δ_b^{13}	^{13}C discrimination of plants
Δ_{b-C_3}	^{13}C discrimination of C_3 plants
Δ_{b-C_4}	^{13}C discrimination of C_4 plants
$\Delta_{f,bur}$	^{18}O discrimination of burning processes
Δ_{inv}	apparent ^{18}O discrimination of invasion (abiotic) CO_2 flux
Δ_{oa}	^{18}O discrimination of CO_2 leaving the ocean to the atmosphere
Δ_R	^{18}O discrimination of respiration
Δ_{strat}	^{18}O discrimination of stratosphere–troposphere exchange
ε	fractionation
ε_{ab}	^{13}C atmosphere–plant equilibrium fractionation
ε_{ao}	^{13}C fractionation of CO_2 entering the ocean
ε_{eq}	^{18}O fractionation effect of CO_2 equilibrating isotopically with water
ε_k^w	kinetic fractionation of water during evaporation
ε_l	^{18}O diffusion fractionation of CO_2 through the stomata
ε_l^w	equilibrium fractionation between water and water vapor
ε_o	^{18}O diffusion fractionation of CO_2 at air–sea interface
ε_{oa}	^{13}C sea-to-air fractionation
ε_s	^{18}O diffusion fractionation of soil respiration
Θ	fraction of CO_2 hydrated in leaf water
τ	turnover time of leaf water corrected for isotope fractionation
ζ	$= (1-h)(\varepsilon_l^w + 1)(\varepsilon_k^w + 1)$

References

Andreae M. O. (1991) Biomass burning: its history, use, and distribution and its impact on environmental quality and global climate. In *Global Biomass Burning: Atmospheric, Climatic and Biospheric Implications* (J. S. Levine, ed.). MIT Press, Cambridge.

Balesdent J. and Recous S. (1997) Les temps de résidence du carbone et le potentiel de stockage de carbone dans quelques sols cultivés français. *Can J Soil Sci* **77**: 187–193.

Bakwin P., Tans P., White J. W. C. and Andres R. J. (1998) Determination of the isotopic ($^{13}C/^{12}C$) discrimination by terrestrial biology from a global network of observations. *Global Biogeochem Cycles* **13**(3): 555–562.

Barbosa P. M., Stroppiana D. and Gregoire J. M. (1999) An assessment of vegetation fire in Africa (1981–1991): burned areas, burned biomass, and atmospheric emissions. *Global Biogeochem Cycles* **13**: 933–950.

Barbour M. M., Schurr U., Henry B. K., Chin Wong S. and Farquhar G. D. (2000) Variation in the oxygen isotope ratio of phloem sap sucrose from castor Bean. Evidence in support of the Péclet effect. *Plant Physiol* **123**: 671–680.

Bariac T. (1988) Les isotopes stables (^{18}O, ^{2}H) de l'eau dans le continuum sol–plante–atmosphere: Conséquence pour la vapeur d'eau atmosphérique. Doctorate, state thesis (in French), University of Paris, VI.

Bariac T., Klamecki A., Jusserand C. and Letolle R. (1987) Evolution de la composition isotopique de l'eau (^{18}O) dans le continuum sol–plante–atmosphere (exemple d'une parcelle cultivée en blé, Versailles, France, Juin (1984))—Part A. *Catena* 14: 55–72.

Bariac T., Jusserand C. and Mariotti A. (1990) Evolution spatio-temporelle de la composition isotopique de l'eau dans le continuum sol–plante–atmosphère. *Geochim Cosmochim Acta* **54**: 413–424.

Battle M., Bender M. L., Tans P. P., White J. W. C., Ellis J. T., Conway T. and Francey R. J. (2000) Global carbon sinks and their variability inferred from atmospheric O_2 and $\delta^{13}C$. *Science* **287**: 2467–2470.

Bender M., Sowers T. and Labeyrie L. (1994) The Dole effect and its variations during the last 130 000 years as measured in the Vostok ice core. *Global Biogeochem Cycles* **8**(3): 363–376.

Bergamaschi P., Hein R., Brenninkmeijer C. A. M. and Crutzen P. J. (2000b) Inverse modeling of global CO cycle—2, Inversion of $^{13}C/^{12}C$ and $^{18}O/^{16}O$ isotopes ratios. *J Geophysical Res* **105**: 1929–1945.

Bousquet P. (1999a) Inverse modeling of annual atmospheric CO_2 sources and sinks—Part 1: Method and control inversion. *J Geophysical Res* **104**(D121): 26161–26178.

Bousquet P. (1999b) Inverse modeling of annual atmospheric CO_2 sources and sinks—Part 2: Sensitivity study. *J Geophysical Res* **104**(D121): 26179–26193.

Bowling D., Baldocchi D. D. and Monson R. K. (1999) Dynamics of isotopic exchange of carbon dioxide in a Tennessee deciduous forest. *Global Biogeochem Cycles* **13**(4): 903–922.

Bowling D., Tans P. P. and Monson R. K. (2001) Partitioning net ecosystem carbon exchange with isotopic fluxes of CO_2. *Global Change Biol* **7**: 127–145.

Brenninkmeier C. A. M., Kraft P. and Mook W. G. (1983) Oxygen isotope fractionation between CO_2 and H_2O. *Isotope Geosci* **1**: 181–190.

Chen J., Chen W., Liu J. and Cihlar J. (2000) Annual carbon balance of Canada's forests during 1895–1996. *Global Biogeochem Cycles* **14**: 839–849.

Ciais P., Tans P. P., Trolier M., White J. W. C. and Francey R. J. (1995a) A large northern hemisphere terrestrial CO_2 sink indicated by the $^{13}C/^{12}C$ ratio of atmospheric CO_2. *Science* **269**: 1098–1102.

Ciais P., Tans P. P., White J. W. C., Trolier M., Francey R. J., Berry J. A., Randall D. A., Sellers P. J., Collatz J. G. and Schimel D. S. (1995b) Partitioning of ocean and land uptake of CO_2 as inferred by $\delta^{13}C$ measurements from the NOAA Climate Monitoring and Diagnostic Laboratory Global Air Sampling Network [CO_2-$\delta^{13}C$]. *J Geophysical Res* **100**(D3): 5051–5070.

Ciais P., Denning A. S., Tans P. P., Berry J. A., Randall D. A., Collatz J. G., Sellers P. J. *et al.* (1997a) A three dimensional synthesis study of $\delta^{18}O$ in atmospheric CO_2. Part 1. Surface fluxes. *J Geophys Res* **102**(D15): 5857–5871.

Ciais P., Tans P. P., Denning A. S., Francey R. J., Trolier M., Meijer H. J., White J. W. C. *et al.* (1997b) A three dimensional synthesis study of $\delta^{18}O$ in atmospheric CO_2. Part 2. Simulations with the TM2 transport model. *J Geophysical Res* **102**(D5): 5873–5883.

Ciais P., Friedlingstein P., Schimel D. S. and Tans P. P. (1999) A Global calculation of the $\delta^{13}C$ of soil respired carbon: Implications for the biospheric uptake of anthropogenic CO_2. *Global Biogeochem Cycles* **13**(2): 519–530.

Cochrane M. A. and Laurence W. F. (2002) Fire as a large scale edge effect in Amazonian forests. *J Tropical Ecol* **18**: 311–325.

Conard S. G. and Ivanova G. A. (1997) Wildfire in Russian boreal forests—potential impacts of fire regime characteristics on emissions and global carbon balance estimates. *Environ Pollut* **98**: 305–313.

Craig H. and Gordon A. (1965) Deuterium and Oxygen-18 variations in the ocean and the marine atmosphere. In *Stable isotopes in Oceanic Studies and Paleotemperatures* (L.o.g.a.N. Science, ed.) pp. 9–130. Pisa.

Cramer W., Kicklighter D. W., Bondeau A., Moore B., III, Churkina G., Nemry B., Ruimy A. and Schloss A. L. (1999) and the Participants of the potsdam NPP Model Intercomparaison. Comparing global models of terrestrial net primary productivity (NPP): Overview and key results. *Global Change Biol* **5**(S1): 1–15.

Cramer W., Bondeau A., Woodward F. I., Prentice C., Betts R. A., Brovkin V., Cox P. M. *et al.* (2001) Global response of terrestrial ecosytem structure and function to CO_2 and climate change: Results from six dynamic global vegetation models. *Global Change Biol* **7**(4): 357–374.

Cuntz M., Ciais P., Hoffmann G., Allison C. E., Francey R. J., Knorr W., Tans P. P., White J. W. C. and Levin I. (2003a) A comprehensive global 3D model of $\delta^{18}O$ in atmospheric CO_2—Part 2. Mapping the atmospheric signal. *J Geophysical Res* **108** doi: 10.1029/2002 JD003154.

Cuntz M., Ciais P., Hoffmann G. and Knorr W. (2003b) A comprehensive global 3D model of $\delta^{18}O$ in atmospheric CO_2—Part 1. Validation of surface processes. *J Geophysical Res* **108** doi: 10.1029/2002 JD003153.

Ekblad A. and Högberg P. (2001) Natural abundance of ^{13}C in CO_2 respired from forest soils reveals speed of link between tree photosynthesis and root respirations. *Oecologia* **127**: 305–308.

Farquhar G. D. and Lloyd J. (1993) Carbon and oxygen isotope effects in the exchange of carbon dioxide between terrestrial plants and the atmosphere. In *Stables Isotopes and Plant Carbon—Water Relations.* (J. R. Ehleringer, A. E. Hall, and G. D. Farquhar, eds) pp. 47–70. Academic Press, New York.

Farquhar G. D., Lloyd J., Taylor J. A., Flanagan L. B., Syvertsen J. P., Hubick K. T., Wong S. C. and Ehleringer R. (1993) Vegetation effects on the isotope composition of oxygen in atmospheric CO_2. *Nature* **363**: 439–443.

Förstel H., Putral A., Schleser G. and Leith H. (1975) The world pattern of oxygen-18 in rainwater and its importance in understanding the biogeochemical oxygen cycle. In *Isotope Ratios as Pollutant Source and Behavior Indicators* (IAEA, ed.). IAEA, Vienna.

Francey R. J. and Tans P. P. (1987) Latitudinal variation in oxygen-18 of atmospheric CO_2. *Nature* **327**: 495–497.

Francey R. J., Tans P. P., Allison C. E., Enting I. G., White J. W. C. and Trolier M. (1995) Changes in oceanic and carbon uptake since (1982). *Nature* **373**: 326–330.

Fung I. Y., Field C. B., Berry J. A., Thompson M. V., Randerson J. T., Malmström C. M., Vitousek P. M. *et al.* (1997) Carbon-13 exchanges between the atmosphere and biosphere. *Global Biogeochem Cycles* **11**(4): 507–533.

Gamo T., Sutsumi M., Sakai H., Nakazawa T., Tanaka M., Honda H., Kubo H. and Itoh T. (1989) Carbon and oxygen isotopic ratios of carbon dioxide of a stratospheric profile over Japan. *Tellus* **41**B: 127–133.

Gat J. R. (2000) Atmospheric water balance—the isotopic perspective. *Hydrol Proc* **14**: 1357–1369.

Gaudinski J. B., Trumbore S. E., Davidson E. A. and Zheng S. (2000) Soil carbon cycling in a temperate forest: Radiocarbon-based estimates of residence times, sequestration rates and partitioning of fluxes. *Biogeochemistry* **51**: 33–69.

Gill A. M., Moore P. H. R., Mc Carthy M. A. and Lang S. (1997) *Contemporary Fire Regimes in the Forests of Southwestern Australia.* CSIRO, Australia Environment, Canberra.

Gillon J. and Yakir D. (2000a) Internal conductance to CO_2 diffusion and $C^{18}O$ discrimination in C_3 leaves—Part A. *Plant Physiol* **123**: 201–213.

Gillon J. and Yakir D. (2000b) Naturally low carbonic anhydrase activity in C_4 and C_3 plants limits discrimintation against $C^{18}OO$ during photosynthesis—Part B. *Plant Cell Environ* **23**: 903–915.

Gillon J. and Yakir D. (2001) Influence of carbonic anhydrase activity in terrestrial vegetation on the ^{18}O content of atmospheric CO_2. *Science* **291**: 2584–2587.

Gitz V. and Ciais P. (2003) Land use amplifying effects on future CO_2 levels. *Global Biogeochem Cycles* **17**(1): 1024.

Gruber N. and Keeling C. D. (2001) An improved estimate of the isotopic air–sea disequilibrium of CO_2: Implications for the oceanic uptake of anthropogenic CO_2. *Geophys Res Lett* **28**(3): 555–558.

Gurney K. R., Law R. M., Denning A. S., Rayner P. J., Baker D., Bousquet P., Bruhwiller L. *et al.* (2002) Towards robust regional estimates of CO_2 sources and sinks using atmospheric transport models. *Nature* **415**: 626–630.

Hao W. M., Liu M. H. and Crutzen P. J. (1990) Estimates of annual and regional release of CO_2 and other trace gases to the atmosphere from fires in the Tropics, based on the FAO statistics for the period 1975–1980. In *Fires in the Tropical Biota: Ecosystem Processes and Global Challenges* (G. J. Goldammer, ed.) pp. 440–462, Springer-Verlag, Berlin.

Hao W. M. and Liu M.-H. (1994) Spatial and temporal distribution of tropical biomass burning. *Global Biogeochem Cycles* **8**: 495–503.

Heimann M. and Keeling C. D. (1989) A three-dimentional model of atmospheric CO_2 transport based on observed winds: 2. Model description and simulated tracer experiments. In *Aspects of Climate Variability in the Pacific and the Western Americas* (D. H. Peterson, ed.) pp. 237–276. AGU, Washington, USA.

Heimann M. and Maier-Reimer E. (1996) On the relations between the oceanic uptake of CO_2 and its carbon isotopes. *Global Biogeochem Cycles* **10**(1): 89–110.

Hesshaimer V. (1997) Tracing the Global Carbon Cycle with Bomb Radiocarbon. PhD Thesis (in German), University of Heidelberg.

Hesterberg R. and Siegenthaler U. (1991) Production and stable isotopic composition of CO_2 in a soil near Bern, Switzerland. *Tellus* **43**B: 197–205.

Houghton R. A. (1999) The annual net flux of carbon to the atmosphere from changes in land use 1850–1990. *Tellus* **51**B: 298–313.

Ito A. and Oikawa T. (2002) A simulation model of the carbon cycle in land ecosystems (Sim-CYCLE): A description based on dry-matter production theory and plot- scale validation. *Ecol Modelling* **151**(2–3): 143–176.

Joos F. and Bruno M. (1998) Long-term variability of the terrestrial and oceanic carbon sinks and the budgets of the carbon isotopes ^{13}C and ^{14}C. *Global Biogeochem Cycles* **12**(2): 277–295.

Joos F., Meyer R., Bruno M. and Leuenberger M. (1999) The variability into the carbon sinks as reconstructed for the last 1000 years. *Geophys Res Lett* **26**: 1437–1441.

Kaplan J. O., Prentice C. and Buchmann N. (2002) The stable isotope composition of the terrestrial biosphere: Modelling at scales from the leaf to the globe. *Global Biogeochem Cycles* **16**(4): 1060–1068.

Keeling C. D. (1995) Interannual extremes in the rate of rise of atmospheric carbon dioxide since 1980. *Nature* **375**: 666–670.

Kroopnick P. and Craig H. (1972) Atmospheric oxygen: Isotopic composition and solubility fractionation. *Science* **175**: 54–55.

Langendörfer U., Cuntz M., Ciais P., Peylin P., Bariac T., Milyukova I., Kolle O., Naegler T. and Levin I. (2002) Modelling of biospheric CO_2 gross fluxes via oxygen isotopes in a spruce forest canopy: A ^{222}Rn calibrated box model approach. *Tellus* **54**B: 476–496.

Leenhouts B. (1998) Assessment for biomass burning in the conterminous United States. *Conservation Ecology [online]* **2**(1): 1–25.

Levin I. (1994) *The Recent State of Carbon Cycling Through the Atmosphere.* Springer-Verlag, Heidelberg.

Lloyd J. and Farquhar G. D. (1994) ^{13}C discrimination during CO_2 assimilation by the terrestrial biosphere. *Oecologia* **99**: 201–215.

Mathieu R. and Bariac T. (1996a) An isotopic study of water movements in clayey soils under a semiarid climate. *Water Resources Res* **32**: 779–789.

Mathieu R. and Bariac T. (1996b) A numerical model for the simulation of stable isotope profiles in drying soils. *J Geophys Res* **101**: 12685–12696.

Melayah A., Bruckler L. and Bariac T. (1996) Modeling the transport of water stable isotope in unsaturated soils under natural conditions. 2 Comparison with field experiments. *Water Resources Res* **32**: 2055–2065.

Merlivat L. and Jouzel J. (1979) Global climatic interpretation of the deuterium—oxygen 18 relationship for precipitation. *J Geophys Res* **84**: 5029–5033.

Miller J. B., Yakir D., White J. W. C. and Tans P. P. (1999) Measurement of $^{18}O^{16}O$ in the soil—atmosphere CO_2 flux. *Global Biogeochem Cycles* **13**: 761–774.

Mouillot F., Rambal S. and Joffre R. (2002) Simulating the effects of climate change on fire frequency and the dynamics of a Mediterranean Maquis woodland. *Global Change Biol* **8**: 423–437.

Murname R. J. and Sarmiento J. L. (2000) Roles of biology and gas exchange in determinig the δ^{13}C distribution in the ocean and the pre-industrial gradient in atmospheric δ^{13}C. *Global Biogeochem Cycles* **14**: 389–405.

Nemani R. R., Keeling C. D., Hashimoto H., Jolly W. M., Piper S. C., Tucker C. J., Myneni R. B. and Running S. W. (2003) Climate-driven increases in global terrestrial net primary production from 1982 to 1999. *Science* **300**: 1560–1563.

Novelli P. C., Conway T. J., Dlugokencky E. J. and Tans P. P. (1995) Recent changes of carbon dioxide, carbon monoxide, and methane in the troposphere and implications for global climate change. *World Meteorol Bull* **44**: 32–37.

Ogée J., Peylin P., Ciais P., Bariac T., Brunet Y., Berbigier P., Roche C., Richard P., Bardoux G. and Bonnefond J.-M. (2003) Partioning net ecosystem carbon exchange into net assimilation and respiration using $^{13}CO_2$ measurements: A cost-effective sampling strategy. *Global Change Biogeochem Cycles* **17**: doi: 10.1029/2002/GB001995.

Peylin P. (1999) The Composition of ^{18}O in Atmospheric CO_2: A New Tracer to Estimate Global Photosynthesis. PhD Thesis (in French), University of Paris, VI.

Peylin P., Ciais P., Tans P. P., Six K. D., Berry J. A. and Denning A. S. (1997) ^{18}O in atmospheric CO_2 simulated by a 3-D transport model: Sensitivity study to vegetation and soil fractionation factors. *Phys Chem Earth* **21**: 463–469.

Potter C., Brooks-Genovese V., Klooster S., Bobo M. and Torregrosa A. (2001) Biomass burning losses of carbon estimated from ecosystem modelling and satellite data analysis for the Brazilian Amazon region. *Atmos Environ* **35**: 1773–1781.

Raich J. W. and Potter C. S. (1995) Global patterns of carbon dioxide emissions from soils. *Global Biogeochem Cycles* **9**(1): 23–36.

Randerson J. T., Thompson M. V. and Field C. B. (1998) Linking ^{13}C based estimates of land and ocean sinks with predictions of carbon storage from CO_2 fertilization of plant growth. *Tellus* **51**B: 668–678.

Randerson J. T., Collatz G. J., Fessenden J. E., Munoz A. D., Still C. J., Berry J. A., Fung I. Y., Suits N. and Denning A. S. (2001) A possible global covariance between terrestrial gross primary production and ^{13}C discrimination: Consequences for the atmospheric 13C budget and its response to ENSO. *Global Biogeochem Cycles* **16**(4): doi: 10.1029/2001/GB001845.

Randerson J. T., Still C. J., Ballé J. J., Fung I. Y., Donney S. C., Tans P. P., Conway T. J., White J. W. C., Waughn B., Suits N. and Denning A. S. (2002b) Carbon isotope discrimination of arctic and boreal biomes inferred from remote atmsopheric measurements and a biosphere–atmosphere model. *Global Biogeochem Cycles* **16**(3): doi: 10.1029/2002/GB001435.

Randerson J. T., Van der Werf G. R., Collatz G. J., Giglio L., Still C. J., Kasischke E. S., Kasibhatla P., Defries R. S. and Tucker C. J. (2004) Interannual variability in fire emissions from C3 and C4 ecosystems. *Global Change Biol*—Submitted.

Rayner P. J., Enting I. G., Francey R. J. and Langenfelds R. (1999) Reconstructing the recent carbon cycle form atmospheric CO_2, $\delta^{13}C$ and O_2/N_2 observations. *Tellus* **51**B: 213–232.

Riley W. J., Still C. J., Torn M. S. and Berry J. A. (2002) A mechanistic model of $H_2^{18}O$ and $C^{18}OO$ fluxes between ecosystems ant the atmosphere: Model description and sensitivity analyses. *Global Biogeochem Cycles* **16**: doi: 10.1029/2002/GB001878.

Röckmann T., Brenninkmeijer C. A. M., Saueressig G., Bergamaschi P., Crowley J., Fischer H. and Crutzen P. J. (1998) Mass independent fractionation of oxygen isotopes in atmospheric CO due to the reaction CO + OH. *Science* **281**: 544–546.

Roden J. S. and Ehleringer J. R. (1999) Observations of hydrogen and oxygen isotopes in leaf water confirm the Craig–Gordon model under wide–ranging environmental conditions. *Plant Physiol* **120**: 1165–1173.

Schimel D. S., Braswell B. H., Holland B. A., McKeown R., Ojima D. S., Painter T. H., Parton W. J. and Townsend A. R. (1994) Climatic, edaphic and biotic controls over the storage and turnover of carbon in soils. *Global Biogeochem Cycles* **8**: 279–293.

Scholze M., Kaplan J. O., Knorr W. and Heimann M. (2003) Climate and interannual variability of the atmosphere–biosphere $^{13}CO_2$ flux. *Geophys Res Lett* **30**(2): 1097.

Seiler W. and Crutzen P. J. (1980) Estimates of gross and net fluxes of carbon between the biosphere and the atmosphere from biomass burning. *Climatic Change* **2**: 207–247.

Shvidenko A. and Nilsson S. (2000) Fire and the carbon budget of Russian forests. In *Fire, Climate Change and Carbon Cycling in the Boreal Forest* (E. S. Kasischke and B. J. Stocks, eds) doi: 10.1029/2001/GL013454.

Stern L. A., Amundson R. and Baisden W. T. (2001) Influence of soils on oxygen isotope ratio of atmospheric CO_2. *Global Biogeochem Cycles* **15**(3): 753–760.

Tans P. P. (1998) Oxygen isotopic equilibrium between dioxide and water in soils. *Tellus* **50**B: 163–178; Erratum: *Tellus* **50**B: 400.

Tans P. P., Berry J. A. and Keeling R. (1993) Oceanic ^{13}C data: A new window on CO_2 uptake by the oceans. *Global Biogeochem Cycles* **7**: 353–368.

Thiemens M. H. (1999) Mass-independant isotope effects in planetary atmospheres and the early solar system. *Science* **283**: 341–345.

Thompson M. V., Randerson J. T., Malmström C. M. and Field C. B. (1996) Change in net primary production and heterotrophic respiration : How much is necessary to sustain the terrestrial carbon sink? *Global Biogeochem Cycles* **10**(4): 711–726.

Townsend A. R., Asner G. P., White J. W. C. and Tans P. P. (2002) Land use effects on atmospheric ^{13}C imply a sizable terrestrial carbon sink in tropical latitudes. *Geophys Res Lett* **29**(10): pp. 289–311. Springer-Verlag, New York.

Townsend A. R., Vitousek P. M., DesMarais D. and Tharpe A. (1997) The effects of temperature and soil carbon pool structure on CO_2 and $^{13}CO_2$ fluxes from five Hawaiian soils. *Biogeochemistry* **38**: 1–17.

Trabaud L. (1994) Widland fire cycles and history in central southern France. In *International Conference on Forest Fire Research*, pp. 545–556. Coimbra, Portugal.

Trudinger C. M., Enting I. G., Francey R. J., Etheridge D. M. and Rayner P. J. (1999) Long-term variability in the global carbon cycle inferred from a high-precision CO_2 and $\delta^{13}C$ ice-core record. *Tellus* **51**B: pp. 233–248.

Valentini R., Matteucci G., Doman A. J., Schulze E.-D., Rebmann C., Moors E. J., Granier A. *et al.* (2000) Respiration as the main determinant of carbon balance in European forests. *Nature* **404**: 861–865.

Vogel J. C., Grootes P. M. and Mook W. G. (1970) Isotopic fractionation between gaseous and dissolved carbon dioxide. *Z Physik* **230**: 225–238.

Wang C., Gower S. T., Wang Y., Zhao H., Yan P. and Bond-Lamberty B. P. (2001) The influence of fire on carbon distribution and net primary production of boreal Larix gmelinii forests in north-eastern China. *Global Change Biol* **7**: 719–730.

Wanninkhof R. (2002) *Gas Transfer at Water Surfaces*. AGU, Washington DC.

Wirth C., Czimczik C. I. and Schulze E.-D. (2002) Beyond annual budgets: carbon flux at different temporal scales in fire-prone Siberian Scots pine forests. *Tellus* **54**B: 611–641.

Yakir D. and Sternberg L. S. L. (2000) The use of stable isotopes to study ecosystem gas exchange. *Oecologia* **123**: 297–311.

Yakir D. and Wang X. F. (1996) Fluxes of CO_2 and water between terrestrial vegetation and the atmosphere estimated from isotope measurements. *Nature* **380**: 515–517.

Yurtsever Y. and Gat J. R. (1981) *Atmospheric Waters* pp. 103–142. IAEA, Vienna.

Zahn A., Neubert R. and Platt U. (2000) Fate of long-lived trace species near the northern hemispheric tropopause: Isotopic composition of carbon dioxide. *J Geophys Res* **105**(D5): 6719–6735.

Zahn A., Neubert R., Maiss M. and Platt U. (1999) Fate of long-lived trace species near the northern hemispheric tropopause: Carbon dioxide, methane, ozone, and sulfur hexafluoride. *J Geophys Res* **104**(D11): 13923–13942.

15

Factors Influencing the Stable Isotopic Content of Atmospheric N_2O

Thom Rahn

Introduction

Solar radiation reaching the Earth's surface as visible and ultraviolet radiation is re-emitted as long wave (infra-red/IR) radiation that can be absorbed by gases in the atmosphere, thus trapping this energy and warming the surface. The most important of these 'greenhouse' gases are water (H_2O) and carbon dioxide (CO_2) but a number of other trace gases have been shown to be very effective at trapping radiation in important windows of the IR spectrum. Chief among these are the naturally occurring gases, methane (CH_4), ozone (O_3), and nitrous oxide (N_2O), and the man-made chlorofluorocarbons (CFCs). Nitrous oxide and the CFCs also have the unique property of their stratospheric reaction products participating in the catalytic destruction of ozone. In the case of N_2O, the increase in globally averaged radiative forcing is estimated at $0.15\,W/m^2$ (Ehhalt *et al.*, 2001) and accounts for $\sim 5\%$ of the total forcing due to all of the recognized greenhouse gases (Rodhe, 1990). Because of these direct and indirect influences on Earth's chemistry and radiation budget, it is imperative that we attempt to understand how human-induced perturbations may affect the global budgets of N_2O.

Early measurements of the N_2O molar mixing ratio in the troposphere employed a variety of techniques yielding a wide range of observed values and resulting in a wide range of predicted lifetimes (Pierotti and Rasmussen, 1977). Short-term variability has since been shown to be small with seasonal amplitudes typically less than measurement precision ($\sim 0.5\,\mu mol/mol$) (Weiss, 1981; Prinn *et al.*, 1990, 2000) with the exception of a $\leq 1\,\mu mol/mol$ seasonal variation in the higher northern latitudes (Weiss, 1981; Prinn *et al.*, 2000; Levin *et al.*, 2002). The global average molar mixing ratio as of 1998 was $\sim 314\,nmol/mol$ with northern hemisphere values averaging $0.8\,nmol/mol$

greater than the southern hemisphere (Ehhalt *et al.*, 2001). This limited variability is one line of evidence (Jobson *et al.*, 1999) indicating that the atmospheric lifetime of N_2O is on the order of c. 114 years (Ehhalt *et al.*, 2001). Because of its considerably long lifetime, perturbations to the N_2O budget will be very slow in responding to remediation efforts, if and when they are implemented.

Although short-term variability is small, trapped gases in ice cores show that the preindustrial N_2O molar mixing ratio was ~265–275 nmol/mol (Machida *et al.*, 1995; Fluckiger *et al.*, 2002) and high-precision records beginning in the latter half of the twentieth century show that there is an ongoing rate of increase of ~0.25%/yr (Weiss, 1981; Khalil and Rasmussen, 1992; Ehhalt *et al.*, 2001). A large proportion of atmospheric N_2O is produced during microbial energy-exchange reactions involving both reduced (NH_3) and oxidized (NO_3^-) forms of nitrogen, both in soils and in aquatic environments. The atmospheric increase is considered to be due to an amplification of these processes arising from application of fertilizers to cultivated soils, but other perturbations to the natural N cycle including production of N_2 fixing crops, animal waste, and increased aerosol deposition of fixed nitrogen as well as biomass burning (natural and deliberate), fuel combustion, and industrial processes also contribute (Bouwman *et al.*, 1995; Kroeze *et al.*, 1999; Ehhalt *et al.*, 2001). The most recent best estimates of the Intergovernmental Panel on Climate Change for the individual N_2O sources are provided in Table 15.1. Once introduced to the troposphere, N_2O is chemically inert until it is delivered to the stratosphere where it is destroyed via photolysis or reaction with excited atomic oxygen, $O(^1D)$ (Minschwaner *et al.*, 1993). Although recent work has shown that the strengths of the known sources are sufficient to balance the well-defined sinks (Mosier *et al.*, 1998; Kroeze *et al.*, 1999), how these sources are spatially distributed, how they have changed over time, and their potential for change in the future are still matters of active debate.

Isotopic Considerations

The stable isotopic composition of atmospheric trace gases provides information about their origin and fate that cannot be determined from mixing ratio measurements alone. Biological source and loss processes (e.g., bacterial production of N_2O or CH_4 and photosynthetic processing of CO_2) are typically accompanied by isotopic selectivity associated with the kinetics of bond formation and destruction (Rosenfeld and Silverman, 1959; Whiticar *et al.*, 1986; Farquhar *et al.*, 1989, 1993). Thermodynamic considerations predict isotopic differentiation between phases and/or reactants under equilibrium conditions (Urey, 1947; Richet *et al.*, 1977).

Table 15.1 Best Estimates of the Annual Mass Fluxes of the Various Known N_2O Sources and Sinks[a]

	Mosier *et al.* (1999), Kroeze *et al.* (1999)		Olivier *et al.* (1998)			
Base year	1994	Range	1990	Range	SAR* 1980s	TAR† 1990s
Sources						
Ocean	3.0	1–5	3.6	2.8–5.7	3	
Atmosphere (*in situ*)	0.6	0.3–1.2	0.6	0.3–1.2		
Tropical soils						
Wet forest	3.0	2.2–3.7			3	
Dry savannas	1.0	0.5–2.0			1	
Temperate soils						
Forests	1.0	0.1–2.0			1	
Grasslands	1.0	0.5–2.0			1	
All soils			6.6	3.3–9.9		
Natural sub-total	9.6	4.6–15.9	10.8	6.4–16.8	9	
Agricultural soils	4.2	0.6–14.8	1.9	0.7–4.3	3.5	
Biomass burning	0.5	0.2–1.0	0.5	0.2–0.8	0.5	
Industrial sources	1.3	0.7–1.8	0.7	0.2–1.1	1.3	
Cattle and feedlots	2.1	0.6–3.1	1.0	0.2–2.0	0.4	
Anthropogenic sub-total	8.1	2.1–20.7	4.1	1.3–7.7	5.7	6.9
Total sources	17.7	6.7–36.6	14.9	7.7–24.5	14.7	
Trend	**3.9**	**3.1–4.7**			**3.9**	**3.8**
Total sinks (stratosphere)	12.3	9–16			12.3	12.6
Implied total source	16.2				16.2	16.4

[a] From Ehhalt *et al.*, 2001.
*Intergovernmental Panel on Climate Change, Second Assessment Report.
†Intergovernmental Panel on Climate Change, Third Assessment Report.

Photochemical reactions and dissociation via photolysis have also been shown to be a source of isotopic fractionation that can be useful for determining different aspects of trace-gas budgets (Rahn *et al.*, 1998; Tyler *et al.*, 2000; Röckmann *et al.*, 2001b; Saueressig *et al.*, 2001; Kaiser *et al.*, 2002a).

Of the three important biologically mediated greenhouse gases, our understanding of the isotopic budget of N_2O lags far behind that of CO_2 and CH_4. The reasons for this are several-fold and include, but are not limited to, the following. The long lifetime of N_2O in the atmosphere means that it is not only well mixed in terms of its mixing ratio but isotopically; therefore, the seasonal and spatial gradients that lend insight into the isotopic budgets of these other trace gases (Mook *et al.*, 1983; Keeling *et al.*, 1984, 1996; Levin *et al.*, 2002; Miller *et al.*, 2002) are not apparent in N_2O; the temporal and spatial isotopic variability of the individual sources is quite large (as will be discussed below), making it difficult to assign unique values for modeling purposes; and, the low mixing ratio leads to inherent difficulties in collection, extraction, and analysis that, in the past, has limited sample throughput as well as the overall precision of the measurements although recent advances have improved on this situation (Röckmann *et al.*, 2003b).

The isotopic signature of N_2O in the free troposphere has been reported by a number of investigators since the mid-1970s (Moore, 1974; Yoshida and Matsuo, 1983; Wahlen and Yoshinari, 1985; Yoshinari and Wahlen, 1985; Kim and Craig, 1990; Rahn and Wahlen, 1997; Cliff *et al.*, 1999; Yoshida and Toyoda, 2000; Kaiser *et al.*, 2003c) with inter-laboratory averages, and in some cases, intra-laboratory ranges varying by several per mil. It is important to note that these studies employed different methodologies and/or reference materials with no inter-laboratory comparison and therefore, the true variability of $^{15}N/^{14}N$ and $^{18}O/^{16}O$ in N_2O remained uncertain. More recently, a study of the variability in the stable isotopic content of tropospheric N_2O at six European stations over a 2-year period has shown that the average $\delta^{15}N$ and $\delta^{18}O$ are $6.72 \pm 0.12‰$ (relative to atmospheric N_2) and $44.62 \pm 0.21‰$ (relative to standard mean ocean water, or SMOW), respectively, with no temporal (over the period of measurement) or spatial trends within the precision of the measurements taken (Kaiser *et al.*, 2003c). These relatively constant values are as would be expected given the atmospheric lifetime discussed above and lend credence to the conclusion that earlier reported variabilities were the result of differences in reference material and lower precision with earlier methodologies.

Isotopomers and Isotopologues

In addition to the ^{18}O and bulk ^{15}N content, there are two additional measures of N_2O isotopic content that are reported in the literature, the '^{15}N position' and the 'mass independent' oxygen isotope anomaly (the deviation of ^{17}O from a strict covariation relative to ^{18}O). Because N_2O is linearly asymmetric, N isotopic substitution can be at either the central or terminal position, leading to what is referred to as ^{15}N position dependent enrichment. Analytical techniques employ either traditional mass spectrometry with the twist that one not only measures the isotopic content of the ionized parent molecule (N_2O^+, masses 44, 45, 46) but the ionic fragment, NO^+ (masses 30 and 31) (Brenninkmeijer and Röckmann, 1999; Toyoda and Yoshida, 1999), or Fourier transform infra-red spectroscopy (FTIR), which is used to compare the absorption spectra of the three different N isotopomers, $^{14}N^{14}N^{16}O$, $^{15}N^{14}N^{16}O$ and $^{14}N^{15}N^{16}O$ (Esler *et al.*, 2000; Griffith *et al.*, 2000; Turatti *et al.*, 2000; Zhang *et al.*,

2000). The former requires that a correction be applied for the recombination of free atomic N and O to form NO and that there be a reference gas for which the distribution of ^{15}N is very precisely known. The FTIR requires a large sample size, which renders the method prohibitive for routine atmospheric sampling. The lack of a common reference gas and inter-laboratory calibration is undoubtedly responsible for the large discrepancies between published data sets of ^{15}N position dependence in the free troposphere (compare $\delta^{15}N^{\beta}_{air}$ of Yoshida and Toyoda [2000] with $^{1}\delta^{15}N$ of Kaiser *et al.* [2003c]). While there is potential for this measurement to provide additional insight into the budget of N_2O, until there is an established universal standard it will apparently not be possible to make meaningful comparisons of reported data. There have been several different systems of nomenclature that have been developed to report the ^{15}N position dependence; (cf. Yung and Miller, 1997; Brenninkmeijer and Röckmann, 1999; Toyoda and Yoshida, 1999).

The other isotopologue that has received some attention is $^{14}N^{14}N^{17}O$, which is usually compared to $^{14}N^{14}N^{18}O$ and is commonly reported as $\Delta^{17}O = \delta^{17}O - 0.52\delta^{18}O$ (although a more rigorous definition, derived without approximation can be found [cf. Kaiser *et al.*, 2003c]). The $\Delta^{17}O$ anomaly is the result of photochemical processes and will be discussed in more detail in the 'Other N_2O Sources' and 'N_2O Loss Processes' sections.

In the following pages I will discuss the mass flux terms that determine the isotopic content of atmospheric N_2O and how they may have varied over time. The discussion will concentrate on the bulk $^{15}N/^{14}N$ and the $^{18}O/^{16}O$ content. Isotopic notation is given in the standard form where $\delta_{samp} = (R_{samp}/R_{std} - 1)$, where R is the heavy to light isotopic ratio of the sample or standard, results being expressed in units of per mil (‰). In the case of N, atmospheric N_2 is the customary standard; in the case of O, I will use the reference standard mean ocean water (SMOW) although the reader should be aware that in the literature, $N_2\ ^{18}O$ results are sometimes also reported relative to atmospheric O_2. I will also describe fractionation and enrichment factors that relate reaction rates (k) and absorption cross-sections (σ) of heavy to light isotopes (e.g., k' to k, respectively); this convention has been adopted in much of the geochemical literature although it is inverse to that recommended by IUPAC as well as to that used in some of the literature dealing with N_2O. For a thorough discussion, the reader is referred to Kaiser *et al.* (2002a, 2003c). Given the problems with ^{15}N position standardization described above, I refer the reader to details of the mass spectrometric methodology and standardization (Brenninkmeijer and Röckmann, 1999; Toyoda and Yoshida, 1999) and to Kaiser *et al.* (2003c) for a more complete discussion. There are also multiply substituted species (e.g., $^{14}N^{15}N^{18}O$, $^{15}N^{14}N^{18}O$, $^{15}N^{15}N^{16}O$) that have yet to be measured and are beyond the scope of this work but that are certainly fodder for future studies (Kaiser *et al.*, 2003a).

N_2O from the Terrestrial Biosphere

The microbial processes responsible for N_2O formation in soils are discussed in more detail elsewhere in this volume (Perez, Chapter 5). In general, N_2O can be formed as a by-product during oxidation of NH_3 (nitrification) or during dissimilatory reduction of NO_3^- (denitrification). Variations on these pathways have been observed as in the case of nitrifier denitrification and methanotrophic nitrification (see Sutka *et al.*, 2003 for summary and additional references). Nitrous oxide itself can serve as an electron acceptor under certain conditions and thus can be reduced before escaping from the soil. The amount of N_2O that does escape to the atmosphere from a given soil is determined by several factors including porosity/permeability, soil water content, substrate availability, and the microbial community itself. It is estimated by the Intergovernmental Panel on Climate Change (IPCC) that natural soils contribute 6 Tg N/yr to the atmosphere as N_2O, while agricultural soils and livestock contribute approximately 4 Tg N/yr (Ehhalt *et al.*, 2001).

Several studies have been conducted to determine the fractionation during production/consumption of N_2O by specific nitrifying (Mariotti *et al.*, 1981; Yoshida, 1988; Webster and Hopkins, 1996; Sutka *et al.*, 2003), denitrifying (Mariotti *et al.*, 1981; Webster and Hopkins, 1996; Barford *et al.*, 1999), and nitrogen-fixing bacterial cultures (Yamazaki *et al.*, 1987) and, collectively, have led to the conclusion that denitrification produces N_2O with greater ^{15}N enrichment than does nitrification, and that subsequent reduction of N_2O, either during denitrification or N_2O fixation, serves to further enrich residual N_2O in both ^{15}N and ^{18}O. In theory, these results could be used to decipher the relative contributions of the various processes in a given ecosystem if the initial conditions are known. In practice, however, there are many more bacteria than those that have been investigated, so the modeling of such systems requires the assumption that fractionations remain constant for a given pathway, regardless of mediating organism. In addition, the isotopic content of substrates is highly variable and often not known and the complications of multiple fractionating processes occurring simultaneously make it extremely difficult to apply these factors quantitatively to field data. A number of studies have described the isotopic content of N_2O actually emitted from terrestrial systems, both natural and agricultural; a thorough summary of these surveys is provided in Chapter 5 (Perez) of this volume. In general, it can be said that the measured isotopic signature, both for ^{15}N and ^{18}O, of N_2O emitted from terrestrial environments is depleted relative to the tropospheric average although anomalously enriched N_2O produced within soils has been described (Mandernack *et al.*, 2000).

N_2O from the Oceans and Other Aquatic Regions

According to the IPCC (Ehhalt *et al.*, 2001), the annual mass flux of N_2O from the oceans is estimated to be of the order of 3 Tg N/yr although variability reported in the literature ranges from 1 to 5.8 Tg N/yr (Nevison *et al.*, 1995, 2003 Mosier *et al.*, 1998; Olivier *et al.*, 1998; Kroeze *et al.*, 1999). It has also been noted that river, estuarine, and coastal marine sources of N_2O may have increased over time due to dissolved inorganic nitrogen loading resulting from fertilizer N runoff, sewage discharge, and atmospheric deposition; indeed, it has been estimated to account for as much as 20% of the current annual anthropogenic production (Seitzinger and Kroeze, 1998; Seitzinger *et al.*, 2002).

A number of studies have been undertaken to investigate the isotopic signature of these N_2O sources with results that are nearly as variable as those from terrestrial soils. Unlike soils, however, the surface oceans are sometimes enriched in both or either ^{15}N and/or ^{18}O relative to tropospheric N_2O (Fig. 15.1). While some isotopic data (Kim and Craig, 1990; Dore *et al.*, 1998; Toyoda *et al.*, 2002) and evidence such as the correlation/anti-correlation of N_2O with dissolved NO_3^-/O_2 (Kim and Craig, 1990) and N_2O/O_2 emission ratios (Lueker *et al.*, 2003) suggest that nitrification contributes significantly to the N_2O flux from a variety of oceanic localities, it is also well established that denitrification plays an important role in oceanic N cycling (Yoshida *et al.*, 1989; Yoshinari *et al.*, 1997; Naqvi *et al.*, 1998; Codispoti *et al.*, 2001). In fact, N_2 production can exceed NO_3^- consumption by as much as ~35% under certain circumstances (Codispoti *et al.*, 2001). In this case, it is obvious that reduction of N_2O will compete significantly with its production and any depletion of ^{15}N or ^{18}O content during production through either nitrification or denitrification will be attenuated by subsequent N_2O reduction. Indeed, it is likely that such processes account for the minimal overlap of the terrestrial and oceanic fields in Fig. 15.1 even though ostensibly the same pathways of nitrification and denitrification are responsible for N_2O production; whereas in the terrestrial environment where soil gas diffusivities can be $\sim 10^5$ greater than diffusion in H_2O, gases produced in-situ escape readily to the atmosphere while, in the oceans, the flux to the atmosphere is dependent in large part on seasonal and/or ephemeral mechanical mixing and N_2O produced in-situ is long-lived enough to be recycled biologically within the system. It is also important to point out that, as with terrestrial

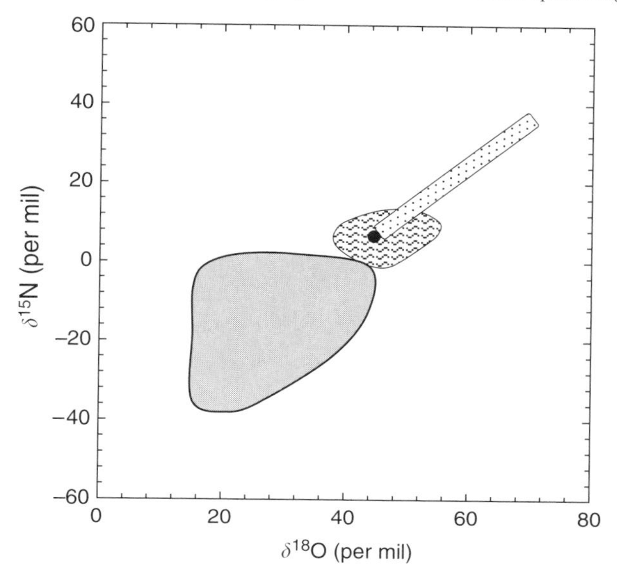

Figure 15.1 A plot of $\delta^{15}N$ vs $\delta^{18}O$ of N_2O in different environments. The gray field represents the range that has been observed emitting from terrestrial ecosystems, both natural and human perturbed. The hatched field is the range of values reported for N_2O dissolved in surface oceans. The stippled field represents stratospheric measurements. The filled circle is the tropospheric average.

systems, quantitative estimates of isotopic fluxes between the various N and O reservoirs are dependent on knowledge of the isotopic composition of the substrates undergoing transformation and such details are not available for most of the oceanic N_2O isotope data reported to date. Finally, an additional factor that must be kept in mind when considering N_2O sea-to-air transfer is that there is fractionation associated with this flux that has been measured as -0.7 and $-1.9‰$, for N and O, respectively (Inoue and Mook, 1994), the heavier isotopes preferentially remaining in solution.

The isotopic content of N_2O emitted from non-ocean aquatic sources has had limited investigation. One study of the Bang Nara River in Thailand has shown that dissolved N_2O is highly variable and seasonal with maximums of $\Delta N_2O = 131$ nM (~'20 times saturation') occurring at the onset of the rainy season. Even with these elevated levels, $\delta^{15}N$ and $\delta^{18}O$ were depleted by only ~7 and ~8‰ respectively, relative to tropospheric N_2O (Boontanon *et al.*, 2000), indicating that apparently in this system as well, N_2O is actively recycled. Again, however, isotopic analysis of substrate was not reported meaning conclusive statements regarding production pathway(s) could not be made.

Other N_2O Sources

The other sources that contribute significantly to atmospheric N_2O are biomass burning (0.5 Tg N/yr) and human industry (nylon and nitric acid production, fossil fuel combustion, 1.3 Tg N/yr) (Ehhalt *et al.*, 2001). Very little work has been performed to determine the N and O isotopic content produced from these sources, leading to the postulation (Rahn and Wahlen, 2000) that it is reasonably similar to atmospheric N_2 and O_2 (or $\delta^{18}O_{SMOW}$ ~22.5‰) in isotopic composition. One sample with ~12 nmol/mol excess N_2O purported to be due to a pollution point source(s), was depleted in both ^{15}N and ^{18}O (Kaiser *et al.*, 2003c). Extrapolation of this result in a two-end-member mixing model would indicate that it arises from a source with $\delta^{15}N = -20‰$ and $\delta^{18}O = -31‰$, which is more in the range of N_2O produced in agricultural systems. The error in this two-point analysis is quite large, however, and it cannot be ruled out that the result does not arise from some source of contamination other than industrial pollution.

One additional source of N_2O in the atmosphere is in-situ production via the reaction

$$NO_2 + NH_2 \rightarrow N_2O + H_2O \tag{15.1}$$

and may account for as much as 5% of the total N_2O sources (Dentener and Crutzen, 1994; Kohlmann and Poppe, 1999). Odd nitrogen (defined as the total of the reactive, oxygenated nitrogen species, e.g., NO, NO_2, NO_3, nitric acid, etc., gas phase species not to be confused with the dissolved ionic species) is actively involved in the catalytic cycling of ozone via the following reactions

$$NO + O_3 \rightarrow NO_2 + O_2 \tag{15.2}$$

$$NO_2 + h\nu \rightarrow NO + O \tag{15.3}$$

$$O + O_2 + M \rightarrow O_3 \tag{15.4}$$

and since ozone is known to have excess ^{17}O, it has been proposed (Röckmann *et al.*, 2001c) that N_2O production from Eq. 15.1 might account for the small but extant $\Delta^{17}O$ anomaly that has been observed in atmospheric N_2O samples (Cliff *et al.*, 1999; Röckmann *et al.*, 2001c). Calculations have shown that a source of ~0.4 Tg N/yr from Eq. 15.1 may be sufficient to account for the ~1‰ $\Delta^{17}O$ anomaly (Röckmann *et al.*, 2001c). It should also be noted that the catalytic cycle that enriches ozone and the nitrogen oxides in ^{17}O and ^{18}O can result in the production of atmospheric NO_3 which becomes a sink from the odd nitrogen cycle when it is aerosolized and deposited at Earth's surface. Nitrate deposited in this manner is then available for consumption by denitrifying bacteria and any $\Delta^{17}O$ anomaly that is present in the NO_3 deposited should be at least partially conserved in N_2O produced (Michalski *et al.*, 2003). It has not yet been definitively shown which of these two pathways is responsible for the observed atmospheric $\Delta^{17}O$ anomaly or whether it may be some combination of the two.

Other gas phase reaction sources of N_2O have been suggested (Zipf and Prasad, 1998; Estupinan *et al.*, 2002) but they are expected to be minor in terms of their mass flux and the details of their isotopic systematics have yet to be elucidated.

N_2O Loss Processes

The destruction of atmospheric N_2O occurs in the stratosphere where it is either photolysed

$$N_2O + h\nu \rightarrow NO + O(^1D) \tag{15.5}$$

or oxidized as in

$$N_2O + O(^1D) \rightarrow 2NO \tag{15.6a}$$

$$\rightarrow N_2 + O_2 \tag{15.6b}$$

These processes account for the loss of ~12.6 Tg N/yr (Ehhalt *et al.*, 2001) with Eq. 15.5 accounting for approximately 90% of the loss, Eq. 15.6a ~6%, and Eq. 15.6b ~4% (Eq. 15.6a is also the source of odd nitrogen in the stratosphere that initiates the catalytic cycling of O_3 as illustrated in Eqs 15.2–15.4). An early enigma of the isotopic budget of N_2O was the enrichment of ^{15}N and ^{18}O in tropospheric samples relative to any other known values other than deep oceans (Yoshida and Matsuo, 1983; Yoshida *et al.*, 1984; Wahlen and Yoshinari, 1985; Yoshinari and Wahlen, 1985; Yoshida, 1988, 1989; Kim and Craig, 1990).

The concept of a stratosphere to troposphere return flux counterbalancing the depleted terrestrial and oceanic source terms was first proposed in 1983 (Yoshida and Matsuo, 1983) and the discovery of a significant enrichment of both ^{15}N and ^{18}O in stratospheric N_2O apparently verified this hypothesis but it was ultimately concluded that the effect was to overcompensate for the depleted sources (Kim and Craig, 1993). Furthermore, it was concluded that photolysis and photooxidation (Eqs 15.5 and 15.6) could not account for the observed stratospheric enrichments (Johnston *et al.*, 1995) indicating that the details of the global N_2O budget, in particular its isotopic systematics, were poorly understood. An explanation for the stratospheric enrichment was provided when Yung and Miller (1997) theorized that photolysis could indeed significantly fractionate N_2O while

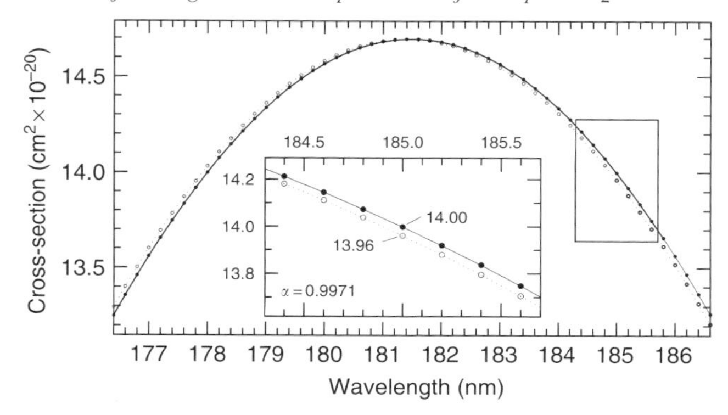

Figure 15.2 Representation of the theoretical shift in cross-section with ^{18}O substitution according to Yung and Miller (1997). The inset plot is a detail of the highlighted section between 184 and 187 nm. From Rahn and Wahlen (1999).

not violating the results of Johnston *et al.* (1995). Their model proposed a wavelength-dependent mechanism for the photolytic fractionation of N_2O based on subtle shifts in the absorption spectrum of the different isotopologues due to variations in zero point energy. This principle is demonstrated graphically in Fig. 15.2 using the recommended absorption cross-section spectral function. The curve representing the ^{18}O-substituted species (dashed curve) is slightly blue shifted, by $-27.5\,cm^{-1}$ (Yung and Miller, 1997), relative to the normal curve (both calculated at 300 K). Cross-sections are equal where the two curves cross near the absorption peaks but a clear separation is observed on both shoulders. The inset plot of Fig. 15.2 details the highlighted section on the higher wavelength shoulder. Illustrated are the cross-sections at $\lambda = 185\,nm$ for the normal curve $(14.00 \times 10^{-20}\,cm^2)$ and the blue shifted $N_2\,^{18}O$ curve $(13.96 \times 10^{-20}\,cm^2)$. Analogous to determining the kinetic fractionation for a chemical reaction, the photolytic fractionation factor will be equal to the ratio of the heavy to light cross-sections or $\sigma_{^{18}O}/\sigma_{^{16}O}(185\,nm) = \alpha_{185} = 0.9971$. When expressed as an enrichment factor, where $\varepsilon_\lambda = 1000(\alpha_\lambda - 1)$, $\varepsilon_{185} = -2.9‰$.

In reality, the situation is complicated even further by vibrational structure in the absorption continuum. If we calculate the enrichment factors empirically using published cross-sectional data for isotopically substituted species (Selwyn and Johnston, 1981), we see that the cross-section curves cross several times, causing ε_λ to change sign several times over the spectrum (Fig. 15.3). The first experiments to confirm this fractionation at discrete wavelengths and isotopic natural abundance levels (Rahn *et al.*, 1998) are included in Fig. 15.3. The results compare favorably with the predicted ε_λ value at 193 nm but the cross-sectional data do not extend to wavelengths greater than 197 nm, making a comparison of the 207 nm

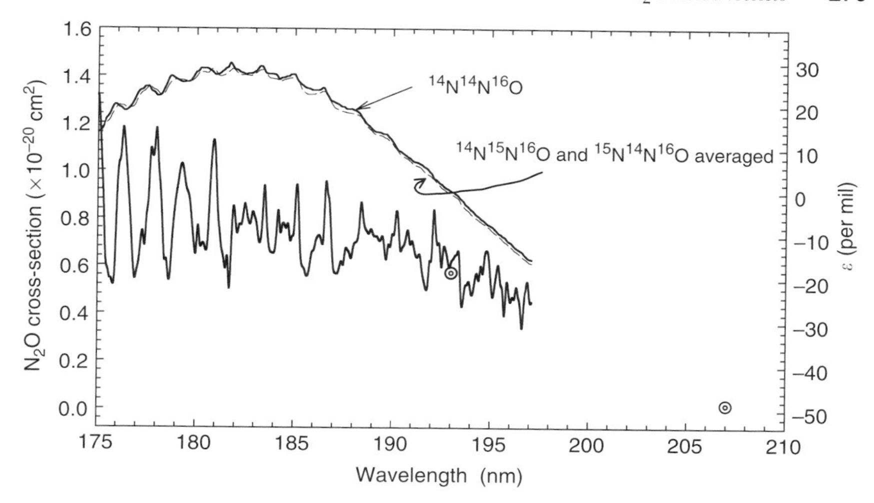

Figure 15.3 Top two curves show absorption cross-sections for the N_2O species as indicated (reproduced from Selwyn and Johnston, 1981). Bottom curve indicates the spectral enrichment factor calculated as described in the text. The two symbols at 193 and 207 m are the laboratory results from Rahn *et al.* (1998). From Rahn and Wahlen (1999).

data impossible. It is important to note here that the data presented in Fig. 15.3 represent the total ^{15}N enrichment such that the cross-section data are the average spectra of pure $^{14}N^{15}NO$ and $^{15}N^{14}NO$ while the data at discrete wavelengths are total $\delta^{15}N$, which is not necessarily the same as $\frac{1}{2}(\delta^{14}N^{15}NO + \delta^{15}N^{14}NO)$. For a complete discussion on the consequences of linear combinations of δ values see Kaiser *et al.* (2003c).

The discrete wavelength data represented in Fig. 15.3 have since been verified by a number of studies that have expanded and improved our knowledge of the photolytic fractionation effects that influence stratospheric N_2O (Umemoto, 1999; Röckmann *et al.*, 2000; Turatti *et al.*, 2000; Zhang *et al.*, 2000; Röckmann *et al.*, 2001a,b; Kaiser *et al.*, 2002b; Kaiser *et al.*, 2003a,b). Perhaps most notable has been the work of Kaiser *et al.* (2003b), which has demonstrated the influence of the vibrational structure near the peak of the absorption continuum by showing simultaneous enrichment of $^{14}N^{15}N^{16}O$ and depletion of $^{15}N^{14}N^{16}O$ and $^{14}N_2{}^{18}O$ in residual N_2O during photolysis at 185 nm.

All of these studies demonstrate more than adequate agreement and the results have revealed fractionations that are approximately double those predicted by the theory of Yung and Miller (1997). Several theoretical studies have subsequently been completed to account for this discrepancy (Johnson *et al.*, 2001; Blake *et al.*, 2003) and have met with reasonable success when applied in chemical modeling of the stratosphere (Blake *et al.*, 2003; McLinden *et al.*, 2003). Furthermore, McLinden *et al.* (2003) and

Blake *et al.* (2003) note that the predicted fractionation due to photolysis is not strictly mass dependent, particularly at longer wavelengths, and may account for as much as half of the observed atmospheric $\Delta^{17}O$ anomaly discussed earlier.

The 10% loss of N_2O due to Eqs 15.6a and 15.6b must also be taken into account when considering the isotopic budget. An early study of the $O(^1D)$ oxidation pathway determined that fractionation for oxygen isotope substitution was $^{18}\varepsilon_{O(^1D)} = 1000[(k(N_2{}^{18}O)/k(N_2{}^{16}O) - 1] = -6‰$ (Johnston *et al.*, 1995). A subsequent study which investigated fractionation for both N and O isotopologues has determined that the O fractionation for Eq. 15.6 is $^{18}\varepsilon_{O(^1D)} = -12.2‰$ and that the total N fractionation is $^{15}\varepsilon_{O(^1D)} = -5.5‰$ (Kaiser *et al.*, 2002a). The discrepancy in the O isotope fractionation is unclear although it is noted that the latter study was conducted under a wide range of conditions (i.e., varying temperature and pressure regimes in a variety of reactors) with no significant variation (Kaiser *et al.*, 2002a).

A final note on the photochemical sinks of N_2O. It has been noted that the ratios of the fractionation factors for the various pathways (e.g., $^{15}\varepsilon_\lambda/^{18}\varepsilon_\lambda$ for photolysis) have unique values which lend insight into processes dominating in the stratosphere (Rahn and Wahlen, 1999; Röckmann *et al.*, 2001b; Kaiser *et al.*, 2002a). The relationships for the different ^{15}N isotopomers in photolysis (Röckmann *et al.*, 2000) and the difference between different fractionation ratios for photolysis and photooxidation (Kaiser *et al.*, 2002a) show particular promise for elucidating stratospheric transport processes.

The N_2O Global Isotopic Budget

Recent analyses that include the full impact of agriculture on the nitrous oxide budget (Mosier *et al.*, 1998; Kroeze *et al.*, 1999) bring the total mass balance of N_2O fluxes and atmospheric accumulation into much better agreement than earlier IPCC assessments (Prather *et al.*, 1995) which estimated that there was an unaccounted for source of ~1.5 Tg N/yr. The latter, along with the poorly understood isotopic evidence found in the 1990s, such as the observed ^{17}O excess (Cliff and Thiemens, 1997; Cliff *et al.*, 1999; Röckmann *et al.*, 2001c) and the stratospheric enrichments of ^{15}N and ^{18}O (Kim and Craig, 1993; Rahn and Wahlen, 1997), led to numerous efforts to find a 'missing source' of nitrous oxide. As described above, the stratospheric enrichments are now well understood in terms of the fractionations associated with photolysis and photooxidation (Johnston *et al.*, 1995; Yung and Miller, 1997; Rahn *et al.*, 1998; Umemoto, 1999; Röckmann *et al.*, 2000, 2001b; Johnson *et al.*, 2001; Kaiser *et al.*, 2002a; Blake *et al.*, 2003) and there are several candidate mechanisms for the observed

^{17}O excess (Röckmann *et al.*, 2001c; Estupinan *et al.*, 2002; Blake *et al.*, 2003; McLinden *et al.*, 2003; Michalski *et al.*, 2003; Kaiser *et al.*, 2004). The laboratory work investigating the fractionation resulting from nitrous oxide photochemistry has led to the unusual circumstance that, in the N_2O system, the isotopic systematics of the stratosphere are far better understood than those influencing the terrestrial and oceanic biospheres. Continued studies of stratospheric N_2O have further refined our understanding of its photochemical fractionation and show potential to lend insight into cross-tropopause transport and stratospheric mixing (Griffith *et al.*, 2000; Toyoda *et al.*, 2001; Parks *et al.*, 2004).

The temporal and spatial variability of the natural and anthropogenic sources of N_2O combined with the wide range of their reported isotopic content (Fig. 15.1), make it practically impossible to assign specific values for these terms in a forward model. Rahn and Wahlen (2000) circumvented this problem when trying to develop a global model by employing the tactic of utilizing the better known terms of the budget to solve for the isotopic content of the terrestrial source. They concluded that although additional sources of N_2O could not be discounted, such sources are not necessary for closure of the isotopic budget within the constraints of the known parameters. While this may not be the most satisfying result, it may well be that a tightly closed isotopic budget of N_2O is an unrealistic goal. This is not to say that measurements of atmospheric N_2O are a futile exercise; on the contrary, isotopic measurements of N_2O in discrete environments will always provide details of the processes affecting its production and loss. For instance, when combined with isotopic measurements of N substrates, N_2O isotopes can reveal temporal details of how nutrients cycle in different ecosystems under varying conditions (Perez *et al.*, 2001), and in the stratosphere, the details of how mixing processes take place can be elucidated (Rahn *et al.*, 1998; Griffith *et al.*, 2000; Kaiser *et al.*, 2002a). A common reference material with absolute calibration of the intramolecular distribution of N isotopes in particular will go a long way toward advancing these goals.

Finally, results of N_2O isotopic analysis from air occluded in ice cores have the potential to reveal details of N cycling during glacial and interglacial times. Results of analyses of air in unconsolidated polar snow (firn) (Sowers *et al.*, 2002; Röckmann *et al.*, 2003a) have already shown that the pre-anthropogenic values of $\delta^{15}N$ and $\delta^{18}O$ were enriched by $\sim$2‰ and 1‰ respectively relative to present times, verifying the temporal decrease attributed to isotopically light agricultural emissions first predicted by Rahn and Wahlen (2000). Subsequent measurements of 24 samples spanning the past 35 k year in ice from Taylor Dome, Antarctica have led to the conclusion that, with the exception of the Younger Dryas, the ratio of terrestrial to marine N_2O emissions has remained relatively constant over that time (Sowers *et al.*, 2003). As methodologies improve and more records become available (including ^{15}N position measurements), there will be much to learn about N_2O and the cycling of nitrogen.

References

Barford C. C., Montoya J. P., Altabet M. A. and Mitchell R. (1999) Steady-state nitrogen isotope effects of N_2 and N_2O production in *Paracoccus denitrificans*. *Appl Environ Microbiol* **65**: 989–994.

Blake G. A., Liang M. C., Morgan C. G. and Yung L. Y. (2003) A Born–Oppenheimer photolysis model of N_2O fractionation. *Geophys Res Lett* **30**: doi: 10.1029/2003GL016932.

Boontanon N., Ueda S., Kanatharana P. and Wada E. (2000) Intramolecular stable isotope ratios of N_2O in the tropical swamp forest in Thailand. *Naturwissenschaften* **87**: 188–192.

Bouwman A. F., VanderHoek K. W. and Olivier J. G. J. (1995) Uncertainties in the global source distribution of nitrous oxide. *J Geophys Res* **100**: 2785–2800.

Brenninkmeijer C. A. M. and Röckmann T. (1999) Mass spectrometry of the intramolecular nitrogen isotope distribution of environmental nitrous oxide using fragmentation analysis. *Rapid Commun Mass Spectrometry* **13**: 2028–2033.

Cliff S. and Thiemens M. (1997) The $^{18}O/^{16}O$ and $^{17}O/^{16}O$ ratios in atmospheric nitrous oxide: A mass-independent anomaly. *Science* **278**: 1774–1776.

Cliff S. S., Brenninkmeijer C. A. M. and Thiemens H M. (1999) First measurement of the $^{18}O/^{16}O$ and $^{17}O/^{16}O$ ratios in stratospheric nitrous oxide: A mass-independent anomaly. *J Geophys Res* **104**: 16171–16175.

Codispoti L. A., Brandes J. A., Christensen J. P., Devol A. H., Naqvi S. W. A., Paerl H. W. and Yoshinari T. (2001) The oceanic fixed nitrogen and nitrous oxide budgets: Moving targets as we enter the anthropocene? *Scientia Marina* **65**: 85–105.

Dentener F. J. and Crutzen J. P. (1994) A 3-dimensional model of the global ammonia cycle. *J Atmos Chem* **19**: 331–369.

Dore J. E., Popp B. N., Karl D. M. and Sansone J. F. (1998) A large source of atmospheric nitrous oxide from subtropical North Pacific surface waters. *Nature* **396**: 63–66.

Ehhalt D., Prather M., Dentener F., Derwent R., Dlugokencky E., Holland E., Isaksen I., Katima J. *et al.* (2001) Atmospheric chemistry and greenhouse gases. In *Climate Change 2001: The Scientific Basis.* Contribution of Working Group 1 to the Third Assessment Report of the Intergovernmental Panel on Climate Change. Cambridge University Press, Cambridge, UK and New York, NY.

Esler M. B., Griffith D. W. T., Wilson S. R. and Steele P. L. (2000) Precision trace gas analysis by FT-IR spectroscopy. 1. Simultaneous analysis of CO_2, CH_4, N_2O, and CO in air. *Anal Chem* **72**: 206–215.

Estupinan E., Nicovich J., Li J., Cunnold D. and Wine P. (2002) Investigation of N_2O production from 266 and 532 nm laser flash photolysis of $O_3/N_2/O_2$ mixtures. *J Phys Chemis* **106**A: 5880–5890.

Farquhar G. D., Ehleringer J. R. and Hubick T. K. (1989) Carbon isotope discrimination and photosynthesis. *Annu Rev Plant Physio Plant Mol Biol* **40**: 503–537.

Farquhar G. D., Lloyd J., Taylor J. A., Flanagan L. B., Syvertsen J. P., Hubick T. K., Wong S. C. and Ehleringer R. J. (1993) Vegetation effects on the istope composition of oxygen in atmospheric CO_2. *Nature* **363**: 439–443.

Fluckiger J., Monnin E., Stauffer B., Schwander J., Stocker T., Chappellaz J., Raynaud D. and Barnola J. (2002) High-resolution Holocene N_2O ice core record and its relationship with CH_4 and CO_2. *Global Biogeochem Cycles* **16**: 1010.

Griffith D. W. T., Toon G. C., Sen B., Blavier J. F. and Toth R. A. (2000) Vertical profiles of nitrous oxide isotopomer fractionation measured in the stratosphere. *Geophys Res Lett* **27**: 2485–2488.

Inoue H. and Mook W. (1994) Equilibrium and kinetic nitrogen and oxygen isotope fractionations between dissolved ad gaseous N_2O. *Chem Geol* **113**: 135–148.

Jobson B., McKeen S., Parrish D., Fehsenfeld F., Blake D., Goldstein A., Schauffler S. and Elkins J. (1999) Trace gas mixing ratio variability versus lifetime in the troposphere and stratosphere: Observations. *J Geophys Res* **104**: 16091–16113.

Johnson M., Billing G., Gruodis A. and Janssen M. (2001) Photolysis of nitrous oxide isotopomers studied by time-dependent hermite propagation. *J Phys Chem* **105**A: 8672–8680.

Johnston J. C., Cliff S. S. and Thiemens M. (1995) Measurement of multioxygen isotopic ($\delta^{18}O$ and $\delta^{17}O$) fractionation factors in the stratospheric sink reactions of nitrous oxide. *J Geophys Res* **100**: 16801–16804.

Kaiser J., Brenninkmeijer C. and Röckmann T. (2002a) Intramolecular ^{15}N and ^{18}O fractionation in the reaction of N_2O with $O(^1D)$ and its implications for the stratospheric N_2O isotope signature. *J Geophys Res* **107**: 4214.

Kaiser J., Röckmann T. and Brenninkmeijer C. (2002b) Temperature dependence of isotope fractionation in N_2O photolysis. *Phys Chem Chem Phys* **4**: 4420–4430.

Kaiser J., Röckmann T. and Brenninkmeijer C. (2003a) Assessment of (NNO) $^{15}N^{15}N^{16}O$ as a tracer of stratospheric processes. *Geophys Res Lett* **30**: 1046.

Kaiser J., Röckmann T., Brenninkmeijer C. and Crutzen P. (2003b) Wavelength dependence of isotope fractionation in N_2O photolysis. *Atmospheric Chem Phys* **3**: 303–313.

Kaiser J., Röckmann T. and Brenninkmeijer C. A. M. (2003c) Complete and accurate mass-spectrometric analysis of tropospheric nitrous oxide. *J Geophys Res* **108**: doi: 10.1029/2003JD003613.

Kaiser J., Röckmann T. and Brenninkmeijer C. A. M. (2004) Contribution of mass-dependent fractionation to the oxygen isotope anomaly of atmospheric nitrous oxide. *J Geophys Res* **109** doi: 10.1029/2003 JD00 4088.

Keeling C. D., Carter A. F. and Mook G W. (1984) Seasonal, latitudinal, and secular variations in the abundance and isotopic ratios of atmospheric carbon dioxide. 1. Results from oceanographic cruises in the tropical Pacific Ocean. *J Geophys Res Atmos* **89**: 4615–4628.

Keeling C., Chin J. and Whorf T. (1996) Increased activity of northern vegetation inferred from atmospheric CO_2 measurements. *Nature* **382**: 146–149.

Khalil M. A. K. and Rasmussen R. A. (1992) The global sources of nitrous oxide. *J Geophys Res* **97**: 14651–14660.

Kim K. R. and Craig H. (1990) Two-isotope characterization of N_2O in the Pacific Ocean and constraints on its origin in deep-water. *Nature* **347**: 58–61.

Kim K. R. and Craig H. (1993) ^{15}N and ^{18}O characteristics of nitrous oxide: A global perspective. *Science* **262**: 1855–1857.

Kohlmann J. and Poppe D. (1999) The tropospheric gas-phase degradation of NH_3 and its impact on the formation of N_2O and NO_x. *J Atmos Chem* **32**: 397–415.

Kroeze C., Mosier A. and Bouwman L. (1999) Closing the global N_2O budget: A retrospective analysis 1500–1994 *Global Biogeochem Cycles* **13**: 1–8.

Levin I., Ciais P., Langenfelds R., Schmidt M., Ramonet M., Sidorov K., Tchebakova N. *et al.* (2002) Three years of trace gas observations over the EuroSiberian domain derived from aircraft sampling—a concerted action. *Tellus* **54**: 696–712.

Lueker T., Walker S., Vollmer M., Keeling R., Nevison C., Weiss R. and Garcia H. (2003) Coastal upwelling air-sea fluxes revealed in atmospheric observations of O-2/N-2, CO_2 and N_2O—art. no. 1292 *Geophys Res Lett* **30**: 1292.

Machida T., Nakazawa T., Fujii Y., Aoki S. and Watanabe O. (1995) Increase in the atmospheric nitrous oxide concentration during the last 250 years. *Geophy Res Lett* **22**: 2921–2924.

Mandernack K. W., Rahn T., Kinney C. and Wahlen M. (2000) The biogeochemical controls of the $\delta^{15}N$ and $\delta^{18}O$ of N_2O produced in landfill cover soils. *J Geophys Res Atmos* **105**: 17709–17720.

Mariotti A., Germon J. C., Hubert P., Kaiser P., Letolle R., Tardieux A. and Tardieux P. (1981) Experimental determination of nitrogen kinetic isotope fractionation: Some principles; illustration for the denitrification and nitrification processes. *Plant Soil* **62**: 413–430.

McLinden C., Prather M. and Johnson M. (2003) Global modeling of the isotopic analogues of N_2O: Stratospheric distributions, budgets, and the $^{17}O/^{18}O$ mass-independent anomaly. *J Geophys Res Atmos* **108**: 4233.

Michalski G., Scott Z., Kabiling M. and Thiemens M. (2003) First measurements and modeling of $\Delta^{17}O$ in atmospheric nitrate. *Geophys Res Lett* **30**: 1870.

Miller J., Mack K., Dissly R., White J., Dlugokencky E. and Tans P. (2002) Development of analytical methods and measurements of $^{13}C/^{12}C$ in atmospheric CH_4 from the NOAA Climate Monitoring and Diagnostics Laboratory global air sampling network. *J Geophys Res* **107**: 4178.

Minschwaner K., Salawitch R. J. and McElroy B. M. (1993) Absorption of solar radiation by O_2: Implications for O_3 and lifetimes of N_2O, $CFCl_3$, and CF_2Cl_2. *J Geophys Res* **98**: 10543–10561.

Mook W. G., Koopmans M., Carter A. F. and Keeling D. C. (1983) Seasonal, latitudinal, and secular variations in the abundance and isotopic ratios of atmospheric carbon dioxide. 1. Results from land stations. *J Geophys Res* **88**: 10915–10933.

Moore H. (1974) Isotopic measurement of atmospheric nitrogen compounds. *Tellus* **26**: 169–174.

Mosier A., Kroeze C., Nevison C., Oenema O., Seitzinger S. and vanCleemput O. (1998) Closing the global N_2O budget: nitrous oxide emissions through the agricultural nitrogen cycle—OECD/IPCC/IEA phase II development of IPCC guidelines for national greenhouse gas inventory methodology. *Nutrient Cycling Agroecosystems* **52**: 225–248.

Naqvi S., Yoshinari T., Jayakumar D., Altabet M., Narvekar P., Devol A., Brandes J. and Codispoti L. (1998) Budgetary and biogeochemical implications of N_2O isotope signatures in the Arabian Sea. *Nature* **394**: 462–464.

Nevison C., Butler J. and Elkins J. (2003) Global distribution of N_2O and the Delta N_2O-AOU yield in the subsurface ocean. *Global Biogeochem Cycles* **17**: 1119.

Nevison C. D., Weiss R. F. and Erickson D. J. (1995) Global oceanic emissions of nitrous oxide. *J Geophys Res* **100**: 15809–15820.

Olivier J., Bouwman A., VanderHoek K. and Berdowski J. (1998) Global air emission inventories for anthropogenic sources of NO_x, NH_3 and N_2O in (1990) *Environmental Pollution: 1st International Nitrogen Conference 1998* **102**: 135–148.

Parks S., Atlas E. L. and Boering A. K. (2004) Measurements of N_2O isotopologues in the stratosphere: Influence of transport on the apparent enrichment factors and the isotope fluxes to the troposphere. *J Geophys Res* **109**: doi: 10.1029/2003JD003731.

Perez T., Trumbore S. E., Tyler S. C., Matson P. A., Ortiz-Monasterio I., Rahn T. and Griffith D. W. T. (2001) Identifying the agricultural imprint on the global N_2O budget using stable isotopes. *J Geophys Res* **106**: 9869–9878.

Pierotti D. and Rasmussen R. A. (1977) Atmospheric distribution of nitrous oxide. *J Geophys Res* **82**: 5823–5832.

Prather M., Derwent R., Ehhalt D., Fraser P., Sanhueza E. and Zhou X. (1995) Other trace gases and atmospheric chemistry. In *Climate Change 1994: Radiative Forcing* and *An Evaluation of the IPCC IS92 Emission Scenarios.* Cambridge University Press, Cambridge, UK and New York, NY.

Prinn R., Cunnold D., Rasmussen R. A., Simmonds P. G., Alyea F., Crawford A., Fraser P. and Rosen R. (1990) Atmospheric emissions and trends of nitrous oxide deduced from 10 years of ALE-GAGE data. *J Geophys Res* **95**: 18369–18385.

Prinn R., Weiss R., Fraser P., Simmonds P., Cunnold D., Alyea F., O'Doherty S. *et al.* (2000) A history of chemically and radiatively important gases in air deduced from ALE/GAGE/AGAGE. *J Geophys Res Atmos* **105**: 17751–17792.

Rahn T. and Wahlen M. (1997) Stable isotope enrichment in stratospheric nitrous oxide. *Science* **278**: 1776–1778.

Rahn T., Zhang H., Wahlen M. and Blake A. G. (1998) Stable isotope fractionation during ultraviolet photolysis of N_2O. *Geophys Res Lett* **25**: 4489–4492.

Rahn T. and Wahlen M. (1999) Nitrous oxide loss processes. IGACtivities: Newsletter of the International Global Atmospheric Chemistry Project **16**: http://www.igac.noaa.gov/newsletter/16/.

Rahn T. and Wahlen M. (2000) A reassessment of the global isotopic budget of atmospheric nitrous oxide. *Global Biogeochem Cycles* **14**: 537–543.

Richet P., Bottinga Y. and Javoy M. (1977) A review of hydrogen, carbon, nitrogen, oxygen, sulphur, and chlorine isotope fractionation among gaseous molecules. *Annu Rev Earth Planetary Sci* **5**: 65–110.

Röckmann T., Brenninkmeijer C. A. M., Wollenhaupt M., Crowley J. N. and Crutzen J. P. (2000) Measurement of the isotopic fractionation of (NNO) $^{15}N^{14}N$, (NNO) $^{14}N^{15}N$, (NNO) $^{14}N^{14}N^{18}O$ in the UV photolysis of nitrous oxide. *Geophys Res Lett* **27**: 1399–1402.

Röckmann T., Kaiser J., Brenninkmeijer C., Brand W., Borchers R. and Crowley J. (2001a) The position dependent ^{15}N enrichment of nitrous oxide in the stratosphere. *Isotopes Environ Health Stud* **37**: 91–95.

Röckmann T., Kaiser J., Brenninkmeijer C. A. M., Crowley J. N., Borchers R., Brand W. A. and Crutzen J. P. (2001b) Isotopic enrichment of nitrous oxide ((NNO) $^{15}N^{14}N$, (NNO) $^{14}N^{15}N$, (NNO) $^{14}N^{14}N^{18}O$) in the stratosphere and in the laboratory. *J Geophys Res* **106**: 10403–10410.

Röckmann T., Kaiser J., Crowley J. N., Brenninkmeijer C. A. M. and Crutzen P. J. (2001c) The origin of the anomalous or 'mass-independent' oxygen isotope fractionation in tropospheric N_2O. *Geophys Res Lett* **28**: 503–506.

Röckmann T., Kaiser J. and Brenninkmeijer C. (2003a) The isotopic fingerprint of the pre-industrial and the anthropogenic N_2O source. *Atmos Chem Phys* **3**: 315–323.

Röckmann T., Kaiser J., Brenninkmeijer C. and Brand W. (2003b) Gas chromatography/isotope-ratio mass spectrometry method for high-precision position-dependent ^{15}N and ^{18}O measurements of atmospheric nitrous oxide. *Rapid Commun Mass Spectrometry* **17**: 1897–1908.

Rodhe H. (1990) A comparison of the contribution of various gases to the greenhouse-effect. *Science* **248**: 1217–1219.

Rosenfeld W. D. and Silverman S. R.(1959) Carbon isotope fractionation in bacterial production of methane. *Science* **130**: 1658–1659.

Saueressig G., Crowley J. N., Bergamaschi P., Bruhl C., Brenninkmeijer C. A. M. and Fischer H. (2001) Carbon 13 and D kinetic isotope effects in the reactions of CH_4 with $O(^1D)$ and OH: New laboratory measurements and their implications for the isotopic composition of stratospheric methane. *J Geophys Res* **106**: 23127–23138.

Seitzinger S. and Kroeze C. (1998) Global distribution of nitrous oxide production and N inputs in freshwater and coastal marine ecosystems. *Global Biogeochem Cycles* **12**: 93–113.

Seitzinger S., Kroeze C., Bouwman A., Caraco N., Dentener F. and Styles R. (2002) Global patterns of dissolved inorganic and particulate nitrogen inputs to coastal systems: Recent conditions and future projections. *Estuaries* **25**: 640–655.

Selwyn G. and Johnston H. S. (1981) Ultraviolet absorption cross section of nitrous oxide as a function of temperature and isotopic substitution. *J Chem Phys* **74**: 3791–3803.

Sowers T., Alley R. and Jubenville J. (2003) Ice core records of atmospheric N_2O covering the last 106,000 years. *Science* **301**: 945–948.

Sowers T., Rodebaugh A., Yoshida N. and Toyoda S. (2002) Extending records of the isotopic composition of atmospheric N_2O back to 1800 AD from air trapped in snow at the South Pole and the Greenland Ice Sheet Project II ice core. *Global Biogeochem Cycles* **16**: 1129.

Sutka R., Ostrom N., Ostrom P., Gandhi H. and Breznak J. (2003) Nitrogen isotopomer site preference of N_2O produced by *Nitrosomonas europaea* and *Methylococcus capsulatus* Bath. *Rapid Commun Mass Spectrometry* **17**: 738–745.

Toyoda S. and Yoshida N. (1999) Determination of nitrogen isotopomers of nitrous oxide on a modified isotope ratio mass spectrometer. *Anal Chem* **71**: 4711–4718.

Toyoda S., Yoshida N., Miwa T., Matsui Y., Yamagishi H., Tsunogai U., Nojiri Y. and Tsurushima N. (2002) Production mechanism and global budget of N_2O inferred from its isotopomers in the western North Pacific—art. no. 1037 *Geophys Res Lett* **29**: 1037.

Toyoda S., Yoshida N., Urabe T., Aoki S., Nakazawa T., Sugawara S. and Honda H. (2001) Fractionation of N_2O isotopomers in the stratosphere. *J Geophys Res* **106**: 7515–7522.

Turatti F., Griffith D. W. T., Wilson S. R., Esler M. B., Rahn T., Zhang H. and Blake G. A. (2000) Positionally dependent ^{15}N fractionation factors in the UV photolysis of N_2O determined by high resolution FTIR spectroscopy. *Geophys Res Lett* **27**: 2489–2492.

Tyler S. C., Ajie H. O., Rice A. L., Cicerone R. J. and Tuazon E. C. (2000) Experimentally determined kinetic isotope effects in the reaction of CH_4 with Cl: Implications for atmospheric CH_4. *Geophys Res Lett* **27**: 1715–1718.

Umemoto H. (1999) $^{14}N/^{15}N$ isotope effect in the UV photodissociation of N_2O. *Chem Phys Lett* **314**: 267–272.

Urey, H. C. (1947) The thermodynamics of isotopic substances. *J Chem Soc* **May**: 562–581.

Wahlen M. and Yoshinari T. (1985) Oxygen isotope ratios in N_2O from different environments. *Nature* **313**: 780–782.

Webster E. and Hopkins D. (1996) Nitrogen and oxygen isotope ratios of nitrous oxide emitted from soil and produced by nitrifying and denitrifying bacteria. *Biol Fertil Soils* **22**: 326–330.

Weiss, R. F. (1981) The temporal and spatial distribution of tropospheric nitrous oxide. *J Geophys Res* **86**: 7185–7195.

Whiticar M. J., Faber E. and Schoell M. (1986) Biogenic methane formation in marine and freshwater environments: CO_2 reduction vs. acetate fermentation. *Geochim Cosmochim Acta* **50**: 693–709.

Yamazaki T., Yoshida N., Wada E. and Matsuo S. (1987) N_2O reduction by *Azotobacter-vinelandii* with emphasis on kinetic nitrogen isotope effects. *Plant Cell Physiol* **28**: 263–271.

Yoshida N. (1988) ^{15}N-depleted N_2O as a product of nitrification. *Nature* **335**: 528–529.

Yoshida N. and Matsuo S. (1983) Nitrogen isotope ratio of atmospheric N_2O as a key to the global cycle of N_2O. *Geochem J* **17**: 231–239.

Yoshida N., Hattori A., Saino T., Matsuo S. and Wada E. (1984) $^{15}N/^{14}N$ ratio of dissolved N_2O in the Eastern Tropical Pacific Ocean. *Nature* **307**: 442–444.

Yoshida N., Morimoto H., Hirano M., Koike I., Matsuo S., Wada E., Saino T. and Hattori A. (1989) Nitrification rates and ^{15}N abundances of N_2O and NO_3^- in the Western North Pacific. *Nature* **342**: 895–897.

Yoshida N. and Toyoda S. (2000) Constraining the atmospheric N_2O budget from intramolecular site preference in N_2O isotopomers. *Nature* **405**: 330–334.

Yoshinari T. and Wahlen M. (1985) Oxygen isotope ratios in N_2O from nitrification at a wastewater treatment facility. *Nature* **317**: 349–350.

Yoshinari T., Altabet M., Naqvi S., Codispoti L., Jayakumar A., Kuhland M. and Devol A. (1997) Nitrogen and oxygen isotopic composition of N_2O from suboxic waters of the eastern tropical North Pacific and the Arabian Sea—Measurement by continuous-flow isotope-ratio monitoring. *Marine Chem* **56**: 253–264.

Yung Y. L. and Miller E. C. (1997) Isotopic fractionation of stratospheric nitrous oxide. *Science* **278**: 1778–1780.

Zhang H., Wennberg P. O., Wu V. H. and Blake A G. (2000) Fractionation of $^{14}N^{15}N^{16}O$ and $^{15}N^{14}N^{16}O$ during photolysis at 213 nm. *Geophys Res Lett* **27**: 2481–2484.

Zipf E. and Prasad S. (1998) Experimental evidence that excited ozone is a source of nitrous oxide. *Geophys Res Lett* **25**: 4333–4336.

16

The Carbon Isotopic Composition of Atmospheric Methane and its Constraint on the Global Methane Budget

John B. Miller

Introduction

Chapter Overview

In this chapter I will develop the context of how the carbon isotopic composition of atmospheric methane can be used to estimate source and sink fluxes. First, I will describe the theory behind using atmospheric measurements to constrain surface sources and sinks and then apply this specifically to isotopic ratios in the atmosphere. I will then review the history of atmospheric $\delta^{13}C$ measurements. In the next section I will use these measurements to estimate sources and sinks and examine the sensitivity of our estimates to uncertainties in our model's parameters. I will end with a brief discussion of how other isotopic species like $^{14}CH_4$ and CH_3D might add to the information contained within the $\delta^{13}C$ data.

The Importance of CH_4 in the Atmosphere

Atmospheric CH_4 is an important chemical component of both the stratosphere and troposphere and is a major contributor to the enhanced greenhouse effect. In the stratosphere, methane is a major source of water vapor (Jones and Pyle, 1984), and is the primary sink for chlorine radicals (Cicerone and Oremland, 1988), and thus plays an important role in the regulation of stratospheric ozone levels. In the troposphere, CH_4 consumes about 25% of all hydroxyl radicals, and as a result is an in-situ source of CO and O_3 (Thompson, 1992). Methane strongly absorbs outgoing longwave radiation (OLR) around 7.7 μm, a 'window' of the earth's emitted infra-red spectrum where neither water nor carbon dioxide absorbs strongly. Models indicate that methane's contribution to greenhouse warming is twenty times that of CO_2 on a per molecule basis (Lashof and Ahuja, 1990). It is estimated that methane has accounted for approximately 20% of the increase in radiative forcing by trace gases since the onset of the industrial era (Lelieveld *et al.*, 1998). Moreover, because the present-day methane budget is close to being in balance, the opportunity exists to actually *reduce* atmospheric CH_4 concentrations by reducing anthropogenic sources, and thus reduce the concentration of an important greenhouse gas (Hansen and Sato, 2001).

The mole fraction of methane in the atmosphere has more than doubled in the last 150 years (Etheridge *et al.*, 1992, 1998) and over that time has been highly correlated with human population growth (Blunier *et al.*, 1993). The growth rate of methane in the atmosphere has averaged nearly %/yr over the last 40 years but has decreased substantially in the last 8 years (Dlugokencky *et al.*, 1998). Neither the rapid increase nor the recent slow down is clearly understood, and this is directly related to the large uncertainties in the magnitudes and spatial distribution of identified methane sources. Estimates of the magnitudes of various sources have been based upon scaled-up field measurements and forward (Fung *et al.*, 1991) and inverse (Brown, 1993; Hein *et al.*, 1997; Houweling *et al.*, 1999) modeling approaches based on atmospheric measurements. Nonetheless, considerable uncertainties remain in the estimates of source magnitudes.

The Methane Budget

The Top–Down Approach Understanding the variability of the CH_4 growth rate is a prerequisite to predicting future atmospheric concentrations. The basic approach in understanding concentrations and trends in concentrations of well-mixed atmospheric gases is to construct a budget for these gases. The budget expresses the imbalance between sources and sinks to the atmosphere, such that the difference between sources and sinks is equal to the growth rate in the atmosphere. If something is known about atmospheric circulation, spatial patterns in atmospheric concentration can impose significant constraints on the geographic location of sources and sinks (e.g., Tans *et al.*, 1990; Fung *et al.*, 1991). Seasonal variations in concentration are often a direct result of seasonal source and sink activity and are yet another constraint on source and sink activity (Dlugokencky *et al.*, 1997; Randerson *et al.*, 1997). However, with all the ways that concentration data can be analyzed, the budget of CH_4 is still underconstrained.

The Bottom–Up Approach It is also possible to estimate the sources of CH_4 independent of any atmospheric information. Measurements of CH_4 fluxes made at the local scale from wetlands, animals, or biomass burning can be scaled up in space and time (e.g., Aselmann and Crutzen, 1989; Hao and Ward, 1993). Process-based models of emissions can also be used to estimate fluxes from wetlands (Walter and Heimann, 2000; Cao *et al.*, 1996) and rice paddies (Cao *et al.*, 1995). Economic and fossil fuel production statistics can also used to estimate fluxes of CH_4 from fossil fuel related activities (Olivier *et al.*, 1999; van Aardenne *et al.*, 2001). On the sink side, OH concentrations can be estimated from CH_3CCl_3 (Prinn *et al.*, 1995; Montzka *et al.*, 2000), and when combined with laboratory-based estimates of the rate coefficient of $CH_4 + OH$ (Vaghjiani and Ravishankara, 1991), can be used to estimate the predominant CH_4 sink. Process models have also been used to estimate the magnitude of the soil sink (Ridgwell *et al.*, 1999). Table 16.1 shows bottom–up estimates of CH_4 sources and sinks based on the work of Lelieveld *et al.* (1998). If the total amount of methane consumed in a year is known well, the sum of all bottom–up estimates can be checked against atmospheric measurements. This is because this sum must equal the observed annual increase of atmospheric methane (Eq. 16.1), which can be measured very accurately.

Table 16.1a Estimates of Emissions and Isotopic Signature for Most Sources Grouped by Estimated $\delta^{13}C$ Signature

Source	Strength[1] (Tg/yr)	$\delta^{13}C$ (‰)	References
Wetlands[2]	150	−60[3]	(Quay *et al.*, 1988; Stevens, 1993)
Rice	80	−64[4]	(Stevens, 1993; Tyler *et al.*, 1994)
Ruminants	85	−60[5]	(Stevens, 1993)
Termites	20	−57	(Tyler, 1986; Gupta *et al.*, 1996)
Landfills	40	−52[6]	(Tyler, 1986; Levin *et al.*, 1993)
Wastewater	25	−54[6]	(Levin *et al.*, 1999; Lowry *et al.*, 2001)
Animal waste	30	−52	(Tyler, 1986)
Ocean flux[7]	10	−43[6]	(Holmes *et al.*, 2000; Sansone *et al.*, 2001)
Gas and Coal	110	−40[8]	(Schoell, 1980; Deines, 1980)
Biomass burning	40	−25[9]	(Craig *et al.*, 1988)
	Total 590	Weighted average −52.9	

[1]Source strengths based on Lelieveld (1998).
[2]Includes tundra and freshwater sources.
[3]Unweighted averages from both review studies.
[4]10‰ seasonal cycle in Tyler (1994).
[5]Average of C_3 and C_4 diets.
[6]Average of two studies.
[7]Hydrates implicitly included.
[8]Unweighted average of coal and gas δ values.
[9]Likely to reflect $\delta^{13}C$ of plant biomass (C_3 or C_4).

Table 16.1b Estimates of Sink Strength and Associated Isotopic Fractionation

Sink	Strength (Tg/yr)	ε (‰)	References
Tropospheric OH[1]	506	$-5.4/-3.9$[2]	(Cantrell *et al.*, 1990; Saueressig *et al.*, 2001)
Soils	30	-21	(King *et al.*, 1989)
Stratosphere	40	-12[3]	(Hein *et al.*, 1997)
	Total 576	Weighted average $-6.7/-5.4$	

[1]From IPCC (2001).
[2]Fractionation estimates are statistically different at 95%.
[3]Weighted fractionation of Cl, OH, and O^1D.

Isotopic Budgets

Measuring the $^{13}C/^{12}C$ ratio of CH_4 allows us to use the temporal and spatial patterns of two molecular species, instead of one, to constrain the global CH_4 budget. However, the budget will not be fully constrained. To do that I would need to add an independently behaving isotopic species for every term in the budget. Nonetheless, the more isotopic species I measure, the more aspects of the budget can be constrained. From the atmospheric (or top–down) perspective, I can describe the global CH_4 budget using Eq. 16.1

$$\frac{d[CH_4]}{dt} = Q - \frac{[CH_4]}{\tau} \tag{16.1}$$

where Q is the sum of all methane sources and τ is the lifetime of CH_4 methane in the atmosphere. Simply put, the accumulation of methane in the atmosphere is equal to the difference in source and sink fluxes. I can measure the methane mole fraction and its rate of change, so if I can estimate τ based on the global mean OH concentration (and minor contributions from consumption in soil and destruction by molecules other than OH), then I can determine the total flux of all methane sources, Q.

Before proceeding I will define δ, the isotopic ratio expressed as per mil (‰) or part per thousand:

$$\delta \equiv \left(\frac{R_{sample}}{R_{standard}} - 1\right)1000‰ \tag{16.2}$$

where R = rare isotope/abundant isotope. If I can measure the carbon isotopic ratio of methane in the atmosphere, $\delta^{13}C$, and know the fractionation associated with its destruction by OH and other processes, then I can estimate the flux weighted mean isotopic ratio of all methane sources, δ_q (Lassey *et al.*, 2000)

$$\delta_q = \alpha\delta_a + \varepsilon \tag{16.3}$$

where δ_a is the isotopic ratio of atmospheric methane. For a given sink process, $\varepsilon = (\alpha - 1)1000$, and $\alpha = k_{abundant}/k_{rare}$ where k is the rate coefficient for methane destruction. The values of ε and α in Eq. 16.3 are weighted by the magnitude of each sink process. Equation 16.3 is an approximation that assumes both the methane concentration and isotopic ratio are at steady state in the atmosphere. Knowing Q and δ_q imposes a constraint on the magnitude of individual methane sources or source-types, i.e., those sources with common characteristic isotopic ratios.

In particular, methane sources may be divided into three categories: bacterially produced methane, like that from wetlands or ruminant animals; fossil fuel methane, like that associated with coal and natural gas deposits; and methane produced from biomass burning. Each of these three classes has a fairly distinct isotopic signature, with bacterial methane $\delta^{13}C \cong -60‰$, thermogenic methane $\delta^{13}C \cong -40‰$, and biomass burning methane $\delta^{13}C \cong -25‰$ (e.g., Quay *et al.*, 1991). Individual methane sources may differ significantly from their source type's characteristic signature (Conny and Currie, 1996), but the average values above are probably valid on large spatial scales.

If I split up all methane sources into bacterial, fossil fuel and biomass burning fractions I can write the following equations:

$$Q = Q_{bact} + Q_{ff} + Q_{bmb} \tag{16.4}$$

$$\delta_q Q = \delta_{bact} Q_{bact} + \delta_{ff} Q_{ff} + \delta_{bmb} Q_{bmb} \tag{16.5}$$

where *bact* is the bacterial fraction, *ff* the fossil fuel related fraction, and *bmb* the biomass burning fraction. Now there are two equations with three unknown fluxes (Q_{bact}, Q_{ff}, and Q_{bmb}), so I cannot uniquely partition Q unless I specify a value for the bacterial, fossil fuel, or biomass burning fraction. Later, I will use $^{14}CH_4$ data to specify Q_{ff} and then solve Eqs 16.4 and 16.5. However, even with three unknowns the constraint provided by Eqs 16.4 and 16.5 is powerful, because there is a limited set of source fluxes that will satisfy both equations. This is especially true for the biomass burning fraction, because its source signature of $-25‰$ differs more from δ_q than the other source signatures. Additionally, any bottom–up estimate of these source fractions must also satisfy Eqs 16.4 and 16.5.

$\delta^{13}CH_4$ Observations

Measurement Techniques

Measurements of $\delta^{13}C$ in atmospheric methane are most often made on air that has been sampled into a low pressure (3–10 psig) flask or high pressure cylinder (500–2000 psig). Traditional analysis methods (e.g., Stevens and Rust, 1982) use 15–60 L of air and therefore use air from high pressure or high volume cylinders. In these 'off-line' analyses, air is extracted from the cylinder and the CO is converted to CO_2, which along with the original CO_2 is cryogenically removed from the air sample. After non-methane hydrocarbons are removed, the sample is combusted and the resulting CO_2 produced from methane is separated from the air cryogenically and analyzed on a dual-inlet isotope ratio mass spectrometer (IRMS) where its isotopic ratio is compared to CO_2 with a known carbon isotopic ratio. Typical precision for the traditional 'off-line' $\delta^{13}C$ analysis is 0.02–0.05‰, but the process is labor intensive and analysis time is long.

In contrast to traditional analyses, a gas chromatography–isotope ratio mass spectrometry (GC-IRMS) system can use much less air (10–50 mL) than the traditional system (1–20 L). In high precision GC-IRMS systems for methane $\delta^{13}C$ analysis (Miller *et al.*, 2002; Rice *et al.*, 2001), 10–50 mL of air is entrained in a helium stream and transferred to a $-130\,°C$ substrate, where methane is retained but air (N_2, O_2, and Ar) is vented. The concentrated CH_4 then passes through a chromatographic column where it is separated from residual air as well as CO and CO_2. The CH_4 peak eluting from the column is then combusted to form CO_2, which enters the mass spectrometer. The precision of GC-IRMS systems ranges from 0.04 to 0.20‰, but this sacrifice is small compared to relative sample size requirements, which allows for easier collection of atmospheric samples and shorter analysis times. The GC-IRMS process can also be automated, allowing for a much higher throughput of samples compared to the traditional method (Miller *et al.*, 2002).

History of Atmospheric and Surface $\delta^{13}C$ Measurements

The first sustained atmospheric measurements of $\delta^{13}C$ of CH_4 were by Stevens (1995), who reported 201 measurements, mostly from assorted sites in the continental United States, between 1978 and 1989. Quay *et al.* (1991, 1999, http://cdiac.esd.ornl.gov) reported over 1000 measurements between 1987 and 1996 from bi-weekly sampling at Pt. Barrow, AK, Olympic Peninsula, WA, and Mauna Loa, HI, in addition to less frequent sampling at Cape Grim, Tasmania, Fraserdale, Canada, and the Marshal Islands, and from Pacific Ocean ship transects. $\delta^{13}C$ of methane in the

Southern Hemisphere has been regularly monitored at Baring Head, New Zealand since 1990 (Lowe *et al.*, 1994), and regular measurements of $\delta^{13}C$ have also been made at Niwot Ridge, CO, and Montana de Oro, CA (Gupta *et al.*, 1996; Tyler *et al.*, 1999) since 1989 and 1995, respectively, and Izana, Tennerife since 1997 (Saueressig *et al.*, 2001). More than 2000 measurements of $\delta^{13}C$ in CH_4 have also been made weekly at six of the NOAA/CMDL Cooperative Air Sampling Network sites since January 1998 (Miller *et al.*, 2002). Currently, $\delta^{13}C$ of methane is being measured at thirteen of the NOAA/CMDL network sites (see ftp.cmdl.noaa.gov/ccg/ch4c13). A total of 19 regularly monitored sites exist that have data published in the literature.

In addition to measurements of $\delta^{13}C$ in atmospheric methane, measurements of source isotopic signatures are necessary to apply the isotopic budget approach. Table 16.1 lists references for measurements of methane source and sink signatures. The large variety of biogeochemical processes that produce methane means that no single model can be used to predict the isotopic signature of methane fluxes. Instead, global estimates of isotopic ratios in methane emissions are based mainly on a relatively small number of field measurements (e.g., Conny and Currie, 1996; Tyler, 1986). Recent progress has been made in better understanding the isotopic signature of emissions from wetlands (e.g., Popp *et al.*, 1999), rice paddies (e.g., Chidthaisong *et al.*, 2002; Kruger *et al.*, 2002), and landfills (e.g., Liptay *et al.*, 1998; Chanton and Liptay, 2000). However, the biogeochemical complexity of the processes, and the lack of measurements of variables thought to influence those processes precludes the use of a predictive model for isotopic signatures at this time. This stands in sharp contrast to the case of CO_2 fractionation, where well-tested models exist that predict the fractionation occurring during assimilation of CO_2 by both C_3 and C_4 plants (e.g., Farquhar *et al.*, 1989).

Firn, Ice Core, and Air Archive Measurements

While there have been numerous measurements of the paleo-atmospheric concentration of methane over decadal (Battle *et al.*, 1996), centennial (Nakazawa *et al.*, 1993; Etheridge *et al.*, 1998), and millennial and longer scales (Blunier *et al.*, 1995; Brook *et al.*, 1996; Chappellaz *et al.*, 1997), there have been only a few measurements of $\delta^{13}C$ in atmospheric CH_4 made on pre-1978 air. Craig *et al.* (1988) reported two values of $\delta^{13}C$ from a northern hemisphere ice core with a gas age estimated to be about 200 years before present. Sowers (2000b) recently measured the $\delta^{13}C$ of methane trapped in Antarctic ice spanning the last glacial/inter-glacial transition (33 k–2 kyr before the present) and from Greenland ice during the preindustrial period (20–200 yr before the present) (Sowers, 2000a). More recently, Braunlich *et al.* (2001) presented $\delta^{13}C$ measurements of air trapped in firn as old as

50 years. Francey *et al.* (1999) have also reported a δ^{13}C record derived from both Antarctic firn air and air archived in high pressure cylinders.

Sowers (2000b) found that glacial (e.g., 20 kyr before the present) values of δ^{13}C were 3.4‰ heavier than average preindustrial Holocene (12 kyr before the present) values. Over the same time period, methane concentrations increased from about 380 ppb during the last glacial period to 700 ppb in the preindustrial Holocene (Chappellaz *et al.*, 1990). Sowers attributed these changes to a doubling in methane emissions from northern wetlands following deglaciation. This was based on the idea that the isotopic signature for northern wetland emissions is about −65‰ whereas for tropical emissions it is −53‰ (Stevens, 1993). However, the uncertainty in the isotopic signature of wetlands emissions in one region can be as large as the apparent latitudinal gradient (Conny and Currie, 1996). Using a single value for wetlands emissions of −60‰, these measured changes in CH_4 and δ^{13}C following deglaciation can be explained by a doubling of global wetland emissions from 80 to 160 Tg/yr.

For ice core air with a gas age of 200 years before the present, Sowers measured a δ^{13}C value of −49.2‰ which is similar to that of Craig *et al.* (1988), who measured a value of −49.6‰. The measurements of Craig *et al.* (1988) were not corrected for diffusional fractionation in the firn layer (Francey *et al.*, 1999), which may explain the offset between the two sets of measurements. Both of these values are significantly heavier than previous model estimates (Lassey *et al.*, 2000; Miller, 1999), and could be evidence that estimates of preindustrial biomass burning are too low. Firn gas and air archive measurements do match the model estimates very well, but the measurements extend only as far back as 1950. More measurements of δ^{13}C from firn and recent ice cores are needed to paint a clearer picture of the preindustrial and early industrial methane budget.

Modern Trend, Latitudinal Gradient, and Seasonal Cycles

20th Century Trend The firn and air-archive records show that as methane concentration increased during the second half of the twentieth century, δ^{13}C also increased. While the CH_4 mole fraction increased at a rate of about 10 ppb/yr, δ^{13}C increased at about 0.03‰/yr. These increases are most likely a result of the dramatic increase in global fossil fuel use and an increase in biomass burning. These are the only two source classes that could make the value of atmospheric δ^{13}C more positive.

Starting in the mid-1980s and continuing on until the end of the century, the globally averaged CH_4 growth rate slowed substantially from 13 ppb/yr in 1985 to about 3 ppb/yr in 1999. Likewise, Francey *et al.* (1999) have shown that the growth rate of δ^{13}C also has slowed down, although the slowdown appears to begin in the mid-1990s as opposed to the mid-1980s. What was the cause of the slowdowns in CH_4 and $\delta^{13}C$? Dlugokencky *et al.* (1998) showed that as long as τ was assumed to be constant over time (Prinn *et al.*, 1995), the source term Q (Eq. 16.1) remained nearly constant over this time. The decrease in the growth rate of CH_4 could then be explained best as the approach of the sources and sinks toward steady state, i.e., $d[CH_4]/dt = 0$. The analysis by Francey *et al.* (1999) of $\delta^{13}C$ from the Cape Grim Air Archive also supports the notion that both Q and τ have been nearly constant and that the system is approaching steady state. That the $\delta^{13}C$ growth rate slowed down some years after CH_4 is also consistent with the ideas of Tans (1997), who pointed out that the adjustment time after perturbation is longer for $\delta^{13}C$ than CH_4 concentration.

Most recently, the growth rates for both CH_4 and $\delta^{13}C$ have been nearly flat since 1999. Dlugokencky *et al.* (1998) estimated that steady state for atmospheric CH_4 might be reached in the next 20 years. The current steady state, however, is more likely a result of the decay of the huge pulse of global methane emissions that occurred during 1998 as a result of increased emissions from northern and tropical wetlands and northern biomass burning (Dlugokencky *et al.*, 2001).

Latitudinal Gradient The north–south gradients in methane mole fraction and its isotopic composition are important constraints on the location and strength of methane sources and sinks (Fung *et al.*, 1991). The mole fraction latitudinal gradient is well established (e.g., Steele *et al.*, 1987; Dlugokencky *et al.*, 1994), and Quay *et al.* (1991, 1999) and Miller *et al.* (2002) have reported an annual mean gradient for $\delta^{13}C$. Figure 16.1 shows the annual mean gradients for 1998–2000 between 90° S and 71° N. The average difference between South Pole (SPO) and Barrow, AK, (BRW–SPO) was -0.65 ± 0.1‰ in 1998, -0.48 ± 0.1‰ in 1999, and -0.48 ± 0.1‰ in 2000. Quay *et al.* (1991) reported a mean annual average difference of -0.54 ± 0.05‰ between BRW (71° N) and CGO (41° S) during the years 1989–1995. For the period 1998–2000 the mean hemispheric difference from the six sites in Fig. 16.1 is 0.19‰. It is important to note that two of the northern hemisphere sites, MLO and NWR, are situated at altitudes greater than 3000 m. The methane mole fraction at these sites is substantially less than at sites at comparable latitudes but at sea level. It is likely that the northern hemisphere average $\delta^{13}C$ calculated here is too positive, because $\delta^{13}C$ values increase with height (Tyler *et al.*, 1999). Thus, it is also likely that the interhemispheric difference is underestimated. For the purposes of the model (see below) the data have been corrected for these altitude effects, which the values in Table 16.2 reflect.

The $\delta^{13}C$ values in the southern hemisphere are higher than $\delta^{13}C$ values in the northern hemisphere. This feature is the combined effect of two separate processes. The majority of emissions originate in the northern

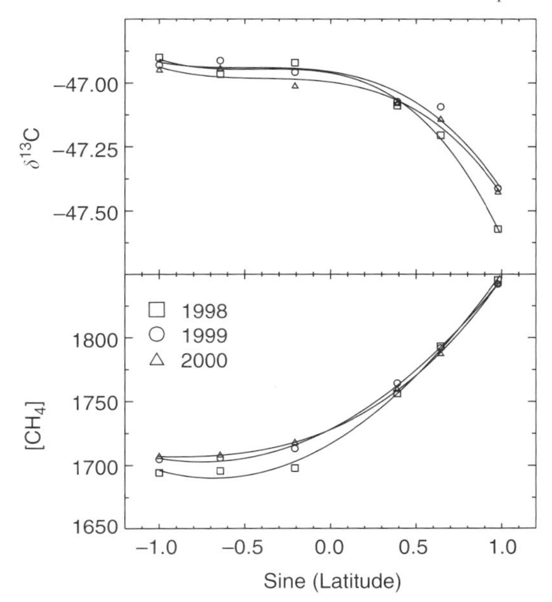

Figure 16.1 Latitudinal gradient of CH_4 and $\delta^{13}C$ during 1998–2000. The gradient is constructed from six NOAA/CMDL sites: (from north to south) Barrow, Alaska 71° N (BRW), Niwot Ridge, Colorado 40° N (NWR), Mauna Loa, Hawaii 20° N (MLO), American Samoa 14° S (SMO), Cape Grim, Tasmania 41° S (CGO), South Pole 90° S (SPO). The symbols are annual mean $\delta^{13}C$ and CH_4 values at each station and the lines are cubic polynomial fits to the annual means. The latitude axis is scaled by sine of latitude to account for the reduction in zonal area with increasing latitude.

hemisphere, and the reaction with OH enriches the ^{13}C content of the methane remaining in the atmosphere. Methane reaching the southern hemisphere has had more time to react with OH than methane in the northern hemisphere, leaving it more enriched in ^{13}C. Another prominent feature is the lack of a significant $\delta^{13}C$ gradient in the southern hemisphere, which can be explained by low emissions and rapid atmospheric mixing (Law *et al.*, 1992).

Southern Hemisphere Seasonal Cycles Southern hemisphere sites SPO (90° S), CGO (41° S), and SMO (14° S) do not exhibit strong seasonal variability in the NOAA/CMDL data set, except during 1998 (Fig. 16.2). During this period substantial decreases in $\delta^{13}C$ values are present during August, September, and October, especially at SMO and CGO. The conventional assumption is that seasonal variations in SH mole fractions are driven mostly by OH oxidation, but the magnitude of the dip in $\delta^{13}C$ values is too large to be explained by OH alone. One possible contribution to the observed dip is the positive 12 Tg/yr anomaly in tropical wetland emissions during 1998

Table 16.2a Input Parameters for Two-Box Model

Units	dX/dt[a] (ppb/yr)	X[a] (ppb)	$d\delta/dt$[b] (‰/yr)	δ[a] (‰)	k_{ex} (1/yr)	k_{12}[c] (1/yr)	ε[d] (‰)	FFP[e] (Tg/yr)	δ_B[f] (‰)	δ_{BMB}[f] (‰)	δ_{FFP}[f] (‰)
N	5.5	1791	0.02 ± 0.02	-47.2	1.0 ± 0.1	0.1071 ± 0.01	-6.4 ± 0.8	124 ± 47	-61 ± 2	-24 ± 2	-43 ± 2
S	10.0	1705	0.02 ± 0.02	-46.9	1.0 ± 0.1	0.1057 ± 0.01	-6.2 ± 0.8	11 ± 4	-61 ± 2	-24 ± 2	-43 ± 2

[a] Average of measured values from NOAA/CMDL global network during 1998–1999.

[b] From Quay *et al.* (1999).

[c] Calculated as $k_{12} = k_{OH} + k_{SOIL} + k_{STRAT}$. k_{OH} is 1/10.5 and is taken from Montzka *et al.* (2000); k_{SOIL} was 1/484.2 and was calculated as a first-order loss assuming a 30 Tg/yr soil sink and a global CH_4 burden of 1750 ppb. We assume that $\frac{2}{3}$ of the soil sink is in the northern hemisphere. k_{STRAT} is 1/110 and is taken from Scientific Assessment of Ozone Depletion: 1998.

[d] $\varepsilon = 1000(\alpha - 1)$. Calculated as $\alpha = (\alpha_{OH}k_{OH} + \alpha_{SOI}k_{SOIL} + \alpha_{STRAT}k_{STRAT})/k_{12}$. α_{OH} is 0.9946 and is taken from Cantrell *et al.* (1990); α_{SOIL} is 0.979 and is taken from King *et al.* (1989); α_{STRAT} is 0.988 and was calculated by weighting α_{OH} and α_{Cl} by the strengths of Cl and OH sinks in the stratosphere according to Hein *et al.* (1997). Errors were determined only by propagating errors in k_{12} and assigning an error to α_{OH} of 0.0009, the error estimate of Cantrell *et al.* (1990).

[e] *FFP* is the sum of the Fung *et al.* (1991) categories: gas venting, gas leaks, coal mining, and landfills. The total of the fossil fuel categories was 100 Tg/yr and was calculated from Quay *et al.* (1999) $^{14}CH_4$ data. Landfill emissions are taken as 35 Tg/yr, which is the average of the Hein *et al.* (1997) and Fung *et al.* (1991) estimates. The north/south division (92%/8%) is based on Table 4 of Fung *et al.* (1991). The error estimates are derived from the range in landfill emission estimates (20 Tg/yr) and the range for fossil fuel emissions of 50% given in Quay *et al.* (1999).

[f] B (bacterial emissions) are defined as the sum of the Fung *et al.* categories: bogs, swamps, tundra, rice, animals, termites, and clathrates; *BMB* is biomass burning; *FFP* defined as above. δ values were calculated using source signatures from Table 1 of Quay *et al.* (1999) with weightings from the global totals of the Fung *et al.* categories listed above. We assume that source signatures are the same for each hemisphere.

Parameter	Units	Northern hemisphere	Southern hemisphere	Comments
dX/dt	(ppb/yr)	5.5	10	Average of measured values from NOAA/CMDL global network during 1998–1999
X	(ppb)	1791	1705	
$d\delta/dt$	(‰/yr)	0.02 ± 0.02	0.02 ± 0.02	From Quay *et al.* (1999)
δ	(‰)	−47.2	−46.9	Average of measured values from NOAA/CMDL global network during 1998–1999
k_{ex}	(1/yr)	1.0 ± 0.1	1.0 ± 0.1	
k_{12}	(1/yr)	0.1071 ± 0.01	0.1057 ± 0.01	Calculated as $k_{12} = k_{OH} + k_{SOIL} + k_{STRAT}$. k_{OH} is 1/10.5 and is taken from Montzka *et al.* (2000); k_{SOIL} was 1/484.2 and was calculated as a first-order loss assuming a 30 Tg/yr soil sink and a global CH_4 burden of 1750 ppb. We assume that $\frac{2}{3}$ of the soil sink is in the northern hemisphere. k_{STRAT} is 1/110 and is taken from Scientific Assessment of Ozone Depletion: 1998.
ε	(‰)	-6.4 ± 0.8	-6.2 ± 0.8	$\varepsilon = 1000(\alpha - 1)$. Calculated as $\alpha = (\alpha_{OH}k_{OH} + \alpha_{SOI}k_{SOIL} + \alpha_{STRAT}k_{STRAT})/k_{12}$. α_{OH} is 0.9946 and is taken from Cantrell *et al.* (1990); α_{SOIL} is 0.979 and is taken from King *et al.* (1989); α_{STRAT} is 0.988 and was calculated by weighting α_{OH} and α_{Cl} by the strengths of Cl and OH sinks in the stratosphere according to Hein *et al.* (1997). Errors were determined only by propagating errors in k_{12} and assigning an error to α_{OH} of 0.0009, the error estimate of Cantrell *et al.* (1990).
FFP	(Tg/yr)	124 ± 47	11 ± 4	*FFP* is the sum of the Fung *et al.* (1991) categories: gas venting, gas leaks, coal mining, and landfills. The total of the fossil fuel categories was 100 Tg/yr and was calculated from Quay *et al.* (1999) $^{14}CH_4$ data. Landfill emissions are taken as 35 Tg/yr, which is the average of the Hein *et al.* (1997) and Fung *et al.* (1991) estimates. The north/south division (92%/8%) is based on Table 4 of Fung *et al.* (1991). The error estimates are derived from the range in landfill emission estimates (20 Tg/yr) and the range for fossil fuel emissions of 50% given in Quay *et al.* (1999).
δ_B	(‰)	-61 ± 2	-61 ± 2	*B* (bacterial emissions) are defined as the sum of the Fung *et al.* categories: bogs, swamps, tundra, rice, animals, termites and clathrates; *BMB* is biomass burning; *FFP* defined as above. δ values were calculated using source signatures from Table 1 of Quay *et al.* (1999) with weightings from the global totals of the Fung *et al.* categories listed above. We assume that source signatures are the same for each hemisphere.
δ_{BMB}	(‰)	-24 ± 2	-24 ± 2	
δ_{FFP}	(‰)	-43 ± 2	-43 ± 2	

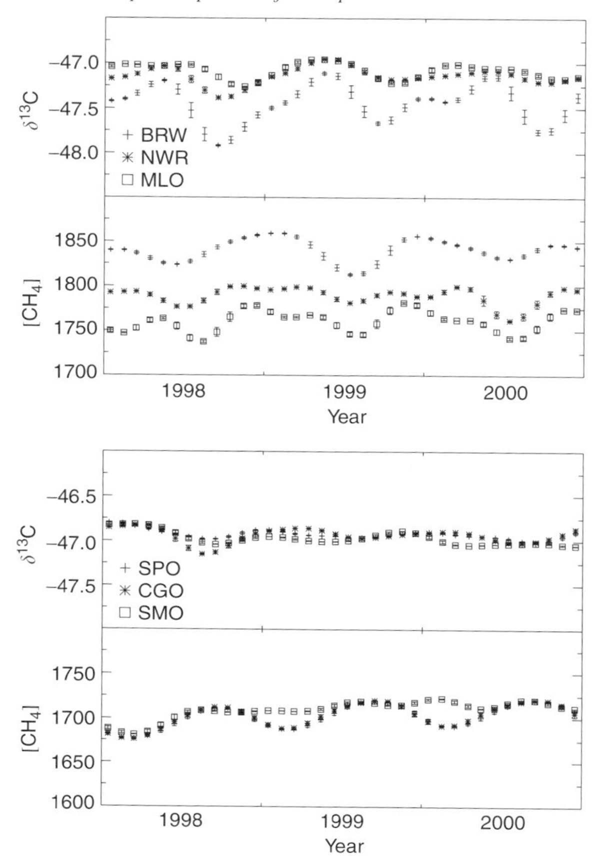

Figure 16.2 Monthly mean seasonal cycles of CH_4 and $\delta^{13}C$ from 1998 to 2000 at six NOAA/CMDL stations. The monthly means are calculated by sampling a curve fit to the raw data. This method accounts for missing data in some months at certain sites. The error bars are the standard deviations of sampled data used to construct the monthly mean and are not related to measurement precision. Station codes are identified in the text and in the caption for Fig. 16.1.

proposed by Dlugokencky *et al.* (2001). A +12 Tg/yr anomaly would result in a −0.13‰ anomaly in the lower southern atmosphere (0–30° S) if the emissions mixed evenly through the entire semi-hemisphere and if the signature of the wetland source were −60‰. The seasonal cycle amplitudes at CGO, based on the smooth curve fit to the data (Miller *et al.*, 2002), were 0.30‰ in 1998, 0.17‰ in 1999, and 0.15‰ in 2000. Large tropical wetland emissions could help to explain the presence of the dip at SMO and CGO in 1998.

Lowe *et al.* (1997) showed distinct seasonal cycles in $\delta^{13}C$ between 1989 and 1997 from air collected at Baring Head. They calculated that the amplitude of the observed seasonal cycle was too large to be explained solely on the basis of OH oxidation. If the methane mole fraction seasonal amplitude were controlled completely by OH destruction (as might be the case at the South Pole), the amplitude of the δ^{13}C signal would be at most 0.1‰ according to the following Rayleigh model of CH_4 consumption.

$$\delta - \delta_o \approx -\varepsilon \frac{\Delta M}{M} \tag{16.6}$$

Here, δ and δ_o are the final, and original isotopic ratios, expressed in δ-notation (‰ units), ε is the kinetic fraction factor due to reaction with OH [$\varepsilon = -5.4$‰ (Cantrell *et al.*, 1990) or $\varepsilon = -3.9$‰ (Saueressig *et al.*, 2001)] and $\Delta M/M$ is the fraction of total methane destroyed ($\Delta M/M =$ 30 ppb/1700 ppb for southern hemisphere sites). The expected amplitude is 0.10‰ (−5.4) or 0.07‰ (−3.9). When the seasonal cycles amplitude exceeds these limits, it is an indication that processes other than destruction by OH are at work.

Northern Hemisphere Seasonal Cycles Seasonal variations of δ^{13}C are more pronounced in the northern hemisphere, because about 75% of methane emissions originate there (Fung *et al.*, 1991). Mean northern hemisphere mole fractions average about 90 ppb higher than in the southern hemisphere, and in 1998–2000 δ^{13}C values were about 0.3‰ lower in the northern hemisphere than in the southern hemisphere. Seasonal variations are most evident at BRW where the seasonal cycle amplitude has averaged about 0.6‰, with the maximum in May and the minimum at the end of September. The timing of the minimum and maximum indicates high northern sources as the primary driver of seasonal variability. δ^{13}C values start to decrease in May and continue through the summer despite the fact that destruction of CH_4 by OH is largest during this time of year. This probably occurs because emissions from isotopically light sources like wetlands are greatest during the summer, and bacterial emissions have 2–3 times the impact on δ^{13}C values than OH for the same change in mole fraction. Seasonal patterns at NWR and MLO are less pronounced than at BRW, but like the southern hemisphere exhibit deeper minima in 1998 (Fig. 16.2).

Calculating CH_4 Sources With the Help of δ^{13}C

Global Means Using Eq. 16.1 and a value of CH_4 lifetime, $\tau = 9.4$ yr, $[CH_4] = 1750$ ppb, $d[CH_4]/dt = 5$ ppb/yr, and a conversion factor of 2.767 ppb/Tg (Fung *et al.*, 1991), we find that $Q = 529$ Tg/yr. Using the exact formulation of Lassey *et al.* (2000), we will calculate δ_q:

$$\delta_q = \alpha\delta_a + \varepsilon - \frac{\varepsilon(1 + \delta_a/1000)}{Q}\frac{d[CH_4]}{dt} + \frac{d\delta_a}{dt}\frac{[CH_4]}{Q} \tag{16.7}$$

Using a global mean value of $\delta_a = -47.1‰$, a value of $\varepsilon = -6.3‰$, and a $\delta^{13}C$ growth rate of 0.01‰/yr, we find that the global, flux weighted isotopic signature of all sources, $\delta_q = -53.2‰$. Now we can use Eqs 16.4 and 16.5 to estimate the annual global CH_4 emissions from bacterial processes and biomass burning after separately calculating fossil fuel emissions using $^{14}CH_4$ data. Based on the data of Quay *et al.* (1999) we find the fossil fuel related flux of CH_4 to be 100 Tg/yr. Assuming values for δ_{ff}, δ_{bmb}, and δ_{bio} of −40, −25, and −60‰, the global value of $Q_{bmb} = 46$ Tg/yr, and $Q_{bio} = 383$ Tg/yr. This simple calculation produces estimates of Q_{bmb} and Q_{bio} that are similar to those in Table 16.1, where $Q_{bio} = 420$ Tg/yr and $Q_{bmb} = 40$ Tg/yr.

Results from an Inverse 2-Box Model The spatial patterns in the $\delta^{13}C$ and CH_4 data, specifically the latitudinal gradient, can be used to infer the spatial distribution of CH_4 sources. In the following example, we use a two-box model of the atmosphere, and therefore use the interhemispheric difference. We again specify the fossil fuel flux based on $^{14}CH_4$ data but now include emissions from landfills because of their similar spatial patterns. Dividing the Earth at the equator, we can solve for the emissions from bacterial processes and biomass burning in each hemisphere, if we know the interhemispheric exchange constant, k_{ex}. In Eqs 16.8–16.11 we have four equations and four unknowns: bacterial and biomass burning emissions for the northern and southern hemispheres. The values of all the other terms are listed in Table 16.2.

$$\dot{X}_N = B_N + BMB_N + FFP_N + k_{12}X_N + k_{ex}(X_N - X_S) \tag{16.8}$$

$$\dot{X}_S = B_S + BMB_S + FFP_s + k_{12}X_S - k_{ex}(X_N - X_S) \tag{16.9}$$

$$\dot{\delta}_N = \frac{B_N}{X_N}(\delta_B - \delta_N) + \frac{BMB_N}{X_N}(\delta_{BMB} - \delta_N) + \frac{FFP_N}{X_N}(\delta_{FFP} - \delta_N) + \varepsilon k_{12} + k_{ex}\frac{X_S}{X_N}(\delta_N - \delta_S) \tag{16.10}$$

$$\dot{\delta}_S = \frac{B_S}{X_S}(\delta_B - \delta_S) + \frac{BMB_N}{X_N}(\delta_{BMB} - \delta_S) + \frac{FFP_S}{X_S}(\delta_{FFP} - \delta_S) + \varepsilon k_{12} - k_{ex}\frac{X_N}{X_S}(\delta_N - \delta_S) \tag{16.11}$$

Here, X represents the mole fraction and the over-dot denotes the time derivative. B is the hemispheric total of bacterial emissions, BMB, of biomass burning emissions, and FFP, of fossil fuel related emissions, plus those from landfills. Fossil fuel and landfill sources were grouped together because both are estimated to be more than 90% in the northern hemisphere.

When we solve these two systems of two linear equations using the best estimates of the terms on the left hand side, the global emission totals are: bacterial $= 355 \pm 48$ Tg/yr; biomass burning $= 56 \pm 37$ Tg/yr. The hemispheric totals are $B_N = 250 \pm 33$ Tg/yr, $B_S = 106 \pm 21$ Tg/yr, $BMB_N = 23 \pm 30$ Tg/yr, and $BMB_S = 31 \pm 10$ Tg/yr. The global ratio of the *B*, *FFP*, and *BMB* emissions is 65/25/10, which is similar to that obtained by Fung *et al.* (1991) of 64/25/11, Crutzen *et al.* (1995) of 72/22/6, Hein *et al.* (1997) of 70/22/7, and Lelieveld *et al.* (1998) of 73/19/7.

Uncertainties and Sensitivities Even though we use a simple two-box model of the atmosphere, this can be an excellent tool to investigate sensitivity of source partitioning to uncertainty in the parameters listed in Table 16.2. A Monte Carlo approach is used in which all parameters are assigned uncertainties, which are listed in Table 16.2. To estimate the error, each known term in Eqs 16.8–16.11 is represented by a normally distributed set of points such that the mean and standard deviation are equal to the mean and uncertainty shown in Table 16.2. Equations 16.8–16.11 are solved 10,000 times, each time randomly choosing values from the distribution of each term. The uncertainties on the fluxes are then simply the mean and standard deviation of the solutions from the 10,000 trials.

Table 16.3 lists the calculated sensitivities of the calculated fluxes to the parameters listed in Table 16.2. The dominant source of uncertainty is the

Table 16.3 Sensitivity of Source Partitioning to Parameter Changes in the Inverse-Box-Model

Parameter	Units	Source	Sensitivity Globally	Sensitivity NH	Sensitivity SH	Total Uncertainty (Tg)[a]
δ_N–δ_S	Tg ‰$^{-1}$	*B*	0	69	−69	0
X_N–X_S	Tg ppb^{-1}	*B*	0	0.91	−0.91	0
%Cl added	Tg %$^{-1}$	*B*	11.2	3.9	7.3	Not estimated
		BMB[b]	−6.3	−2.2	−4.1	Not estimated
δ_B	Tg ‰$^{-1}$	*B*	9.7	7.3	2.4	19.4
δ_{BMB}	Tg ‰$^{-1}$	*B*	1.5	0.95	0.55	3.0
δ_{FFP}	Tg ‰$^{-1}$	*B*	6.7	6.3	0.4	13.4
FFP	Tg Tg^{-1}	*B*	−0.51	−0.47	−0.04	26.0
$d\delta/dt$	Tg ‰$^{-1}$yr	*B*	−131	−67	−64	2.6
dX/dt	Tg ppb^{-1}yr	*B*	1.7	0.9	0.8	3.4

[a]These are global values computed by multiplying the sensitivity by the estimated uncertainty from Table 16.2. The uncertainty for dX/dt is set at 2 ppb/yr.

[b]For all other source sensitivities, *BMB* (biomass burning) is simply of the opposite sign as *B* (bacterial), such that total global emissions remain constant. Adding a Cl sink increases the total emissions in the model, which requires that *BMB* and *B* sensitivities not be of equal magnitude.

fossil fuel emission rate, especially for northern hemisphere sources. Uncertainty in source δ values is the next most important source of error. The derived fluxes are much less sensitive to the isotopic signature of biomass burning fluxes than to changes in the signatures of *FFP* or bacterial emissions. For any source, a 2‰ change in source signature corresponds to an approximate 5% change in the flux of that source. Five percent of the biomass burning source is only about 3 Tg, whereas for bacterial emissions 5% is 18 Tg. Global flux partitioning is therefore not very sensitive to the isotopic signature of biomass burning emissions but is sensitive to the weighted signature of all bacterial emissions. The interhemispheric exchange constant, k_{ex}, does not influence global partitioning, but has a big impact on partitioning of a source between hemispheres. For example, although our global *BMB* value agrees with the earlier estimate of Fung *et al.* (1991), our BMB_S estimate is very different. However, the level of agreement in north/south partitioning is largely a function of our choice of k_{ex}, which is a poorly constrained parameter (Denning *et al.*, 1999).

Interestingly, it is evident that improving the precision and accuracy of our atmospheric measurements will not dramatically alter our ability to partition sources, at least when using annual hemispheric average $\delta^{13}C$ values. Changing the global average $\delta^{13}C$ by 0.1‰ would only alter emissions partitioning by about 1.5 Tg/yr in our model. The biggest improvements in emission partitioning will come from better constraining fossil fuel emissions and by better understanding the isotopic ratio of bacterial emissions and how and why they vary. Recent measurements of ε_{OH} (Saueressig *et al.*, 2001) show a value of −3.9‰, contrasting with the earlier measurement of −5.4‰ (Cantrell *et al.*, 1990). Using the newer value of ε_{OH} results in an increase in calculated biomass burning emissions of about 25 Tg with a concomitant decrease in bacterial emissions. The other important way we can improve the sink side of the equation is by better determining the lifetime of CH_4 in the atmosphere, including the magnitude of the soil sink, and the possible existence of a tropospheric Cl sink.

A Role for Chlorine? What is the extent to which atomic Cl in the marine boundary layer (MBL) consumes CH_4? Several recent studies have suggested the possibility that CH_4 is oxidized by Cl in the marine boundary layer (Gupta *et al.*, 1996; Vogt *et al.*, 1996; Wingenter *et al.*, 1999; Allan *et al.*, 2001a,b). Wingenter *et al.* (1999) estimated that 2% of CH_4 in the MBL is destroyed by Cl, but Singh *et al.* (1996) estimated that no more than 2% of CH_4 is consumed by Cl in the troposphere. Because of the large isotopic fractionation in the reaction $CH_4 + Cl$ (Saueressig *et al.*, 1995), atmospheric $\delta^{13}C$ is a good tracer for the activity of Cl. A model experiment (Miller *et al.*, 2002) suggests an upper limit for Cl of 6% of the total sink. If it were more, the biomass burning source would be less than 20 Tg/yr, which is unlikely from a bottom–up point of view. However, the absence of a chlorine sink is also consistent with the data.

Three-Dimensional Inverse Models The paucity of atmospheric δ^{13}C data has limited its inclusion as input into global inverse models studying the CH_4 budget. The inverse model of Hein *et al.* (1997) included a small amount of δ^{13}C data. Perhaps not surprisingly, it found that including δ^{13}C data did not significantly change the fluxes that were derived solely from CH_4 mole fraction measurements. In this study, δ^{13}C measurements from just three northern hemisphere sites were used as constraints. As more measurements become available it is likely that δ^{13}C measurements will prove more useful as formal constraints in inverse studies.

How Can We Improve the Utility of δ^{13}C Measurements? As we have seen above, CH_4 concentration and δ^{13}C measurements alone cannot uniquely specify methane sources, even when divided into just three broad categories. To arrive at a unique solution, we added information on the fossil fuel fraction that was derived from $^{14}CH_4$ measurements. More generally, δ^{13}C measurements will be most useful when measurements of another methane species are added to the mix. Both $^{14}CH_4$ and CH_3D can serve this role. Measurements of $^{14}CH_4$ have specific use in determining fossil fuel related emissions. Although this is, in principle, a very powerful tracer, there is considerable uncertainty in the $^{14}CH_4$ budget related to emissions of $^{14}CH_4$ from nuclear power plants (Kunz, 1985). The signal of δD in CH_4 may be a strong marker for the latitudinal origin of methane fluxes (Waldron *et al.*, 1999), because of the strong gradient in δD of H_2O of about 100‰ from equator to pole (Dansgaard, 1964). This will be true insofar as the source of hydrogen in bacterially produced methane is water (either directly or indirectly). More importantly, the large difference in CH_3D wetland emissions has also been directly observed (Bellisario *et al.*, 1999; Smith *et al.*, 2000). With the advent of continuous flow measurement techniques for δD of CH_4 (Hilkert *et al.*, 1999; Rice *et al.*, 2001), atmospheric and source δD measurements should become an important complement to the ongoing measurements of δ^{13}C. As we have seen above in the case of δ^{13}C, it will be very important to accurately characterize the deuterium signature of CH_4 emissions. Atmospheric δ^{13}C measurements and, more generally, isotopic measurements will certainly help us to better understand the global methane budget, but we will need to better understand the processes responsible for determining fractionation during both production and consumption of methane.

Up to this point, the full extent of spatial and temporal information contained within the δ^{13}C measurements has not been fully utilized. This will require the merging of data sets between different laboratories using different sample collection, analysis, and standardization procedures. Arriving at a unified data set of atmospheric $\delta^{13}C$ measurements will require common isotopic standards and careful intercomparison of both sample and reference gases. So, it remains to be seen to what extent atmospheric $\delta^{13}C$ measurements will improve our estimates of sources and sinks compared to estimates based on CH_4 alone.

References

Allan W., Lowe D. C. and Cainey J. M. (2001a) Active chlorine in the remote marine boundary layer: Modeling anomalous measurements of delta C-13 in methane. *Geophys Res Lett* **28**: 3239–3242.

Allan W., Manning M. R., Lassey K. R., Lowe D. C. and Gomez A. J. (2001b) Modeling the variation of delta C-13 in atmospheric methane: Phase ellipses and the kinetic isotope effect. *Global Biogeochem Cycles* **15**: 467–481.

Aselmann I. and Crutzen P. J. (1989) Global distribution of natural freshwater wetlands and rice paddies, their net primary productivity, seasonality and possible methane sources. *J Atmos Chem* **8**: 307–358.

Battle M., Bender M., Sowers T., Tans P. P., Butler J. H., Elkins J. W., Ellis J. T. *et al.* (1996) Atmospheric gas concentrations over the past century measured in air from firn at the South Pole. *Nature* **383**: 231–235.

Bellisario L. M., Bubier J. L., Moore T. R. and Chanton J. P. (1999) Controls on CH_4 emissions from a northern peatland. *Global Biogeochem Cycles* **13**: 81–91.

Blunier T., Chappellaz J. A., Schwander J., Barnola J. M., Desperts T., Stauffer B. and Raynaud D. (1993) Atmospheric methane: Record from a Greenland ice core over the last 1000 Years. *Geophys Res Lett* **20**: 2219–2222.

Blunier T., Chappellaz J., Schwander J., Stauffer B. and Raynaud D. (1995) Variations in atmospheric methane concentration during the Holocene epoch. *Nature* **374**: 46–49.

Braunlich M., Aballanin O., Marik T., Jockel P., Brenninkmeijer C. A. M., Chappellaz J., Barnola J. M., Mulvaney R. and Sturges W. T. (2001) Changes in the global atmospheric methane budget over the last decades inferred from C-13 and D isotopic analysis of Antarctic firn air. *J Geophys Res* **106**: 20465–20481.

Brook E. J., Sowers T. and Orchardo J. (1996) Rapid variations in atmospheric methane concentration during the past 110 000 years. *Science* **273**: 1087–1091.

Brown M. (1993) Deduction of emissions of source gases using an objective inversion algorithm and a chemical-transport model. *J Geophys Res* **98**: 12639–12660.

Cantrell C. A., Shetter R. E., McDaniel A. H., Calvert J. G., Davidson J. A., Lowe D. C., Tyler S. C., Cicerone R. J. and Greenberg J. P. (1990) Carbon kinetic isotope effect in the oxidation of methane by the hydroxyl radical. *J Geophys Res* **95**: 22455–22462.

Cao M. K., Dent J. B. and Heal O. W. (1995) Modeling methane emissions from rice paddies. *Global Biogeochem Cycles* **9**: 183–195.

Cao M. K., Marshall S. and Gregson K. (1996) Global carbon exchange and methane emissions from natural wetlands: Application of a process-based model. *J Geophys Res* **101**: 14399–14414.

Chanton J. and Liptay K. (2000) Seasonal variation in methane oxidation in a landfill cover soil as determined by an in situ stable isotope technique. *Global Biogeochem Cycles* **14**: 51–60.

Chappellaz J., Barnola J. M., Raynaud D., Korotkevich Y. S. and Lorius C. (1990) Ice-core record of atmospheric methane over the past 160 000 years. *Nature* **345**: 127–131.

Chappellaz J., Blunier T., Kints S., Dallenbach A., Barnola J. M., Schwander J., Raynaud D. and Stauffer B. (1997) Changes in the atmospheric CH_4 gradient between Greenland and Antarctica during the Holocene. *J Geophys Res* **102**: 15987–15997.

Chidthaisong A., Chin K. J., Valentine D. L. and Tyler S. C. (2002) A comparison of isotope fractionation of carbon and hydrogen from paddy field rice roots and soil bacterial enrichments during CO_2/H-2 methanogenesis. *Geochim Cosmochim Acta* **66**: 983–995.

Cicerone R. J. and Oremland R. S. (1988) Biogeochemical aspects of atmospheric methane. *Global Biogeochem Cycles* **2**: 299–327.

Conny J. M. and Currie L. A. (1996) The isotopic characterization of methane, non-methane hydrocarbons and formaldehyde in the troposphere. *Atmos Environ* **30**: 621–638.

Craig H., Chou C. C., Welhan J. A., Stevens C. M. and Engelkemeir A. (1988) The isotopic composition of methane in polar ice cores. *Science* **242**: 1535–1539.

Crutzen P. J. (1995) The role of methane in atmospheric chemistry and climate. In *Ruminant Physiology: Digestion, Metabolism, Growth and Reproduction: Proceedings of the Eighth International Symposium on Ruminant Physiology* (W. v. Engelhardt, S. Leonhardt-Marek, G. Breves and D. Giesecke, eds) pp. 291–315. Ferdinand Enke Verlag, Stuttgart.

Dansgaard W. (1964) Stable isotopes in precipitation. *Tellus* **16**: 436–468.

Deines P. (1980) The isotopic composition of reduced organic carbon. In *Handbook of Environmental Isotope Geochemistry* (P. Fritz and J. C. Fontes, eds) pp. 329–406. Elsevier Scientific.

Denning A. S., Holzer M., Gurney K. R., Heimann M., Law R. M., Rayner P. J., Fung I. Y. *et al.* (1999) Three-dimensional transport and concentration of SF6–A model intercomparison study (TransCom 2). *Tellus Ser B* **51**: 266–297.

Dlugokencky E. J., Steele L. P., Lang P. M. and Masarie K. A. (1994) The growth-rate and distribution of atmospheric methane. *J Geophys Res* **99**: 17021–17043.

Dlugokencky E. J., Masarie K. A., Tans P. P., Conway T. J. and Xiong X. (1997) Is the amplitude of the methane seasonal cycle changing? *Atmos Environ* **31**: 21–26.

Dlugokencky E. J., Masarie K. A., Lang P. M. and Tans P. P. (1998) Continuing decline in the growth rate of the atmospheric methane burden. *Nature* **393**: 447–450.

Dlugokencky E. J., Walter B. P. and Kaischke E. S. (2001) Measurements of an anomalous global methane increase during 1998. *Geophys Res Lett* **28**: 499–503.

Etheridge D. M., Pearman G. I. and Fraser P. J. (1992) Changes in tropospheric methane between 1841 and 1978 from a high accumulation-rate antarctic ice core. *Tellus Ser B* **44**: 282–294.

Etheridge D. M., Steele L. P., Francey R. J. and Langenfelds R. L. (1998) Atmospheric methane between 1000 AD and present: Evidence of anthropogenic emissions and climatic variability. *J Geophys Res* **103**: 15979–15993.

Farquhar G. D., Ehleringer J. R. and Hubick K. T. (1989) Carbon isotope discrimination during photosynthesis. *Annu Rev Plant Physiol Plant Mol Biol* **9**: 121–137.

Francey R. J., Manning M. R., Allison C. E., Coram S. A., Etheridge D. M., Langenfelds R. L., Lowe D. C. and Steele L. P. (1999) A history of delta C-13 in atmospheric CH_4 from the Cape Grim air archive and antarctic firn air. *J Geophys Res* **104**: 23631–23643.

Fung I., John J., Lerner J., Matthews E., Prather M., Steele L. P. and Fraser P. J. (1991) 3-dimensional model synthesis of the global methane cycle. *J Geophys Res* **96**: 13033–13065.

Gupta M., Tyler S. and Cicerone R. (1996) Modeling atmospheric $\delta^{13}CH_4$ and the causes of recent changes in atmospheric CH_4 amounts. *J Geophys Res* **101**: 22923–22932.

Hansen J. E. and Sato M. (2001) Trends of measured climate forcing agents. *PNAS* **98**: 14778–14783.

Hao W. M. and Ward D. E. (1993) Methane production from global biomass burning. *J Geophys Res* **98**: 20657–20661.

Hein R., Crutzen P. J. and Heimann M. (1997) An inverse modeling approach to investigate the global atmospheric methane cycle. *Global Biogeochem Cycles* **11**: 43–76.

Hilkert A. W., Douthitt C. B., Schluter H. J. and Brand W. A. (1999) Isotope ratio monitoring gas chromatography mass spectrometry of D/H by high temperature conversion isotope ratio mass spectrometry. *Rapid Comm Mass Spectrometry* **13**: 1226–1230.

Holmes M. E., Sansone F. J., Rust T. M. and Popp B. N. (2000) Methane production, consumption, and air-sea exchange in the open ocean: An evaluation based on carbon isotopic ratios. *Global Biogeochem Cycles* **14**: 1–10.

Houweling S., Kaminski T., Dentener F., Lelieveld J. and Heimann M. (1999) Inverse modeling of methane sources and sinks using the adjoint of a global transport model. *J Geophys Res* **104**: 26173–26160.

Jones R. L. and Pyle J. A. (1984) Observations of CH_4 and N_2O by the Nimbus-7 SAMS: A comparison with *in situ* data and two-dimensional numerical model calculations. *J Geophys Res* **89**: 5263–5279.

King S. L., Quay P. D. and Lansdown J. M. (1989) The C-13/C-12 kinetic isotope effect for soil oxidation of methane at ambient atmospheric concentrations. *J Geophys Res* **94**: 18273–18277.

Kruger M., Eller G., Conrad R. and Frenzel P. (2002) Seasonal variation in pathways of CH_4 production and in CH_4 oxidation in rice fields determined by stable carbon isotopes and specific inhibitors. *Global Change Biol* **8**: 265–280.

Kunz C. (1985) Carbon-14 discharge at three light water reactors. *Health Phys* **49**: 25–35.

Lashof D. A. and Ahuja D. R. (1990) Relative contributions of greenhouse gas emissions to global warming. *Nature* **344**: 529–531.

Lassey K. R., Lowe D. C. and Manning M. R. (2000) The trend in atmospheric methane delta C-13 implications for isotopic constraints on the global methane budget. *Global Biogeochem Cycles* **14**: 41–49.

Law R., Simmonds I. and Budd W. F. (1992) Application of an atmospheric tracer model to high southern latitudes. *Tellus* **44**: 358–370.

Lelieveld J., Crutzen P. J. and Dentener F. J. (1998) Changing concentration, lifetime and climate forcing of atmospheric methane. *Tellus Ser B* **50**: 128–150.

Levin I., Bergamaschi P., Dorr H. and Trapp D. (1993) Stable isotopic signature of methane from major sources in Germany. *Chemosphere* **26**: 161–177.

Levin I., Glatzel-Mattheier H., Marik T., Cuntz M., Schmidt M. and Worthy D. E. (1999) Verification of German methane emission inventories and their recent changes based on atmospheric observations. *J Geophys Res* **104**: 3447–3456.

Liptay K., Chanton J., Czepiel P. and Mosher B. (1998) Use of stable isotopes to determine methane oxidation in landfill cover soils. *J Geophys Res* **103**: 8243–8250.

Lowe D. C., Brenninkmeijer C. A. M., Brailsford G. W., Lassey K. R., Gomez A. J. and Nisbet E. G. (1994) Concentration and C-13 records of atmospheric methane in New Zealand and Antarctica–evidence for changes in methane sources. *J Geophys Res* **99**: 16913–16925.

Lowe D. C., Manning M. R., Brailsford G. W. and Bromley A. M. (1997) The 1991–1992 atmospheric methane anomaly: Southern hemisphere C-13 decrease and growth rate fluctuations. *Geophys Res Lett* **24**: 857–860.

Lowry D., Holmes C. W., Rata N. D., O'Brien P. and Nisbet E. G. (2001) London methane emissions: Use of diurnal changes in concentration and delta C-13 to identify urban sources and verify inventories. *J Geophys Res* **106**: 7427–7448.

Miller J. B. (1999) Application of Gas Chromatography Isotope Ratio Mass Spectrometry (GC-IRMS) to Atmospheric Budgets of $C^{18}OO$ and $^{13}CH_4$. PhD thesis, University of Colorado, Boulder.

Miller J. B., Mack K. A., Dissly R., White J. W. C., Dlugokencky E. J. and Tans P. P. (2002) Development of analytical methods and measurements of $^{13}C/^{12}C$ in atmospheric CH_4 from the NOAA/CMDL global air sampling network. *J Geophys Res* **107**: doi: 10.1029/2001JD000630.

Montzka S. A., Spivakovsky C. M., Butler J. H., Elkins J. W., Lock L. T. and Mondeel D. J. (2000) New observational constraints for atmospheric hydroxyl on global and hemispheric scales. *Science* **288**: 500–503.

Nakazawa T., Machida T., Tanaka M., Fujii Y., Aoki S. and Watanabe O. (1993) Differences of the atmospheric CH_4 concentration between the arctic and antarctic regions in the pre-industrial/pre-agricultural era. *Geophys Res Lett* **20**: 943–946.

Olivier J. G. J., Bouwman A. F., Berdowski J. J. M., Veldt C., Bloos J. P. J., Visschedijk A. H. J., van de Maas C. W. M. and Zandweld P. Y. J. (1999) Sectoral emission inventories of greenhouse gases for 1990 on per country basis as well as on 10×10. *Environ Sci Policy* **2**: 241–264.

Popp T. J., Chanton J. P., Whiting G. J. and Grant N. (1999) Methane stable isotope distribution at a Carex dominated fen in north central Alberta. *Global Biogeochem Cycles* **13**: 1063–1077.

Prinn R. G., Weiss R. F., Miller B. R., Huang J., Alyea F. N., Cunnold D. M., Fraser P. J., Hartley D. E. and Simmonds P. G. (1995) Atmospheric trends and lifetime of CH_3CCl_3 and global OH concentrations. *Science* **269**: 187–192.

Quay P. D., King S. L., Landsdown J. M. and Wilbur D. O. (1988) Isotopic composition of methane released from wetlands: Implications for the increase in atmospheric methane. *Global Biogeochem Cycles* **2**: 385–397.

Quay P. D., King S. L., Stutsman J., Wilbur D. O., Steele L. P., Fung I., Gammon R. H. *et al.* (1991) Carbon isotopic composition of atmospheric CH_4: Fossil and biomass burning source strengths. *Global Biogeochem Cycles* **5**: 25–47.

Quay P., Stutsman J., Wilbur D., Snover A., Dlugokencky E. and Brown T. (1999) The isotopic composition of atmospheric methane. *Global Biogeochem Cycles* **13**: 445–461.

Randerson J. T., Thompson M. V., Conway T. J., Fung I. Y. and Field C. B. (1997) The contribution of terrestrial sources and sinks to trends in the seasonal cycle of atmospheric carbon dioxide. *Global Biogeochem Cycles* **11**: 535–560.

Rice A. L., Gotoh A. A., Ajie H. O. and Tyler S. C. (2001) High-precision continuous-flow measurement of delta C-13 and delta D of atmospheric CH_4. *Anal Chem* **73**: 4104–4110.

Ridgwell A. J., Marshall S. J. and Gregson K. (1999) Consumption of atmospheric methane by soils: A process-based model. *Global Biogeochem Cycles* **13**: 59–70.

Sansone F. J., Popp B. N., Gasc A., Graham A. W. and Rust T. M. (2001) Highly elevated methane in the eastern tropical North Pacific and associated isotopically enriched fluxes to the atmosphere. *Geophys Res Lett* **28**: 4567–4570.

Saueressig G., Bergamaschi P., Crowley J. N., Fischer H. and Harris G. W. (1995) Carbon kinetic isotope effect in the reaction of CH_4 with Cl Atoms. *Geophys Res Lett* **22**: 1225–1228.

Saueressig G., Crowley J. N., Bergamaschi P., Bruhl C., Brenninkmeijer C. A. M. and Fischer H. (2001) Carbon 13 and D kinetic isotope effects in the reactions of CH_4 with O(D-1) and OH: New laboratory measurements and their implications for the isotopic composition of stratospheric methane. *J Geophys Res* **106**: 23127–23138.

Schoell M. (1980) The hydrogen and carbon isotopic composition of methane from natural gases of various origins. *Geochim Cosmochim Acta* **44**: 649–661.

Singh H. B., Thakur A. N., Chen Y. E. and Kanakidou M. (1996) Tetrachloroethylene as an indicator of low Cl atom concentrations in the troposphere. *Geophys Res Lett* **23**: 1529–1532.

Smith L. K., Lewis W. M., Chanton J. P., Cronin G. and Hamilton S. K. (2000) Methane emissions from the Orinoco River floodplain, Venezuela. *Biogeochemistry* **51**: 113–140.

Sowers T. A. (2000a) The delta ^{13}C of Atmospheric CH_4 over the last 200 years as recorded in the GISP II ice core. *Eos Trans AGU* **81**: S20.

Sowers T. A. (2000b) The delta ^{13}C of atmospheric CH_4 over the past 30,000 years as recorded in the Taylor Dome ice core. *Eos Trans AGU* **81**: F78.

Steele L. P., Fraser P. J., Rasmussen R. A., Khalil M. A. K., Conway T. J., Crawford A. J., Gammon R. H., Masarie K. A. and Thoning K. W. (1987) The global distribution of methane in the troposphere. *J Atmos Chem* **5**: 125–171.

Stevens C. M. (1993) Isotopic abundances in the atmosphere and sources. In *Atmospheric Methane: Sources, Sinks and Role in Global Change* (M. A. K. Khalil, ed.) pp. 11–21. Springer-Verlag, Berlin.

Stevens C. M. (1995) Carbon-13 isotopic abundance and concentration of atmospheric methane for background air in the southern and northern hemispheres from 1978 to 1989. **Publication 4388** Environmental Science Division, U.S. Department of Energy.

Stevens C. M. and Rust F. E. (1982) The carbon isotopic composition of atmospheric methane. *J Geophys Res* **87**: 4879–4882.

Tans P. P. (1997) A note on isotopic ratios and the global atmospheric methane budget. *Global Biogeochem Cycles* **11**: 77–81.

Tans P. P., Fung I. Y. and Takahashi T. (1990) Observational constraints on the global atmospheric CO_2 budget. *Science* **247**: 1431–1438.

Thompson A. M. (1992) The oxidizing capacity of the earth's atmosphere: probable past and future changes. *Science* **256**: 1157–1165.

Tyler S. C. (1986) Stable carbon isotope ratios in atmospheric methane and some of its sources. *J Geophys Res* **91**: 13332–12338.

Tyler S. C., Ajie H. O., Gupta M. L., Cicerone R. J., Blake D. R. and Dlugokencky E. J. (1999) Stable carbon isotopic composition of atmospheric methane: A comparison of surface level and free tropospheric air. *J Geophys Res* **104**: 13895–13910.

Tyler S. C., Brailsford G. W., Yagi K., Minami K. and Cicerone R. J. (1994) Seasonal variations in methane flux and delta(CH_4)C-13 values for rice paddies in Japan and their implications. *Global Biogeochem Cycles* **8**: 1–12.

Vaghjiani G. L. and Ravishankara A. R. (1991) New measurement of the rate coefficient for the reaction of OH with methane. *Nature* **350**: 406–409.

van Aardenne J. A., Dentener F. J., Olivier J. G. J., Goldewijk C. and Lelieveld J. (2001) A 1 degree×1 degree resolution data set of historical anthropogenic trace gas emissions for the period 1890–1990. *Global Biogeochem Cycles* **15**: 909–928.

Vogt R., Crutzen P. J. and Sander R. (1996) A mechanism for halogen release from sea-salt aerosol in the remote marine boundary layer. *Nature* **383**: 327–330.

Waldron S., Lansdown J. M., Scott E. M., Fallick A. E. and Hall A. J. (1999) The global influence of the hydrogen isotope composition of water on that of bacteriogenic methane from shallow freshwater environments. *Geochim Cosmochim Acta* **63**: 2237–2245.

Walter B. P. and Heimann M. (2000) A process-based, climate-sensitive model to derive methane emissions from natural wetlands: Application to five wetland sites, sensitivity to model parameters, and climate. *Global Biogeochem Cycles* **14**: 745–765.

Wingenter O. W., Blake D. R., Blake N. J., Sive B. C., Rowland F. S., Atlas E. and Flocke F. (1999) Tropospheric hydroxyl radical and atomic chlorine concentrations, and mixing timescales determined from hydrocarbon and halocarbon measurements made over the Southern Ocean. *J Geophys Res* **104**: 21819–21828.

Physiological Ecology

A Series of Monographs, Texts, and Treatises

T. T. KOZLOWSKI. Growth and Development of Trees, Volumes I and II, 1971

D. HILLEL. Soil and Water: Physical Principles and Processes, 1971

V. B. YOUNGER and C. M. McKELL (Eds.). The Biology and Utilization of Grasses, 1972

J. B. MUDD and T. T. KOZLOWSKI (Eds.). Responses of Plants to Air Pollution, 1975

R. DAUBENMIRE. Plant Geography, 1978

J. LEVITT. Responses of Plants to Environmental Stresses, Second Edition
Volume I: Chilling, Freezing, and High Temperature Stresses, 1980
Volume II: Water, Radiation, Salt, and Other Stresses, 1980

J. A. LARSEN (Ed.). The Boreal Ecosystem, 1980

S. A. GAUTHREAUX, JR. (Ed.). Animal Migration, Orientation, and Navigation, 1981

F. J. VERNBERG and W. B. VERNBERG (Eds.). Functional Adaptations of Marine Organisms, 1981

R. D. DURBIN (Ed.). Toxins in Plant Disease, 1981

C. P. LYMAN, J. S. WILLIS, A. MALAN, and L. C. H. WANG. Hibernation and Torpor in Mammals and Birds, 1982

T. T. KOZLOWSKI (Ed.). Flooding and Plant Growth, 1984

E. I. RICE. Allelopathy, Second Edition, 1984

M. L. CODY. Habitat Selection in Birds, 1985

R. J. HAYNES, K. C. CAMERON, K. M. COH, and K. K. SHERLOCK (Eds.). Mineral Nitrogen in the Plant–Soil System, 1986

T. T. KOZLOWSKI, P. J. KRAMER, and S. G. PALLARDY. The Physiological Ecology of Woody Plants, 1991

H. A. MOONEY, W. E. WINNER, and E. J. PELL (Eds.). Response of Plants to Multiple Stresses, 1991

F. S. CHAPIN III, R. L. JEFFERIES, J. F. REYNOLDS, G. R. SHAVER, and J. SVOBODA (Eds.). Arctic Ecosystems in a Changing Climate: An Ecophysiological Perspective , 1991

T. D. SHARKEY, E. A. HOLLAND, and H. A. MOONEY (Eds.). Trace Gas Emissions by Plants, 1991

U. SEELIGER (Ed.). Coastal Plant Communities of Latin America, 1992

JAMES R. EHLERINGER and CHRISTOPHER B. FIELD (Eds.). Scaling Physiological Processes: Leaf to Globe, 1993

JAMES R. EHLERINGER, ANTHONY E. HALL, and GRAHAM D. FARQUHAR (Eds.). Stable Isotopes and Plant Carbon–Water Relations, 1993

E. -D. SCHULZE (Ed.). Flux Control in Biological Systems, 1993

MARTYN M. CALDWELL and ROBERT W. PEARCY (Eds.). Exploitation of Environmental Heterogeneity by Plants: Ecophysiological Processes Above- and Belowground, 1994

WILLIAM K. SMITH and THOMAS M. HINCKLEY (Eds.). Resource Physiology of Conifers: Acquisition, Allocation, and Utilization, 1995.

WILLIAM K. SMITH and THOMAS M. HINCKLEY (Eds.). Economic Physiology of Coniferous Forests, 1995.

MARGARET D. LOWMAN and NALINI M. NADKARNI (Eds.). Forest Canopies, 1995

BARBARA L. GARTNER (Ed.). Plant Stems: Physiology and Functional Morphology, 1995

GEORGE W. KOCH and HAROLD A. MOONEY (Eds.). Carbon Dioxide and Terrestrial Ecosystems, 1996

CHRISTIAN KORNER and FAKHRI A. BAZZAZ (Eds.). Carbon Dioxide, Populations, and Communities, 1996

J. LANDSBERG and S. GOWER. Applications of Physiological Ecology to Forest Management, 1997

THEODORE KOZLAWSKI and STEPHEN PALLARDY. Growth Control in Woody Plants, 1997

FAKHRI BAZZAZ & JOHN GRACE. Plant Resource Allocation, 1997

LOUISE JACKSON. Ecology in Agriculture, 1997

ROWAN SAGE and RUSSELL MONSON C4. Plant Biology, 1998

YIGI LUO and HAROLD MOONEY. Carbon Dioxide and Environmental Stress, 1999

Index